The Cognitive Revolution in Educational Psychology

A volume in
Current Perspectives on Cognition, Learning, and Instruction
Series Editors: James M. Royer

The Cognitive Revolution in Educational Psychology

Edited by

James M. Royer

University of Massachusetts, Amherst

IAP
INFORMATION AGE
PUBLISHING

80 Mason Street • Greenwich, Connecticut 06830 • www.infoagepub.com

Library of Congress Cataloging-in-Publication Data

The impact of the cognitive revolution in educational psychology /
edited by James M. Royer.
p. cm. – (Current perspectives on cognition, learning, and
instruction)
Includes bibliographical references.
ISBN 1-59311-162-2 (pbk.) – ISBN 1-59311-163-0 (hardcover)
1. Educational psychology–History. 2. Cognitive science–History.
I. Royer, James M., 1941- II. Series.
BF81.I47 2005
370.15–dc22

2005005129

Printed in the United States of America

CONTENTS

CHAPTER 1

THE COGNITIVE REVOLUTION IN EDUCATIONAL PSYCHOLOGY

James M. Royer
University of Massachusetts, Amherst

ABSTRACT

This chapter serves as a foreword to the remaining chapters in the book. The chapter provides a description of the cognitive revolution, which began in the 1950s and reached full fruition in the late 1960s. The term "cognitive revolution" began to be used to take advantage of an analysis of scientific revolutions in general that was developed by Thomas Kuhn. The next section of the chapter describes how some aspects of the cognitive revolution seem to fit Kuhn's analytic framework, and others do not. Following this analysis, the chapter turns to examining the impact of the cognitive revolution in educational psychology as illustrated by the remaining chapters in the book. Each of the chapters is briefly described and the end of the chapter comments on the likely direction the revolution will take in the future.

The cognitive revolution began on September 11, 1956. That at least is the date picked by George Miller, one of the principal instigators of the revolution (Miller, 2003). The occasion was a symposium at the Massachusetts

The Impact of the Cognitive Revolution on Educational Psychology, pages 1–12

Institute of Technology concerned with information theory. Miller remembers papers that day by Newell and Simon on their "logic machine," by Nate Rochester from IBM on his work with the then largest computer in the world that was designed to test Donald Hebb's theory of cell assemblies, and several papers on language that was highlighted by Noam Chomsky's presentation of his new theory of syntax. The day's session ended with papers by G. C. Szikali on perceptual recognition, and by Swetts and Birdsall on the use of signal detection theory to explain aspects of perceptual recognition.

Miller remembers his feelings of that day as follows:

> I left the symposium with a conviction, more intuitive than rational, that experimental psychology, theoretical linguistics, and the computer simulation of cognitive processes were all pieces from a larger whole and that the future would see a progressive elaboration and coordination of their shared concerns. (2003, p. 143)

The revolution that was initiated that day, or at least in that time period, involved a clear paradigm shift in psychology. It involved the replacement of a psychological science dominated by the view that the correct subject of psychological investigation was observable behavior, with one dominated by the view that psychology should be about explaining the internal workings of mind. And, as Miller foresaw, very early on the new science morphed into cognitive science, an umbrella that encompassed psychologists, linguists, computer scientists, and later yet, neuroscientists.

DOES THE TERM "COGNITIVE REVOLUTION" FIT?

The term "cognitive revolution" didn't come into vogue until after Thomas Kuhn published his classic book on scientific revolutions (Kuhn, 1962). Thereafter, the overthrow of behavioral views with cognitive views was referred to as a revolution to take advantage of Kuhn's analysis of how new scientific ideas come to dominate older ones. However, there are aspects of the cognitive revolution that do not fit Kuhn's analysis. Kuhn talked about how paradigm shifts from one scientific view to another were driven by discovery and invention, with discovery being the discovery of new facts that did not fit well into the old paradigm, and invention being the development of new theory that challenged the important aspects of the old theory. As time passes, Kuhn explained, more factual anomalies develop and new theories that are better able to explain the data emerge, and eventually more scientists come to believe in the newer perspectives than believe in the older perspectives. In effect, the revolution is a numerical weighting process where the number of new believers becomes greater than the number of old believers.

The old believers do not go quietly into the night, however, in Kuhn's story. Many of them remain in the field and continue to defend the older views against the upstart encroachers until their numbers dwindle to the point where they have little influence on the community of scientists.

The parts of the cognitive revolution that do not seem to fit Kuhn's story have to do with the discovery of new and anomalous facts and the development of competing theory. As Miller (2003) describes in his remembrance, there were few facts around that were directly contrary to behavioral principles (whatever those were), and there were few (if any) explanatory systems around that were grand enough to be described as competitors to behavioral views. Rather, what seemed to happen was that psychologists of Miller's ilk simply found the ideas being entertained under the cognitive umbrella much more interesting than those they entertained under the behavioral umbrella. For instance, before his conversion to cognitivism, Miller describes his wrestling with writing a book on the psychology of language (Miller, 1951) while trying to remain true to the tenants of behaviorism.

This analysis of the cognitive revolution suggests that the revolution was based not so much on competition between anomalous facts and competing theories, but rather on the fact that many researchers and theorists simply became bored with the constraints imposed by behavioral views. Under the new umbrella of cognitivism scientists could entertain new and exciting ideas and new and exciting research agendas. Moreover, they could engage in discourse with researchers and theoreticians from other disciplines while pursuing their own interests.

Kuhn (1962) described how holders of older views did not quietly leave the field and that part of the analogy did seem to fit the view that psychology underwent a cognitive revolution. B. F. Skinner, in particular, was not pleased about the emergent, and soon dominant, cognitive perspective. In a 1984 article, Skinner argued that American education was in a shameful state and there were two principle "roadblocks" that contributed to this state: (1) psychological roadblocks in the form of cognitive psychology and (2) educational roadblocks promulgated by Schools of Education that turned out teachers totally unprepared to utilize scientifically proven techniques (he meant, of course, behavioral procedures) for promoting academic learning. Rather than being an advance in scientific reasoning, Skinner argued that cognitive thinking was a retreat to the nonscientific days of psychological speculation:

> Cognitive psychology is frequently presented as a revolt against behaviorism, but it is not a revolt; it is a retreat. Everyday English is full of terms derived

from ancient explanations of human behavior. We spoke that language when we were young. When we went out into the world and became psychologists, we learned to speak in other ways but made mistakes for which we were punished. But now we can relax. Cognitive psychology is Old Home Week. We are back among friends speaking the language we spoke when we were growing up. We can talk about love and will and ideas and memories and feelings and states of mind, and no one will ask what we mean; no one will raise an eyebrow. (pp. 949–950)

THE IMPACT OF THE COGNITIVE REVOLUTION ON EDUCATIONAL PSYCHOLOGY

Psychologists who studied human learning were the first to begin to march to the beat of the cognitive revolution. One of the noteworthy early accomplishments was the publication of the Atkinson and Shiffrin (1968) model that described the architecture of the human cognitive system along with the control processes that operated within the framework of that architecture. This seminal article spawned a considerable amount of research that examined the implications of a human memory system that contained hierarchically organized processing components and that could control its own functioning. The volitional control of learning processes was a far cry from behavioral perspectives where human learning was governed by stimulus events and consequences to responses made in the presence of those events.

Educational psychologists were a little late in joining the revolution. The field was still very much dominated by behavioral perspectives in the 1960s. For example, Skinner's summary statement on the application of behavioral psychology to education (titled *The Technology of Teaching)* was published in 1968. The contents of the present book, however, indicate how that was soon to change, and in some cases, the chapters illustrate how the ideas in the cognitive revolution were foreshadowed by ideas presented many years before.

The first chapter in the book is a marvelous example of showing that the roots of the cognitive revolution could be traced back to differing perspectives on epistemology as reflected in the thinking of eminent philosophers. In their chapter, Ralph Reynolds and Gale Sinatra review the historical record of three views of human knowledge that they call *experience-centered, mind-centered,* and a third perspective that involves an *interaction* between the other two. Reynolds and Sinatra then trace the impact of these perspectives on modern psychology, beginning with their role in shaping the thinking of early European psychologists and continuing into an analysis of American behaviorism and the subsequent development of cognitive perspectives.

After showing how epistemological issues contributed to the development of American psychological perspectives, Reynolds and Sinatra turn their attention to tracing the impact of these ideas in research on reading. They begin with a brief review of behavioral approaches to reading research and then turn to the impact of the cognitive revolution on the way in which reading was viewed and examined in the research literature. Their review of cognitive approaches to reading research begins with an analysis of basic reading issues like the nature of the memory representation that is produced by the process of learning to read, and then moves to higher-level concerns like the nature of the comprehension process and the impact of teaching reading strategies.

The third chapter in the book examines the history of the cognitive revolution through a lens provided by an examination of the research and writings of Richard C. Anderson, one of the most eminent educational psychologists of his era. James Royer, the author of the chapter, joined Anderson's lab in 1967 when Anderson was still very much a behaviorist. In the years from 1967 to 1970, he watched Anderson change his thinking to incorporate cognitive ideas, and in his chapter Royer notes that he has recently read every publication that Anderson produced up until the early 2000s. During that time Anderson's research and thinking underwent a number of additional changes that mirrored the changes that were occurring in that part of educational psychology that was concerned with learning and thinking. For example, early cognitive perspectives were very much focused on the individual learner. Over time, however, the focus shifted to a concern with the learner in groups, and Anderson's recent contributions to the literature on social cognitive perspectives on learning provide an example of that shifting focus.

Another way of examining the shift from behavioral to cognitive perspectives in educational psychology is to examine the content of popular textbooks and other teaching materials that were used in college courses concerned with theories of learning as taught to students interested in educational psychology. Donald Cunningham has been teaching courses in learning and cognition for 35 years and in Chapter 4 provides an analysis of the books and teaching materials that he used over this 35-year period. Interestingly enough, Cunningham was also a student of Richard Anderson and he provides his own recollections of the shifting theoretical sands as viewed from the perspective of Anderson's changing views of how educationally relevant learning should be explained and researched.

Cunningham then describes his move to Indiana University and the materials he used when teaching his own courses on human learning. He also describes some of the teaching techniques he used in those courses and ties his changing teaching techniques to changing perspectives in the field. Near the end of his chapter, Cunningham describes changes at his university that involve the merging of a number of different academic fields into a common umbrella called the learning sciences. This interdisciplinary approach to research and to graduate training that Cunningham describes is a concrete embodiment of Miller's vision in 1956 of an emerging set of theoretical ideas that could only be examined by viewing them from a variety of disciplinary perspectives.

The next two chapters in the book examine the impact of the cognitive revolution on disciplinary learning. Chapter 5 by Richard Mayer shows how changing views of cognition can contribute to an analysis of how to teach mathematics. Mayer's chapter begins by describing the ongoing debate on how and what to teach in mathematics classrooms—that is, whether to take a basic skills approach or a conceptual understanding approach (this debate has been labeled the math wars in some circles). The chapter then briefly reviews the historical roots of the debate, and goes on to describe three approaches to defining the objectives of mathematics education: one from the perspective of a mathematician who describes objectives in terms of subject-matter knowledge, a second from the perspective of a teacher who defines objectives in terms of instructional activities, and a third view from the perspective of a psychologist who talks about desirable changes in knowledge.

Mayer then describes four major contributions that cognitive science has made to the analysis of mathematics education. These contributions are cognitive task analysis, a description of the types of knowledge that are produced during mathematics learning, a cognitive psychology of learning that is specifically focused on mathematics learning, and a description of how learners build knowledge in their memories. Mayer concludes his chapter by arguing that the cognitive content he has analyzed can help resolve the debate about whether the basic skills approach or the conceptual understanding approach is the best way to teach mathematics.

Chapter 6 provides a comprehensive examination of the impact of the cognitive revolution on science learning and science teaching. The chapter, by Eugenia Etkina, Jose Mestre, and Angela O'Donnel, begins with a description of how science learning and teaching was viewed from a behavioral perspective that focused on observable behaviors. They then describe the fundamental shift that occurred when the field began to view the important outcome of science learning as understanding, an outcome that could not be directly observed. The view that understanding was the critical outcome in science learning put the idea of constructivism in learning to the forefront. Understanding is a process of relating new information to existing knowledge, and the manner in which new knowledge and old knowledge interact to form an amalgam of the two is basically a constructive process.

Etkina and colleagues trace the instructional implications of the constructivist perspective and then examine how research on expertise has shaped our views of science learning outcomes. They then move to an examination of a number of topics that are highly important in science learning including a discussion of the transfer of learning, the role of metacognition and self-reflection in science learning, and how assessment should be conducted in order to advance science learning. Having sketched the foundations of how cognitive thought has contributed to science learning, the authors move to a discussion of the instructional principles that can be derived from their analysis of basic issues. The principles they describe are in the form of a set of activities that should be encouraged and practiced during ideal science instruction.

Having described the principles underlying ideal practice, Etkina and colleagues examine real instructional programs that embody these principles. The chapter ends with a discussion of how teacher training in science should be conducted.

The first six chapters in the book are largely concerned with the impact of the cognitive revolution on learning and thinking, but the revolution had an impact in other areas of educational psychology as well. Chapter 7, by Frank Pajares and Dale Schunk, describes the impact of the revolution on theory and research aimed at understanding academic motivation. Pajares and Schunk take the perspective that modern academic motivation research revolves around processes related to the self. This was not always so, as they describe in the beginning sections of their chapter where they point out that precognitive views of motivation were dominated by three "forces." The first was the behavioral force, which asserted that an internal attribute such as self was anathema because it could not be directly observed. A second force that was relevant to understanding motivation derived from Freudian psychology, which described motivation as deriving from intrapsychic struggles between psychodynamic components such as the id and the superego. The third force was the humanist school championed by, among others, the powerful voice of Abraham Maslow, who insisted that motivation was best described by focusing on inner experiences, internal processes, and self-constructs.

Following the cognitive revolution, a number of lines of inquiry regarding academic motivation rose to prominence and Pajares and Schunk trace two in particular. One focused on constructs concerned with students' perceptions of their own abilities, or competence beliefs, such as self-efficacy, self-concept of ability, self-worth, and helplessness beliefs. Another line explored the role played by achievement goals in human decision making and task selection. These included achievement goal orientations, intrinsic motivation, and cooperative versus competitive goal beliefs and structures. Some lines of inquiry, such as expectancy-value theory and attribution theory, predated the revolution but retained their relevance to the present day. Pajares and Schunk close their chapter with an insightful analysis of what the future of theory and research on academic motivation will look like.

The chapters I have described thus far might be thought of as describing the attempts of educational psychologists to understand learning and motivation, and their attempts to develop procedures for influencing learning and motivation. The next two chapters describe attempts to actually embed learning and motivational principles in an educational context. Chapter 8, by Marcy Driscoll and Kerry Burner, describes changes that occurred in instructional design as a consequence of the cognitive revolution. They begin their analysis by defining instructional technology (it involves how knowledge about human learning can be put to use in the school and in the workplace), and then move to an examination of the body of work generated by Robert Gagné, the grand figure in the instructional design field from the period of the 1950s to the 1990s. Their analysis of Gagné's research and writings, which focuses on looking for evidence of when and how cognitive concepts began to influence his thinking, in a sense parallels the methodology of the analysis in Chapter 3 that examined the impact of the cognitive revolution as viewed from the perspective of the research and writing of Richard Anderson.

Following their analysis of Gagné's work, Driscoll and Burner turn their attention to changes that have been made to the Dick and Cary instructional design model, which is described as being the academic standard for students in the field of instructional design. The model is presented in textbook form and first appeared in 1978. Since then the text has been revised four times, with versions appearing in 1985, 1990, 1996, and the latest appearing in 2002. The Dick and Cary model leads the instructional designer through a series of steps (10 in the most recent version) that are designed to produce efficient and effective instruction. Driscoll and Burner show how the cognitive revolution has resulted in changes in the way the Dick and Cary model is communicated to the instructional designer, and they show how the revolution resulted in changes in the content of the model.

Having completed their analysis of classic perspectives on instructional design, and the changes that occurred in those perspectives as a consequence of the cognitive revolution, Driscoll and Burner turn their attention to recent directions in instructional design. Their analysis of recent directions is framed by the perspective that the field of instructional design can be divided into three worlds (van Merrienboer & Kirschner, 2001): (1) the World of Knowledge, (2) the World of Learning, and (3) the World of Work. Driscoll and Burner point out that the World of Learning and the World of Work perspectives are new directions for the field of instructional design, and their analysis shows how the cognitive revolution has influenced movement in the new directions.

Chapter 9, by Jennifer Wiley, Christopher Sanchez, and Tom Moher, continues with the theme of using psychological principles to directly influence learning, but their focus is on how instructional technology has been used to achieve this goal. The time frame of their analysis of instructional technology begins in the era dominated by behavioral psychology and continues through the recent period of social cognitive and constructivist perspectives. They begin their description during the behavioral era with a discussion of Sidney Pressey's teaching machine that was originally described in a 1926 publication, and then move to describing Skinner's efforts at developing teaching machines. Their analysis then turns to events occurring in the cognitive period that had an impact on the way instructional technology was designed and used. They point out that one of the major shifts that occurred was the replacement of a *transmission* model with a *constructivist* model. The goal of the first is to transmit a body of facts to students, whereas the goal of the later is to have students develop an understanding of the subject matter through a process of exploration and problem solving.

Wiley and colleagues describe how the shift to the constructivist model resulted in four basic implementations of instructional technology: (1) intelligent tutoring systems, (2) simulation or modeling environments, (3) project- or problem-oriented learning environments, and (4) participatory simulations. Following a discussion of each of these implementations, the authors then turn their attention to trends they see in the educational research on instructional technology over the past decade. The first of these trends is a shift from individual to group learning contexts. This trend is consistent with the shift described in a number of other chapters that involves moving from a cognitive perspective to a sociocognitive perspective. Another trend that Wiley and colleagues see is the shift in emphasis on cognitive activity to a shift in emphasis on metacognitive activity. Again, this is a theme that is represented in a number of the other chapters in the volume. Finally, the authors describe a trend away from creating activities to fit around technology, to a movement that embeds technology into meaningful learning activities. In many of the instances where the authors discuss global descriptions of instructional technology, they also provide a real-world context for their discussion by describing specific instructional technologies that embody the qualities they are describing.

The next chapter in the volume examines one of the oldest questions in scientific psychology; specifically, how we are to describe the manner in which early learning influences later learning. This, of course, is the question of how learning transfers, and Gary Phye's Chapter 10 provides a historical analysis of the way transfer has been viewed. Phye's chapter is developed around three themes. The first involves the identification of a philosophy of science that serves to guide the general study of adaptive human behavior, with the issue of the transfer of learning being a special case subservient to the more general issue. Phye's second theme is devoted to telling the story that traces the theoretical models and metaphors that have guided us through three distinctive eras in scientific psychology, beginning from the early 1900s and continuing into modern times. Phye's third theme is integrative and involves his efforts at combining three elements of analysis. First, a philosophy of science of human behavior. Second, competing conceptual models and metaphors that have been used to describe transfer and problem solving. Third, a reconsideration of the materials and experimental research methods in light of the other two analytical frames. The integration that Phye describes involves the combination of experimental and psychometric approaches to the study of transfer and problem solving—an integration that, Phye notes, has in the past been called for by some very eminent psychologists.

The most recent phase of the cognitive revolution, as has been mentioned in several places in this introduction, involves a shift from an analysis of the individual as a being who influences his or her own learning, to an analysis of the individual as part of groups with the groups themselves having their own qualities and characteristics that can influence cognitive performance. It is fitting, therefore, that the book end with an analysis of sociocognitive views of learning. That analysis is provided by Peter Freebody in Chapter 11. Freebody's frame of reference is the conjoined history of psychology, education, and cognitive science, and in his analysis of these three entities, he makes three arguments. First, he points out that the tenants of cognitive psychology were not created whole-cloth to replace the rejected tenants of behaviorism. Rather, some of the ideas in cognitive psychology were foreshadowed by psychologists who predated the behavioral era.

Freebody's second argument revolves around what he calls the moderate and strong use of The Social in cognitive theorizing. The moderate form, as described by Freebody, involves the notion that events external to the individual are "factors" that contribute to the cognitive development of the individual. This view assumes that social context is a kind of "main effect" that can be isolated and studied. The stronger view that Freebody argues for is a much more dynamic perspective where The Social is seen as (1) transformable, (2) both the project and the product of coordinated social actions, and (3) simultaneously and continuously shaping and renewed by group activities.

Near the end of Freebody's chapter he makes his third argument, which is that whereas there is a tendency to view the relationship between science and practice as flowing in one direction, increasingly it is bidirectional.

That is, as a consequence of trying to apply our science to practice, we have necessarily had to reexamine the science, and that, Freebody argues, has pulled us in the direction of paying more attention to social influences in our theorizing.

SPECULATIONS ABOUT FUTURE DIRECTIONS

The story that is told from varying perspectives in the chapters in the book involves moving from the relatively simple, but at the time earthshaking, idea that humans can shape their own learning to perspectives that are increasingly complicated and dynamic. It is clear that increasingly the cognitive revolution is evolving into the social cognitive revolution, or at least it is in some areas of educational psychology. Our theorizing is certainly influenced by the sociocognitive perspectives, as is many aspects of our research. For example, we increasingly see researchers involved with "design experiments" where the researcher is trying to package a whole set of instructional and social manipulations in a dynamic mix that makes it very difficult to sort out individual contributions to cognitive change. The argument is that is the way it should be. As human beings in a real world we soak up knowledge and information in ways that rarely allow an analysis of where particular bits of knowledge and beliefs came from.

However, there are areas of educational psychology where socio-cognitive influences seem to have had little impact. Measurement is one. We have remarkable theoretical claims, sometimes accompanied by some empirical research, about the advances in learning that occur when we view learners as participants in learning communities. However, these claims are difficult to document with tests that meet standard psychometric requirements, and we do not seem to be making much progress in developing tests that evaluate outcomes that theorists tell us are being produced.

Another area where we have difficulty in incorporating sociocognitive perspectives is educational technology. In theory at least it should be possible to use technology to enhance and expand learning communities. But perhaps there is something special about direct human interaction that will not tolerate technological intervention and it remains to be seen whether sociocognitive theorizing will have much of an impact on educational technology.

Another change in our science that is still in progress involves methodology and data analysis. When the cognitive revolution began the gold standard for experimental design was the multifactor experiment that involved random assignment of participants to treatment and control groups. Most researchers nowadays are comfortable with a mix of qualitative, quantitative, and case study procedures that they employ as needed. Moreover, it is increasingly rare to see research reported that involves random assignment of participants to treatments and the utilization of analysis of variance procedures to analyze data.

At this point in time it appears that we are unlikely in the future to see the kind of field-shaking events that occurred when the cognitive revolution began. Rather, it appears that we are in a period where we are steadily and systematically advancing our understanding of psychology related to educational practice, and we are busy applying what we learn to actual educational practice. I could certainly be wrong though, and there may be another revolution just over the horizon.

REFERENCES

Atkinson, R. C., & Shiffrin, R. M. (1968). Human memory: a proposed system and its control processes. In K. W. Spence, and J. T. Spence (Eds.), *The psychology of learning and motivation: Advances in research and theory* (Vol. II, pp. 89–115). New York: Academic Press.

Kuhn, T. (1962). *The structure of scientific revolutions.* Chicago: University of Chicago Press.

Miller, G. A. (2003). The cognitive revolution: A historical perspective. *Trends in Cognitive Sciences, 7,* 141–144.

Pressey, S. L. (1926). A simple apparatus which gives tests and scores - and teaches. School and Society, 23, 373–376.

Skinner, B. F. (1984). The shame of American education. *American Psychologist, 39,* 947–954.

van Merrienboer, J. J. G., & Kirschner, P. A. (2001). Three worlds of instructional design: State of the art and future directions. *Instructional Science, 29,* 429–441.

CHAPTER 2

THE COGNITIVE REVOLUTION IN SCIENTIFIC PSYCHOLOGY

Epistemological Roots and Impact on Reading Research

Ralph E. Reynolds and Gale M. Sinatra
University of Nevada–Las Vegas

ABSTRACT

We examine the epistemological and philosophical roots of modern American psychology in this chapter. In particular, we examine the influence of different epistemological stances on the cognitive revolution in American scientific psychology and on research and theory in reading. We argue that basic process reading research and reading comprehension research have emerged from different epistemological standpoints, leading to a divergence in their research orientations. Finally, we call for the next movement of the cognitive revolution to include concepts such as true conscious control, internally initiated learning, and intentional actions. These constructs are necessary if reading researchers are to understand the multiple levels of the reading process. Tenets from Kantian and Cartesian epistemologies and evolutionary psychology are used to provide theoretical support for the inclusion of intentional concepts into learning research.

The Impact of the Cognitive Revolution on Educational Psychology, pages 13–39

In this chapter, we address and integrate four separate, but related, issues relevant to the development and application of modern American psychology. First, we investigate the impact on psychology of insights gained from work in epistemology (the philosophical study of the origins of and nature of knowledge). In particular, we focus on how psychologists have understood the relative contributions of sensory input and the mind to knowledge acquisition, both historically and currently. We focus on this issue because even though they are seldom acknowledged or even understood, psychology's epistemological roots have at times heavily influenced how we view learning and knowledge. Second, we examine the cognitive revolution in American scientific psychology from the standpoint of epistemology. For example, what appeared to be a dramatic reorientation in American psychology was simply a return to a better warrented epistemological position.

We then examine the impact of epistemological views on research and theory in reading, both before and after the cognitive revolution. It is our view that after the cognitive revolution, research and theory in reading implicitly developed in line with two different epistemological views. One view is more consistent with the research and theory done concerning the acquisition of basic processes in reading, and the other more consistent with research and theory concerning reading comprehension. Our fourth issue involves establishing an epistemological and evolutionary case for why inclusion of concepts such as true conscious control, internally initiated learning, and intentional actions are necessary if reading researchers are to understand the complexity of the reading process.

We begin with a brief review of the epistemological foundations of two psychological meta-theories: behaviorism and cognitivism (Baars, 1986). It seems appropriate to begin with a look at the philosophical study of epistemology because it predates the founding of the discipline of psychology by almost 2,400 years. Epistemologists have the goal of answering a disarmingly simple question—from where does knowledge/understanding come? Two thousand years of philosophical inquiry has demonstrated that the answer to this question is anything but simple. We discuss only three general exemplars proposed by philosophers to answer the knowledge question. This will simplify and reduce what is an enormous body of information. These three approaches are: experience-centered, mind-centered, and interactive—with interactive approaches varying according to the relative contribution of the mind in creating new knowledge (Reynolds, Sinatra, & Jetton, 1996).

Experience-centered approaches promote the idea that the five senses are the only sources of human knowledge. On this view, the mind exists but is seen as totally passive. Learning is seen as externally initiated via contact with the senses. The mind-centered approach suggests that true knowledge is the result of reason and reflection, properties of the mind. Here, the mind is seen as an active participant in learning and the construction of knowledge. Indeed, from this perspective, the mind is seen as the only source of true knowledge. Not surprisingly, the third approach involves the senses and the mind working together to produce knowledge. On this view, learning is still primarily initiated via contact with the senses; however, the mind plays a more active role in the process (Reynolds et al., 1996). Explanations of the exact nature of the role of the mind have varied over time, ranging from a relatively mechanical role suggested by Aristotle to an active, shaping role hypothesized by Kant (1781/1990).

EPISTEMOLOGICAL FOUNDATIONS

Early Epistemologies

The earliest Greek writers held experience-centered views on how human beings acquired knowledge. Their key insight was that all thoughts, concepts, and ideas were ultimately derived from sensory input. It was thought that sensory information was sent to the brain where higher-level knowledge emerged through interaction with the mind/soul, called *Nous*. It was hypothesized that human beings were endowed with a mind/soul at birth. Two questions are implicit in this approach: how does sensory information get to the brain and how do higher-level ideas emerge?

Two early Greek philosophers who wrote on these issues were Alcmaeon (520 B.C.) and Protagoras (485 B.C.–411 B.C.). Alcmaeon proposed that all knowledge came from the senses and traveled through air channels to the brain. It is interesting to note that Alcmaeon had deduced that air was the conveyor of sensory information even though his dissections of animals never revealed any "air tubes." This is neither the first nor last instance of conflicts in understanding based on philosophical or ideological deduction and empirical observation. Later Greeks simplified the system by seeing the nerves and spinal cords as sensory conduits instead of Alcmaeon's "air." The mind/soul somehow allowed human beings to create from these simple sensory inputs higher-level concepts. It is important to note that it was never clarified exactly how the mind contributed to knowledge acquisition. The emphasis was always on sensory input; hence, the designation of this approach as experience-centered.

Protagoras understood the immense philosophical implications of Alcmaeon's proposals. That is, if senses were the only sources of knowledge, then it would be impossible to derive any general, much less absolute, truths because all knowledge would be dependent on each individual's ability to sense accurately/similarly. This is basically the same argument that relativists and postmodernists use today to argue against the existence of objective knowledge. Whether or not humans can ever gain absolute understanding or totally objective understanding has remained an important topic of discussion for philosophers to this day. Protagoras's famous comment, "Man is the measure of all things," was meant to illustrate this point.

The idea that human knowledge was subject to the vagaries and inaccuracies of human sense organs and perceptions was unacceptable to Plato (427 B.C.–346 B.C.). Instead, he suggested that true knowledge consisted solely of concepts and abstractions arrived at through reason. Indeed for Plato, the mind was the only necessary actor in knowledge creation. Plato viewed anything physical, particular, and mortal as by definition flawed. This specifically included all sensory inputs processed by the human brain. In essence, Plato deemed all sensory input as unreliable because it would vary from person to person. The only true knowledge came from reason; hence, it must be wholly derived from the mind. Plato summarized his ideas by noting that both the wise man and the fool perceive the world, but the wise man's judgments frequently had better outcomes. Plato called these ideas the Theory of Forms. The major application to instruction of the Theory of Forms is the Socratic Dialectic. The Socratic Dialectic is a teaching method in which probing questions are used to help learners arrive at knowledge through the application of reason. Plato discusses his epistemology in several dialogues: *Meno, Theaetetus, Apology, Phaedrus, Timaeus, Republic,* and others. The Theory of Forms represented a mind-centered approach to epistemology.

Aristotle (384 B.C.–322 B.C.) studied with Plato for about 20 years and followed him as the leading thinker of the Greek world. Aristotle disagreed with Plato's epistemology. Where Plato had relied on deductive reasoning to come to his epistemological conclusions, Aristotle was the first major empiricist. He observed, collected, experimented, and then induced concepts and principles. Rather than regarding sensations as irrelevant, he regarded them as essential building blocks for knowledge even though they were sometimes flawed. Aristotle suggested that the mind grouped incoming sensory information into general principles, categories, or "universals." This grouping was accomplished rather mechanically on the basis of the temporal and frequency relations among incoming sensations. Universals were the path to higher levels of knowledge and ultimately, wisdom. Finally, Aristotle foreshadowed the idea that incoming sensory information was grouped in terms of simple associations. He stated,

> Whenever we try to recollect something, we experience certain antecedent movements [memories] until finally we come to the one after which comes the one we seek. This is why we hunt up the series, having started in thought either from present intuition or some other, and from something either sim-

> ilar or contrary to what we seek, or else from that which is contiguous to it. (as cited in Hunt, 1993, p. 32)

There are serious conceptual problems with each of these approaches, even beyond the obvious misunderstandings of human physiology (e.g., Aristotle thought the brain was an organ that cooled blood and the heart was the seat of knowledge). Specifically, experience-centered approaches lacked the ability to explain how human beings can process two series of perceptions and conclude that one represents a causal situation while the second is a random series. Also, these approaches cannot adequately explain how higher-level concepts and principles are derived from basic perceptions. Mind-center approaches cannot explain the knowledge we acquire from experimentation and observation. As Reynolds et al. (1996) stated,

> ...if all knowledge is derived (remembered) from the mind, how is it that human beings can discover new facts such as ... knowledge that Io, one of Jupiter's four largest moons, is volcanic; or knowledge that titanium improves the tensile strength of steel. (p. 94)

This type of knowledge certainly was not already in the mind awaiting the probing questions of a Socratic teacher to bring it to awareness. Aristotle's interactive epistemology came closest to beginning to explain how human beings acquired knowledge. Yet, explanations of exactly how the mind created universals and high-level knowledge were neither clear nor explicit. However, we should not be too critical. These early Greek epistemologists defined the terrain for discussion in this area for the next 2.5 millennia.

A period of relative intellectual dormancy followed the flurry of epistemological insights by the early Greeks. Rene Descartes (1596–1650) was the next figure who made major contributions to philosophy, epistemology, and pre-psychology. Many philosophers anoint his work on the mind–body issue as his major accomplishment. Questions about whether and how the mind and the body interact have long been a staple of epistemological discussions. The debate centers on the idea that the body exists in the physical world and the mind exists in the metaphysical world. How then is it possible for them to interact and initiate responses in each other? Clearly, conventional physical concepts such as mass and force are not at work in this interaction.

Using deductive reasoning as his method, Descartes concluded that the mind and the body could and did influence each other. A demonstration of the utility of this conclusion is the practice of modern psychiatry. Psychiatrists can affect mental states and capacities through pharmacological manipulations of the body. Conversely, mental stress has well-documented physical ramifications for some individuals. This insight is of great importance to modern psychology in that it establishes the principle that the mind can and does affect the body. As such, it provides an epistemological foundation for the validity of addressing concepts like internally initiated behaviors, intentional cognitive actions, and volition.

Unfortunately, Descartes's main impact on American psychology came neither from his work on the mind–body issue nor from his epistemological work on derived and innate ideas. Instead, it came from another area of his work: the idea that the human body and mind work like machines. His primary metaphor was that of a clock. And, like a clock, the best way to understand its inner workings was to reduce them to their smallest components. This insight led to the doctrines of mechanism and reductionism, which are still with us today.

British Empiricism and Pre-Psychology

The works of Hobbes (1650), Locke (1690/1959), Berkeley (1709/1957), Hume (1739), Hartley (1749), and Mill (1829) bear most heavily on epistemology and psychology. This group was often called the "British Empiricists" for their penchant for observation and experiment. We focus on John Locke's contributions to epistemology rather than try to do justice to the entire group.

John Locke (1632–1704) has been called the father of British empiricism. He was best known for his work in social and political philosophy. Indeed, John Adams and Thomas Jefferson acknowledged their debt to Locke's writings on democracy and social equality in their own thinking (Ellis, 2000). Locke's only well-known work in epistemology was *An Essay Concerning Human Understanding* (1690/1959). One of his goals was to discover the extent of human understanding. His political philosophy required that he achieve this goal without resorting to the idea that human beings were endowed by God with innate capacities for understanding. Acknowledging innate abilities would have required Locke to entertain ideas that might suggest that some human beings (such as priests and aristocrats) might be endowed with special abilities that others (commoners) did not possess.

The essence of Locke's epistemological thinking was based on three premises: (1) the mind was *not* endowed with innate ideas, (2) the mind worked according to basic mechanical (as opposed to metaphysical) principles, and (3) all knowledge was derived from sensory input. Locke suggested that the human mind had no capabilities other than the ability to combine and compound sensory input. He used the word *reflection* to describe this process; however, he was describing a strictly mechanical process that was not under conscious control—more like apperception than the conscious, self-controlled connotation the word has today.

He supported his notion of the sensory origin of knowledge by proposing that the sum of human knowledge consisted of two types of ideas: simple ideas and complex ideas. Simple ideas were thought to arise from sensory input and perception. These ideas were elemental and arose without conscious awareness or effort. Complex ideas were formed from simple ideas and reflection. Locke proposed that complex ideas were created by associating simple ideas with reflection; however, to Locke this was a matter of mechanics, not inspiration. Locke considered the formation of complex ideas as human learning. He summarized his approach as follows:

> Let us then suppose the mind to be, as we say, white paper, void of all characters:—How comes it to be furnished? Whence comes it by that vast store which the busy and boundless fancy of Man has painted on it with almost endless variety? Whence has it all the materials of reason and knowledge? To this I answer, in one word, EXPERIENCE. In that all of our knowledge is founded; and from that ultimately derives itself. (Locke, 1690/1959, p. 3)

Note that Locke was much more radical in his stance than were his experience-centered forebearers. The early Greeks thought that sensory data was the raw material of knowledge but that the human mind/soul allowed for reason. For Locke, it was simply a matter of the sense organs sending sensations to the brain where they were mechanically combined. This concept of mental combinations of ideas marks the beginning of Associationism.

British Empiricist thought strongly and directly influenced American behaviorism. Hume's (1739) and Hartley's (1749) ideas of contiguity and repetition can be seen in Watson's work (Watson, 1913), while a clear mechanistic worldview is apparent in the work of E. L. Thorndike and B. F. Skinner (Skinner, 1957; Thorndike, 1931). Behaviorists took an even more radical mechanistic approach than their British Empiricist forebearers. Behaviorists suggested that understanding, learning, and the acquisition of knowledge was a relatively simple matter. Empirical work was required to identify the proper organization of stimuli that would maximize observable learning outcomes. Discussion of the role of the mind was deemed metaphysical and hence, inappropriate for real scientific work. Behaviorists removed the mind and all mentalistic terms from the vocabulary of scientific psychology to emphasize this point.

Before leaving the realm of epistemology, it is important to consider the work of Immanuel Kant (1724–1804). Kant's work has been influential in both psychology (as the basis for Gestalt movement in psychology) and linguistics (as one basis for Chomsky's [1959] scathing attack on Skinner's explanation of verbal behavior). Kant took an approach to epistemology that was quite different from Locke's. As he states in his seminal work, the *Critique of Pure Reason*:

> Experience is by no means the only field to which our understanding can be confined. Experience tells us what *is*, but not that it must be necessarily what it is and not otherwise. It therefore, never gives us any really general truths; and our reason, which is particularly anxious for that class of knowledge, is roused by it rather than satisfied. General truths, which at the same time bear the character of an inward necessity, must be independent of experience—clear and certain in themselves. (1781/1990, pp. 182–183)

Kant was suggesting that while experience—externally initiated sensory input—was the raw material of knowledge acquisition and cognition, it was necessary but not sufficient to explain the complexity of human knowledge and learning. Kant suggested that the nature of the human mind was such that it played a role in framing, organizing, and extending that experience. Kant concluded this strand of thought by suggesting that there were some aspects of human knowledge and learning—an understanding of cause and effect relationships, for example—that could not arise through the simple accretion of externally initiated experiences.

Instead, the understanding of cause and effect relationships had to occur through an interaction between elements of sensory experience and the internal framing and organizing aspects of the mind. To Kant, the mind was an active variable in human learning. Human knowledge was more than the simple sum of experiences. Kant could not explain how the mind came to possess innate framing and organizing capacities. He deduced their existence because they were necessary to explain human knowledge. It would have to wait for the advent of evolutionary psychology (Plotkin, 1998) before we could gain some insight as to why the mind possesses these innate traits.

A final aspect of Kant's work warrants mention. Kant investigated the idea of mental representation of knowledge well before the cognitive revolution and modern computers. Evidence of this can be seen in his discussion of the schematization of knowledge. To Kant, schemata were mental constructions that were not, and could not, be derived from sensory data. Instead, they were "...pure conceptions of understanding..." that organized and shaped existing and incoming sensory experiences. To Kant, schemata were not simply items of experience ordered associatively or temporally by a mechanistic mind. They were instead products of the active organizing aspects he suggested were inherent in the human mind. Kant's views were far different from Aristotle's, though both are considered advocates of interactive approaches.

In conclusion, epistemological theories have, over time, oscillated among three approaches: experience-centered, mind-centered, and interactive. The mind/soul played a role in thought and reason in reaction to sensory input for the early Greeks. Descartes and Kant endowed the mind with innate ideas and *a priori* framing and organizing capabilities; thus, giving the mind a necessary, but not sufficient, role in knowledge acquisition and creation. However, by the end of the British Empiricist era, the mind had become a simple machine powered by sensory input and equipped with only mechanistic associative and temporal capacities. It was this view of the mind, rather than the more established (in epistemology) views of Kant and Descartes, that emerged as dominate in American psychology during the Behaviorist era and beyond.

EUROPEAN PSYCHOLOGY AND EPISTEMOLOGY

We now turn our attention to epistemological influences on the new science of psychology. Wilhelm Wundt (1832–1920) founded the science of psychology in 1975 or 1979 (accounts differ). He drew on British Empiricists and German physiologists for his initial thoughts on psychology. He accepted the need for rigorous scientific methodology, empiricism, and aspects of Associationism. However, he also embraced the ideas of Kant (1781/1990) and Mill (1829) in thinking that the mind played an active role in organizing its own content. Wundt declared that the study of consciousness and the contents of the mind were to be the focus of the new field.

Many of Wundt's more cognitive ideas flourished in Europe. For example, Franz Brentano (1838–1917) was a forefather of both Gestalt and humanist psychology. Meanwhile, E. B. Titchener (1867–1927) brought to America only those aspects of Wundt's psychology that emphasized external experience and scientific rigor—though Titchener continued to use mentalistic terms. However, Titchener's definition of consciousness as the sum of our present (sensory) experiences and the mind as the accumulation of our lifetime of (sensory) experiences suggests movement away from the idea of the mind as an independent contributor to learning. Indeed, these positions are strongly reminiscent of the experience-centered epistemologies of Locke and the British Empiricists. American psychology completed its movement away from the idea that the mind had a necessary role in knowledge acquisition with the advent of behaviorism (Watson, 1913).

THE COGNITIVE REVOLUTION IN AMERICAN SCIENTIFIC PSYCHOLOGY

The dates for when the cognitive revolution actually began vary depending on the source. One starting date that is widely discussed is 1959. Noam

Chomsky, a little-known linguist at the time, published his review of Skinner's book *Verbal Behavior* in the journal *Language* that year (Chomsky, 1959; Skinner, 1957). The review devastated behaviorism by showing that the complexities and consistencies in language and language use could not be explained as the sum of sensory experiences. It is important to note that part of Chomsky's argument was based on Kant's notion of innate capacities of the mind.

Others have argued that the founding of the Harvard Center for Cognitive Studies in 1960 by George Miller and Jerome Bruner was the beginning point for the cognitive takeover of psychology. Miller described his feeling at the naming and opening of the center:

> To me, even as late as 1960, using "cognitive" was an act of defiance. It was less outrageous for Jerry [Bruner], of course; social psychologists were never swept away by behaviorism the way experimental psychologists had been. But for someone raised to respect Reductionist science, "Cognitive Psychology" made a definite statement. It meant that I was interested in the mind—I came out of the closet. (cf. Hunt, 1993, p. 513)

However, even in his own words, it is not clear exactly what Miller meant by the term "mind." Was it the passive, mechanical, reactive mind of Locke or the interactive, organizing mind of Descartes and Kant? Even today, the answer one would get to this question might depend on the cognitive psychologist queried.

The cognitive revolution was built on three major insights—none of which were based on epistemological concepts of the mind: (1) computational theory and the computational metaphor, (2) the notion of mental representation, and (3) the rise of the role of theory in scientific psychology. An added benefit was that psychologists were allowed to again use mentalistic terms—mental representation, mind, consciousness, etc.—to describe results of empirical studies.

Computational Theory and Representation

The term "computational theory" represents the mathematical ideas that led to the development of modern computers. In the early 19th century, mathematicians set about clarifying and specifying the foundations of mathematics. One outcome of this effort was the realization that any mathematical system could be expressed via symbolic logic. A second outcome was the understanding that symbolic logic in turn could be represented by electrical circuits (using Boolean set theory). Once it was demonstrated that information could be represented electrically, it became theoretically possible to create machines that could compute any mathematical formula or logical expression (Baars, 1986).

The reality of representation was further enhanced by a thought experiment performed by British mathematician Alan Turing (1950). Turing imagined a simple machine that could make marks on a tape, read them, and erase them. He suggested that with an unlimited supply of tape, this simple machine could compute any conceivable symbolic relationship. In essence, the modern digital computer is an extremely large, fast Turing machine. Taken together, computational theory and Turing's demonstration of representation formed one important basis for the cognitive revolution. Also, they gave cognitivists their signature metaphor for how human beings acquire and represent knowledge—the modern digital computer.

The demonstration that a physical form of mental representation was possible was extremely important for cognitive psychology. It meant that theorists and researchers could now hypothesize different types of mental representations for knowledge and put them to the test. Much of the early empirical work in cognitive psychology was focused on this issue. Also, the idea of representation made possible the creation of explanatory theories that could hypothesize the existence and role of mediating cognitive variables that were neither observable nor directly measurable.

The Resurgence of Theoretical Psychology

Perhaps the most important contribution of the cognitive revolution was that it allowed for the resurgence of theoretical thinking in American experimental psychology. A major tenet of behaviorism was that explanations for data must be made only in terms of observable phenomena. This was very restrictive because it did not allow psychologists to go beyond environmental events (things they saw, heard, touched, etc.) to explain why learning occurred or did not occur. Without permission to go beyond this restrictive view, American learning scholars stagnated theoretically for almost 50 years (see Hull, 1952, for a notable exception). Indeed, American psychologists were considerably less theoretically sophisticated than their European counterparts in the early 1960s when the cognitive revolution began. As Plotkin (1998) suggested in his book, *Evolution in Mind: An Introduction to Evolutionary Psychology*

> ...one of the remarkable features of most, or perhaps all, sciences is that they rely heavily upon causal explanations that invoke causal elements that are not part of normal life and everyday experience. Not only is there no scientist who denies the existence of atoms, but we now invoke the constituent parts of atoms to give us causal explanations for why sodium and chlorine

> combine in the way that they yield ordinary table salt. What behaviourism did was forbid causal explanations if they did not lie within the limits of ordinary everyday experience. Scientifically, this was an extraordinarily bankrupting stance. Cognitivism rescued psychology from this cripplingly narrow vision. It liberated psychologists conceptually and allowed causal powers to be relocated in the mind and the brain, only a very small part of whose workings are visible. (p. 33)

In conclusion, the cognitive revolution changed much in American psychology; however, it turns out that it was not much of a revolution. It was clearly cognitive in the sense that psychologists became free to pursue ideas such as knowing, thinking, perceiving, conceiving, and processing without the requirement that they be observable. Specifically omitted from the new *zeitgeist* was any reference to ideas concerning the effects of emotion, volition, and intention in relation to knowledge acquisition. Baars (1986) amplifies the first point,

> The cognitive revolution was not a spectacular one—no one stormed the Winter Palace, not even metaphorically. Public fireworks were rare. Furthermore, unlike Watson's revolution, the cognitive shift was not self-conscious. No one announced its existence until long after the fact. Experimental psychologists did not set out to make a revolution. Many of them were surprised to find themselves breaking the prohibitions of the behavioristic metatheory, and did so reluctantly. (Baars, 1986, p. 1)

Despite the proliferation of mentalistic terms in the new cognitive literature, most American psychologists continued to implicitly adhere to basic British Empiricist epistemology. In fact, the cognitive revolution was not overly concerned with the role of the mind in an epistemological sense as some assume. To many, knowledge is still thought to originate solely from the senses. Recall that the dominant metaphor for cognitive psychology was and is the digital computer. The computer provided a model from which to study how sensory data passed through a cycle of mechanical processing and emerged as higher-level ideas. Little discussion of the role of the mind was needed because digital computers do not possess one, in anything more than a mechanical sense. However, neither do digital computers possess the innate capacities that Descartes and Kant suggested were necessary to fully understand the nature of human knowledge. Human information processing (HIP) models became the instantiation of the new cognitive orientation toward knowledge.

Human Information Processing

The emergence of HIP models in the late 1960s and early 1970s helped establish the main principles of cognitivism. Virtually all of these processing models can be traced to the early work of Atkinson and Shiffrin (1968); however, many variations and improvements have appeared over the years (see Andre & Phye, 1986). It should be noted that Atkinson and Shiffrin included the concept of executive control in their early model. However, it was not clear how executive control worked or whether it reflected a mind that was logical or mechanical in nature. Indeed, some subsequent modelers removed the concept of executive control altogether. One such early model is presented in Figure 2.1.

The model began with sensory input that was briefly stored in sensory memory. Sensory memory was associated with all of the senses and provided a brief repository for sensory data—just long enough for it to be processed. As the sensory data moved into short-term memory, selective attention was applied and patterns were recognized. Information was further processed to make it ready for long-term memory or a response in short-term memory. Encoding was the process of making information more memorable via elaborative processes, and later relating it to existing knowledge. It is important to note that in these early models, the incoming information was not necessarily changed by existing information, it was just associated with it. Working memory was an aspect of short-term memory in that it involved the conscious processing of information. Long-term mem-

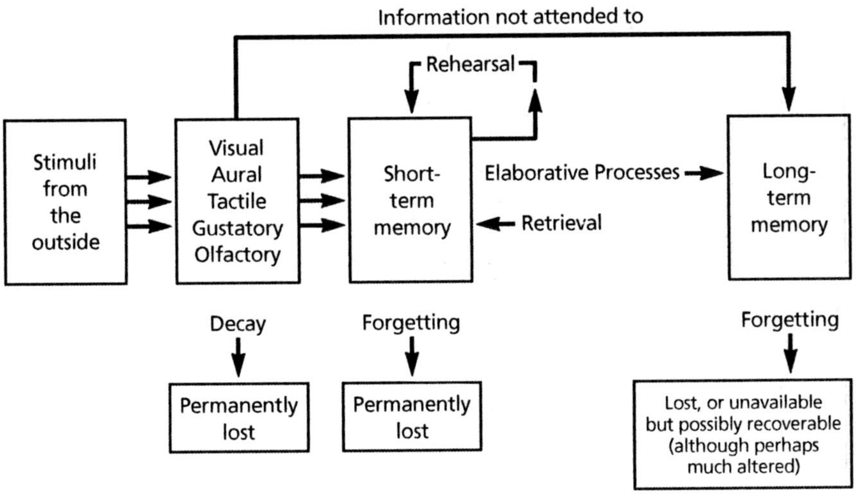

Figure 2.1. Human information model (adapted from Hunt, 1993).

ory was composed of semantic representations of the encoded data.HIP models are almost direct representations of Locke's views; hence, they are modern representations of an experience-centered epistemology. The HIP model gives more detail to how information is processed (e.g., identifying four aspects of memory and the concept of encoding) but allows for little or no role for a Kantian mind. As with the British Empiricists, there is little explanation about how basic sensory input becomes higher-level ideas and concepts.

Even given these limitations, several important ideas have emerged from HIP models that have had an immense impact on how psychologists conceptualize knowledge and knowledge acquisition. Key among these are representation, differentiated memory, encoding, automaticity, selective attention, and the notion of processing information. The idea of automaticity is particularly important.

There is a distinction made in HIP models between automatic and conscious processes. Processes or tasks are said to become automatic when they are overlearned and require very little attention to be engaged. Conscious processes are those that draw increased amounts of attention because of their salience to completing or performing tasks or procedures. Driving a car is the classic example used to delineate these two processes. Driving a car is mostly an automatic task under normal driving conditions and given that one is sufficiently practiced. While driving, a practiced individual can shift gears, engage turn signals, and obey traffic signals while still listening intently to the radio. Recent evidence suggests that the same cannot be said for talking on cell phones. If something occurs that demands the driver's attention (perhaps an accident); the driver's attention will be drawn to the accident. He or she may still be able to shift gears but will likely not be able to listen intently to the radio.

For most HIP psychologists, attention is seen as a fixed and limited cognitive resource (e.g., Kahneman, 1973; Posner & Boise, 1971, Reynolds & Anderson, 1982). Attention is fixed in the sense that it cannot grow to meet additional demands. It is limited in that some tasks may demand more attention than any given learner has available. Selective attention is the ability to attend to some sensory inputs while simultaneously not attending to others. For example, a dedicated heavy metal music aficionado may selectively hear the bass line of a song to the exclusion of the rhythm guitarist or even the lead singer. Behaviorists and many early cognitivists thought that it was some quality of the stimulus or the task demands that oriented attention. To return to the driving example, the suddenness of an accident might reorient the drivers' attention away from the radio. Defined this way, HIP models, which emerged from the cognitive revolution, are very consistent with an experience-centered epistemological view of the mind.

BEHAVIORISM AND READING INSTRUCTION

It is necessary to investigate reading research and reading instruction before the cognitive revolution in order to fully understand its impact on the study of reading. American psychologists have been interested in understanding the nature of the reading process at least since Huey published *The Psychology and Pedagogy of Reading* in 1908. However, the impact of psychological research on reading instruction was initially minimal. This began to change just at the end of the behaviorist era. An interesting irony of behaviorism is that just when it was being replaced by cognitivism as a general approach to understanding learning, it was experiencing a rebirth of sorts in clinics and classrooms (Baars, 1986).

At that time, clinical psychology was dominated by psychoanalytic approaches that required significant investments of time and money on the part of clients, even though successful outcomes occurred only sporadically. A similar situation existed in classrooms. The dominant approaches to teaching reading at the time focused on students' creation of meaning from text rather than the processing of text elements leading to comprehension. However, some critics (Chall, 1967; Flesch, 1955) claimed that the success rate was low and the cost was high. Behavioral instructional methods and intervention methods emerged as attractive alternatives to the dominant approaches because of their stress on scientific rigor and their promise of success verified by observable, testable outcomes. Also, behaviorism had a charismatic proselytizer in B. F. Skinner to deliver the message.

Behavioral instructional methods for reading were centered on reducing to-be-learned tasks to their smallest components, and repeated, active drill and practice on these components until they were learned to mastery. Immediate feedback to errors was given as well as immediate reinforcement for correct responses. The teacher's role in this instructional scheme was to reduce complex lessons to simple component parts, arrange materials so that students learned these small units effectively, and provide feedback and reinforcement. The translation of the behaviorist ideal to reading instruction may have reached its pinnacle with the development of reading programs such as DISTAR and Reading Mastery.

It is important to note that behaviorism was seen to have had highly negative effects on reading instruction by classroom teachers and teacher educators (Sharp, Reynolds, Brown, Morris, & Gunn, 2004). For example, behavioral instructional methods worked well with the small specifiable units inherent in many phonics curricula. They did not work well with the broader, more abstract units needed for a meaning-based reading curriculum. This coincidence linked phonics curricula with behaviorist teaching methods permanently in the minds of many teachers—much to the detriment of phonics curricula. Also, likely linked in teachers' minds because of behaviorism were: (1) scientific psychology with explanations of learning that were simplistic and did not explain higher-level learning, and (2) scientifically-based teaching methods that systematically removed the teacher from curricular and instructional decisions.

If these types of links were indeed made by teachers and teacher educators, it may help explain why the meaning- and student-centered philosophies of Goodman (1971) and Smith (1971) found such a favorable reception with reading teachers. It did not seem to matter that these theories were not validated by scientific research, were based on erroneous assumptions about the reading process (e.g., reading is natural), and were inconsistent with other disciplinary understandings of reading (Stanovich, 2003).

THE EFFECTS OF THE COGNITIVE REVOLUTION ON READING RESEARCH

We discuss the effects of the cognitive revolution on reading research from two perspectives: (1) research and theory concerning the basic processes of reading and (2) research and theory concerning reading comprehension. These two areas of reading research seem to have developed somewhat separately over the years, particularly with reference to the utility of HIP models and implicit conceptions of the role of the mind in knowledge acquisition and creation (Reynolds, 2000). We do not attempt to summarize all of the major research contributions to reading acquisition and reading comprehension. Instead, examples have been selected to illustrate concepts relevant to the present discussion.

Basic Process Reading Research

Research on the basic processes of reading has significantly advanced our understanding of how young readers acquire the necessary skills to read successfully. One reason this has been accomplished is that models of the reading process have been created that are directly based on HIP conceptions. Much of this progress has come since the onset of the cognitive revolution. It is directly related to information-processing theory and reflects substantial use of the computer metaphor. The concepts of automaticity, selective attention, and mental representation have been particularly important to researchers engaged in this work.

Basic processes research has shown that young readers must acquire the ability to automatically identify words if they are to become successful readers (Gough, 1971; Perfetti, 1985; Rayner & Pollatsek, 1989; Stanovich, 1980). Perfetti's verbal efficiency theory (1985) is perhaps the most complete explanation of the need for automaticity in word identification. Like many of his predecessors, Perfetti suggested that attention was a fixed and limited resource. It could be selectively oriented toward stimuli based on their immediate relevance. Attention not needed for initial processing was oriented toward or cascaded onto other attention-demanding tasks.

Using this rationale, Perfetti correctly predicted that inefficient use of attention at the word identification level of reading would lead to inadequate attentional resources remaining for use in the act of comprehension. Hence, good comprehension is less likely to occur when word identification is not automatic. Much empirical work has supported this prediction; however, all of it to date has been correlational in nature. This precludes any cause and effect conclusions about the nature of the relationship between efficient attention use in word identification and improved comprehension (Perfetti, Marron, & Foltz, 1996).

The concept of representation is strongly invoked in basic process reading research. Word identification development has been described in terms of changes in the ways in which words are represented in memory. At first, young children's representations of words are imprecise and incomplete. Representations of words become gradually more precise as children practice. The more precise a word's representation, the less attention must be available to identify it; hence, automaticity occurs when children's representations attain adequate precision for them to be recognized by sight (Adams, 1990; Ehri, 1991, 1998; Gough, 1977; Perfetti, 1985, 1992; Rayner, Foorman, Perfetti, Pesetsky, & Seidenberg, 2001; Share & Stanovich, 1995).

Speculation has been minimal concerning the role of the mind in any of this initial process research. One reason is that the focus of this research is on the initial aspects of acquiring reading skills. The initial processes involved in the acquisition of word identification skill and automaticity are easily modeled and understood within HIP models, information-processing theory, and experience-centered conceptions of the mind. Also on this view, comprehension is often seen as the result of combining word- and sentence-level meanings into higher order concepts by virtue of a somewhat automatic and mechanical combinatorial process. From an epistemological perspective, the emphasis of this work is on the nature and processing of sensory stimuli, not on how sensory data and the human mind interact to create meaning and knowledge. A quintessential example comes from Gough's 1971 model of the reading process. He was so unwilling to address the role the mind played in understanding that he used terms like "Merlin" and "The Place Where Sentences Go When They are Understood" to describe functions critical to comprehension. The deliberate choice of these terms suggests that cognitive psychologists of this time may not have recognized the necessity for describing the role of the mind in knowledge acquisition.

Reading Comprehension Research

Research on the processes involved in reading comprehension did not really begin until well after the cognitive revolution. This is because modern comprehension research could not be done without using mentalistic terms and particularly without relying on the concept of mental representation—both of which were clearly products of the cognitive revolution. Results from reading comprehension research have shown that there are numerous text, task, and reader-based variables that can profoundly influence what readers understand when they read. As suggested earlier, we cannot review all of these findings here (see Pressley, 1998, for a more complete review).

The comprehension research that is most directly inspired by the cognitive revolution is exemplified by the work of Walter Kintsch and his colleagues (Kintsch, 1974, 1977, 1985, 1987, 1998; Kintsch & Keenan, 1973; Kintsch & van Dijk, 1978). Kintsch's model of comprehension has as its foundation a basic HIP model of text processing. It makes explicit use of the computational metaphor and relies on the concepts of metal representation, linguistic parsing, and information processing. It is unabashedly a mechanistic model that relies on cycles of processing and algorithms to move incoming data to higher levels of processing. As Kintsch (1998) states,

> the theory of comprehension presented here is a computational theory. It describes a sequence of steps that, given certain initial conditions, yields outcomes comparable to human comprehension. (p. 8)

Comprehension for Kintsch is similar to the complex ideas that Locke postulated arose from the combination of sensory input and reflection (i.e., the processes of association and apperception to Locke). Kintsch did add the concepts of propositional representation, differentiated memory, spreading activation, and schemata (background knowledge) to his model; however, their inclusion does not change the model's basic epistemological stance. Indeed, Kintsch explicitly makes it clear that he is not entertaining any idea of the mind other than a mechanistic one:

> The mind in this view is not a well-structured, orderly system, but is a little chaotic; being based on perception and experience rather than on logic, being Aristotelian rather than Cartesian. (1998, p. 5)

Schema theory. The advent of schema theory in 1977 provided another theoretical explanation for how one aspect of the comprehension process might work (Anderson, Reynolds, Schallert, & Goetz, 1977; Reynolds, Taylor, Steffensen, Shirey, & Anderson, 1982). Schema theorists suggested that comprehension was the product of an interaction between incoming information and knowledge already stored in memory. Also, schema theory provided a framework for understanding how information might be structured in memory (Rummelhart & Ortony, 1977), how background knowledge interacts with incoming knowledge to produce comprehension and understanding (Anderson et al., 1977; Reynolds et al., 1982), how domain knowledge plays an important role in comprehension and expertise (Alexander & Murphy, 1998), and how background knowledge in the form of epistemological beliefs can influence text comprehension (Kardash & Howell, 2000; Kardash & Scholes, 1996).

While literally hundreds of published empirical studies using subjects from across the age spectrum have supported the idea that background knowledge in all of its forms influences reading comprehension, the exact epistemological stance of schema theory is more difficult to isolate. One strand of schema theory comes from the early work on artificial intelligence by computer and cognitive scientists such as Marvin Minsky and Roger Shank (Minsky, 1975; Shank & Ableson, 1977).

According to these theorists, schema theory is primarily a theory of representation. Schemata are created through the interaction of incoming information and existing knowledge via organizational rules; hence, incoming information can be combined with existing information in relatively mechanical ways. This view of schema theory fits easily with the precepts of the cognitive revolution and experience-centered epistemology. Hence, this view of schema theory could be, and was, successfully used in HIP models of comprehension such as Kintsch (1998) and processing models such as Rayner and Pollatsek (1989).

Conversely, there is a second strand of schema theory that was inspired more by the epistemological ideas of Kant than the computational notions and computer metaphor of the cognitive revolution. On this view, schema theory was seen as a theory of how the innate capacities of the mind shape, organize, and frame incoming and existing knowledge in the form of schemata—"pure conceptions of understanding" that were not, and could not, be derived from sensory data. The primary emphasis was on how existing representations of knowledge (schemata) could shape interpretations of incoming data through the processes of assimilation and accommodation. The nature of this interaction was thought to be complex and internal (not observable in a behaviorist sense).

Anderson and colleagues (1977) designed a study to demonstrate precisely this interaction. Participants were presented with identical stimuli (short stories to read) and given two forms of response opportunities (a free-recall test and a multiple-choice test). The results showed that the identical incoming information was interpreted differently by the two different groups of subjects. The two groups had been selected because they had significantly different existing knowledge concerning the information presented in the stories. This knowledge appears to have influenced subjects' understanding to such a degree that some reported as verbatim phrases that fit with their interpretation but were not in the original story. Others reported never seeing phrases that were in the story. Mental activity was postulated as the mediating cause of the result.

In conclusion, one reason schema theory had such a major impact on psychology and educational psychology was that it appealed to communities of scholars who were (in most cases implicitly) working from quite different epistemological views of the mind. Schema theory contained specific ideas about the nature and impact of the mental representation of knowledge; consequently, it appealed to information-processing scholars. Also, schema theory contained ideas related to the interactivity of knowledge, an increased role for the mind, and the importance of social and cultural knowledge (Reynolds et al., 1982); hence, it appealed to educators.

Strategies instruction. Work on the effects of comprehension strategies instruction on comprehension performance has been another important emphasis in reading comprehension research. Again, literally hundreds of publications have demonstrated that teaching children to use a subset of comprehension strategies has benefits for their overall comprehension success (e.g., Duffy et al., 1984; Pressley, El-Dinera, & Brown, 1992; Pressley & Ghatala, 1990).

The idea that readers can be taught strategies that will improve their reading comprehension performance is so well established in the educational community that it has become a key aspect of most, if not all, research-based reading comprehension curricula in America. A report of the National Reading Panel (2000) identified seven comprehension strategies that have been shown to assist readers' comprehension. They are: comprehension monitoring, cooperative learning, graphic/semantic organizers, generating questions, asking questions, use of story structure, and summarization. The inclusion of comprehension strategies research in this report is impressive because scientific rigor, generalizability, and replicability were the criteria for inclusion in the report.

Even the initial comprehension strategies research was based (perhaps implicitly) on epistemological assumptions about the human mind that went well beyond the experience-centered stipulations of the cognitive revolution and computational theory (Duffy et al., 1984). From the beginning, readers were conceptualized as actively using the contents of their minds (previously learned strategies) to monitor whether or not they were learning, to generate relevant questions about what they understood, and to identify information that was more important than other information and allocate attention to it. The activities of comprehenders and learners engaged in these types of strategic, internally initiated behaviors have yet to be included in the types of processing models used by reading researchers interested in the acquisition of reading skills or many cognitive psychologists interested in learning. Hence, it could be argued that comprehension strategies research seems to owe little directly to the computational metaphor, although a debt to the idea of mental representation and the rise of theory in psychology is clear.

The selective attention strategy. A third area of reading comprehension research involves understanding how attentional resources are used during reading. Results from this work have shown that readers use attention in predictable ways while attempting to comprehend. For the most part, readers attend to text items made important by external task demands. Significantly, this research has shown that a causal relationship exists between the allocation of attention and the text elements that students learn when they read (Reynolds, 1992; Reynolds & Anderson, 1982; Reynolds, Brown, Niederhauser, & Trathen, 2005; Reynolds, Standiford, & Anderson, 1979). In addition, some work has shown that readers may use aspects of their background knowledge to guide their attention (Goetz, Schallert, Reynolds, & Radin, 1983).

The early work in attention and comprehension explained the effects of attention on comprehension via the selective attention hypothesis (Goetz et al., 1983; Reynolds & Anderson, 1982; Reynolds et al., 1979). Following Rothkopf and his colleagues (Rothkopf & Billington, 1979; Rothkopf & Bisbicos, 1967) and information-processing theory, the selective attention hypothesis suggested that information in the text made important by relevant inserted questions would draw the attention of the reader to it. The fact that the questions always referred to the same type of information shaped the readers' attention allocation. In this sense, these studies could be considered mainstream in their adherence to tenets of the cognitive revolution—particularly in their reliance on the notions of text processing, mental representation, and selective attention. Also, they are consistent with the experience-centered epistemology inherent in computational theory and computer models.

Later research into the relationship between attention and comprehension took a rather different approach. The selective attention hypothesis was replaced by a new conception—the selective attention strategy (Reynolds, Wade, Trathen, & Lapan, 1989). This change was significant and heralded a change in the foundational theoretical assumptions being used. As Reynolds and colleagues (1989) stated,

> educators and psychologists have sought to understand how readers learn and recall text information. Early models of the learning process [HIP models] relied on almost mechanical explanations for how prose material was learned and recalled.... the reader was seen as passive and without any real input into

> the learning process.... [others are now beginning to propose] that readers are really active, strategic participants in the learning situation. (p. 159)

Much of the research that followed in this area discussed strategic attention allocation rather than the passive selection of attention by salient stimuli (Reynolds, 1992, 2000).

Just as with schema theory, research on the relationship between attention and comprehension seems to have undergone an epistemological shift and become quite similar to comprehension strategy research. This likely occurred because the theoretical and explanatory tools provided by computational theory and its underlying epistemological stance were becoming inadequate to parsimoniously explain the results that were beginning to surface in educational psychological research.

Just as with the cognitive revolution, there was no loud proclamation that any sort of theoretical shift was occurring. However, gradually comprehension researchers started to move away from relatively passive, mechanical conceptions of learners' minds toward a view of learners' minds as active, causal participants in the learning enterprise. Epistemologically, comprehension researchers appear to have moved away from the experience-centered notions of the mind held by the British empiricists, behaviorists, and many cognitive psychologists. Instead, they implicitly invoke, when they discuss their results, a view of the mind that allows for active mental judgment to influence learning behaviors—reflecting more the views of Kant and Descartes. Ironically, this is somewhat reminiscent of Tolman (1952) suggesting the idea of purposive behavior to account for laboratory results that could not easily be explained inside the confines of the behaviorist meta-theory.

CONCLUSIONS

The impact of the cognitive revolution on reading research is clear but somewhat different depending on the aspect of reading research on which one chooses to focus. It seems that comprehension researchers have moved away from their cognitive revolution roots both theoretically and epistemologically. In contrast, basic processes researchers have found the theoretical precepts of the cognitive revolution sufficient for their explanatory needs. This is certainly a product of the questions being asked and explanations needed to account for results; however, it may also be one reason why it has been suggested that these two areas of research seem to have developed somewhat independently of one another (Reynolds, 2000). Both areas of research were initiated in line with computational principles of the cognitive revolution. However, they may have gradually become less interactive as researchers in the comprehension area engaged different theoretical explanations for their results and different epistemological notions concerning the minds of readers.

The theoretical and epistemological trends apparent in the reading comprehension literature suggest that we might be in the midst of another gradual revolution in psychology. Trends in other areas of psychological research such as conceptual change (Sinatra & Pintrich, 2003) and evolutionary psychology (Plotkin, 1998) suggest the same conclusion. Some psychologists are becoming more comfortable with the idea that the mind adds something to knowledge that cannot be gained from any combination of sensory experiences—associative or otherwise. Also many are beginning to use previously discouraged concepts such as consciousness, volition, intention, and internal initiation to describe research results that cannot easily be explained by current theories.

Completely congruent with these trends are the notions of Kantian and Cartesian epistemology and evolutionary psychology. Indeed, evolutionary psychology provides a reasonable explanation for why the Kantian mind might possess inherent properties that allow us to see cause-and-effect relationships, and organize and shape incoming sensory information. One can infer from Plotkin (1998) that the mind may function as it does because there are many adaptive advantages to doing so. For example, the ability to see cause-and-effect relationships has immense adaptive advantages. It would enable one to accurately anticipate problematic and efficacious situations and take the appropriate actions when needed.

The idea of consciousness control is critical to ideas of volition and intention. Again, Plotkin (1998) has commented on the emergence of consciousness as a critical variable in psychology. He notes that psychologists cannot agree whether or not consciousness is a legitimate area of research or not. He states,

> Banished by the behaviourists, even now there is little consensus as to what consciousness is, how best to think of it, and in some quarters even a sneaking suspicion that behaviourists were right, that it is not a legitimate part of scientific psychology. (p. 265)

Yet, the adaptive aspects of conscious control of behavior and the ability to internally initiate action are obvious. Perhaps the Kantian mind and internal initiation of action couched in the context of evolutionary psychology are ideas whose time has come. The current interests in these ideas may indicate the onset of another period of change in American psychology.

REFERENCES

Adams, M. J. (1990). *Beginning to read.* Cambridge, MA: MIT Press.

Alexander, P. A., Murphy, P. K. (1998). Profiling the differences in students' knowledge, interest, and strategic processing. *Journal of Educational Psychology, 90*(3), 435–447.

Anderson, R. C., Reynolds, R. E., Schallert, D. L., & Goetz, E. T. (1977). Frameworks for the comprehension of discourse. *American Educational Research Journal, 14*(4), 376–381.

Andre, T., & Phye, G.D. (1986). Cognition, learning and education. In G. D. Phye & T. Andre (Eds.), *Cognitive classroom learning* (pp. 1–18). Orlando, FL: Academic Press.

Attkinson, R. C., & Shiffrin, R. M. (1968). Human memory: A proposed system and its control processes. In K. W. Spence & J. T. Spence (Eds.), *The psychology of learning and motivation: Vol. 2, Advances in research and theory* (pp. 89–195). New York: Academic Press.

Baars, B. J. (1986) The cognitive revolution: The rise of a theoretical psychology. In B. J. Baars, *The cognitive revolution in psychology.* New York: Guildford Press.

Berkeley, G. (1709/1957). An essay towards a new theory of vision. In M. W. Calkins (Ed.), *Berkeley: Essay, principles, dialogues* (pp. 1–98). New York: Scribner.

Chall, J. (1967). *Learning to read: The great debate.* New York: McGraw-Hill.

Chomsky, N. (1959). Review of "Verbal Behavior" by B. F. Skinner. *Language 35,* 26–58.

Descartes, R. (1912). *A discourse on method.* London: Dent.

Duffy, G. G., Roehler, L. R., & Mason, L. (1984). *Comprehension instruction: Perspectives and suggestions.* New York: Longman.

Ehri, L. C. (1991). Development of the ability to read words. In R. Barr, M. Kamil, P. Mosenthal, & P. D. Pearson (Eds.), *Handbook of reading research* (Vol. 2, pp. 383–417). New York: Longman.

Ehri, L. C. (1998). Grapheme–phoneme knowledge is essential for learning to read words in English. In J. L. Metsala & L. C. Ehri (Eds.), *Word recognition in beginning literacy.* Mahawah, NJ: Erlbaum.

Ellis, J. (2000). *Founding brothers: The revolutionary generation.* New York: Vintage Books.

Flesch, R. (1955). *Why can't Johnny read?* New York: Harper & Row.

Goetz, E. T., Schallert, D. L., Reynolds, R. E., & Radin, D. I. (1983). Reading in perspective: What real cops and pretend burglars look for in a story. *Journal of Educational Psychology, 75*(4), 500–510.

Goodman, K. S. (1971). Reading: A psycholinguistic guessing game. In H. Singer & R.B. Ruddell (Eds.), *Theoretical models and processes of reading* (pp. 497–508). Newark, DE: International Reading Association.

Gough, P. B. (1971). One second of reading. In H. Singer & R. B. Ruddell (Eds.), *Theoretical models and processes of reading* (pp. 509–535). Newark, DE: International Reading Association.

Hartley, D. (1749). *Observations on man, his frame, his duty, and his expectations.* London: Leake & Frederick.

Hobbes, T. (1650). Human nature. In W. Molesworth (Ed.), *The English works of Thomas Hobbes.* London: John Bohn.

Huey, E.B. (1908). *The psychology and pedagogy of reading.* New York: Macmillon.

Hull, C. L. (1952). *A behavior system.* New Haven, CT: Yale University Press.

Hume, D. (1739). *A treatise of human nature.* London: Noon.

Hunt, M. (1993). *The story of psychology.* New York: Doubleday.

Kahneman, D. (1973). *Attention and effort.* Englewood Cliffs, NJ: Prentice-Hall.

Kant, I. (1781/1990). *Critique of pure reason.* New York: Colonial Press.

Kardash, C. M., & Howell, K. L. (2000). Effects of epistemological beliefs and topic-specific beliefs on undergraduates' cognitive and strategic processing of dual-positional text. *Journal of Educational Psychology, 92,* 524–535.

Kardash, C. M., & Scholes, R. J. (1996). Effects of pre-existing beliefs, epistemological beliefs, and need for cognition on interpretation of controversial issues. *Journal of Educational Psychology, 88,* 26–271.

Kintsch, W. (1974). *The representation of meaning in memory.* Hillsdale, NJ: Erlbaum.

Kintsch, W. (1977). *Memory and cognition.* New York: Wiley.

Kintsch, W. (1985). Text processing: A psychological model. In T. A. van Dijk (Ed.), *Handbook of discourse analysis* (pp. 231–243). London: Academic Press.

Kintsch, W. (1987). Generating scripts from memory. In E. van der Meer &

Kintsch, W. (1998). *Comprehension: A paradigm for cognition.* New York: Cambridge Press.

Kintsch, W., & Keenan, J.M. (1973). Summarizing stories after reading and listening. *Journal of Educational Psychology, 69,* 491–499.

Kintsch, W., & van Dijk, T.A. (1978). Towards a model of text comprehension and production. *Psychological Review, 85,* 36–94.

Locke, J. (1690/1959). *An essay concerning human understanding.* New York: Dover.

Mill, J. (1829). *Analysis of the phenomena of the human mind.* London: Baldwin & Cradock.

Minsky, M.A. (1975). A framework for representing knowledge. In P. H. Winston (Ed.), *The psychology of computer vision.* New York: McGraw-Hill.

National Reading Panel (NRP). (2000). *Teaching children to read: An evidence-based assessment of the scientific research literature on reading and its implications for reading instruction.* Washington, DC: National Institute of Child Health and Human Development.

Perfetti, C. A. (1985). *Reading ability.* New York: Oxford University Press.

Perfetti, C. A. (1992). The representation problem in reading acquisition. In P. B. Gough, L. C. Ehri, & R. Treiman (Eds.), *Reading acquisition* (pp. 145–174). Hillsdale, NJ: Erlbaum.

Perfetti, C. A., Marron, M. A., & Foltz, P. W. (1996). Sources of comprehension failure: Theoretical perspectives and case studies. In C. Cornoldi & J. Oakhill (Eds.), *Reading comprehension difficulties: Processes and intervention* (pp. 137–166). Mahwah, NJ: Erlbaum.

Plotkin, H. (1998). *Evolution in mind: An introduction to evolutionary psychology.* Cambridge, Massachusetts: Harvard University Press.

Posner, M.L., & Boise, S. J. (1971). Components of attention. *Psychological Review, 78,* 391–408.

Pressley, M. (1998). *Reading instruction that works.* New York: The Guilford Press.

Pressley, M., El- Dinera, P. B., & Brown, K. R. (1992). Skilled and not-so-skilled reading: Good information processing and not-so-good information processing. In M. Pressley et al. (Eds.), (1992). *Promoting academic competence and literacy in school* (pp. 91–127). San Diego, CA: Academic Press.

Pressley, M., & Ghatala, E. S. (1990). Self-regulated learning: Monitoring learning from text. *Educational Psychologist* [Special Issue: Self-regulated learning and academic achievement], *25*(1), 19–33.

Rayner, K., Foorman, B. R., Perfetti, C. A., Pesetsky, D., & Seidenberg, M. S. (2001). How psychological science informs the teaching of reading. *Psychological Science in the Public Interest, 2*(2), 31–74.

Rayner, K., & Pollatsek, A. (1989). *The psychology of reading.* Englewood Cliffs, NJ: Prentice Hall.

Reynolds, R. E. (1992). Selective attention and prose learning: Theoretical and empirical research. *Educational Psychology, 4*(4), 1–48.

Reynolds, R. E. (2000). Attentional resource emancipation: Toward understanding the interaction of word identification and comprehension processes in reading. *Scientific Studies of Reading, 4*(3), 169–195.

Reynolds, R. E., & Anderson, R. C. (1982). Influence of questions on the allocation of attention during reading. *Journal of Educational Psychology, 74*(5), 623–633.

Reynolds, R. E., Brown, K. J., Niederhauser, D., & Trathen, W. (2005). The attentional advantage of good comprehenders. Manuscript in preparation.

Reynolds, R. E., Sinatra, G., & Jetton, J. (1996). Views of knowledge acquisition and representation: A continuum from experience-centered to mind-centered. *Educational Psychologist, 31*(2), 93–104.

Reynolds, R. E., Standiford, S. N., & Anderson, R. C. (1979). Distribution of reading time when questions are asked about a restricted category of text information. *Journal of Educational Psychology, 71*(2), 183–190.

Reynolds, R. E., Taylor, M., Steffensen, M. S., Shirey, L. L., & Anderson, R. C. (1982). Cultural schemata and reading comprehension. *Reading Research Quarterly, 17,* 353–366.

Reynolds, R. E., Wade, S. E., Trathen, W., & Lapan, R. (1989). The selective attention strategy and prose learning. In M. Pressley, C. McCormick, & G. Miller (Eds.), *Cognitive strategy research: From basic research to educational applications* (pp. 159–190). New York: Springer-Verlag.

Rothkopf, E. Z., & Billington, M. J. (1979). Goal-guided learning from the text: Inferring a descriptive processing model from inspection times and eye movements. *Journal of Educational Psychology, 71,* 310–327.

Rothkopf, E. Z., & Bisbicos, E. E. (1967). Selective facilitative effects of interspersed questions on learning from written materials. *Journal of Educational Psychology, 58,* 56–61.

Rummelhart, D. E., & Ortony, A. (1977). The representation of knowledge in memory. In R. C. Anderson, R. J. Spiro, & W. E. Montague, (Eds.), *Schooling and the acquisition of knowledge* (pp. 99–135). Hillsdale, NJ: Erlbaum.

Shank, R. C., & Abelson, R. P. (1977). *Scripts, plans, goals, and understanding.* Hillsdale, NJ: Erlbaum.

Share, D. L., & Stanovich, K. E. (1995). Cognitive processes in early reading development: Accommodating individual differences into a model of acquisition. *Issues in Education, 1,* 1–57.

Sharp, A. C., Reynolds, R. E., Brown, K. J., Morris, A., & Gunn, S. (2004). First grade reading instruction empowered through research-based knowledge. *American Reading Forum Online Yearbook.*

Sinatra, G., & Pintrich, P. (2003). The role of intentions in conceptual change learning. In G. Sinatra & P. Pintrich (Eds.), *Intentional conceptual change* (pp. 1–19). Mahwah, NJ: Erlbaum.

Skinner, B. F. (1957). *Verbal behavior.* New York: Appleton.

Smith, F. (1971). *Understanding reading: A psycholinguistic analysis of reading and learning to read.* New York: Holt, Rinehart, and Winston.

Stanovich, K. E. (2003). *Progress in understanding reading.* New York: The Guilford Press.

Stanovich, K. E. (1980). Toward an interactive-compensatory model of individual differences in the development of reading fluency. *Reading Research Quarterly, 16,* 32–71.

Thorndike, E. L. (1931). *Human learning.* New York: Appleton.

Tolman, E. C. (1952). Autobiography. In E. G. Boring, H. S. Langfield, H. Werner, & R. Yerkes (Eds.), *A history of psychology in autobiography* (Vol. 4, pp. 323–339). Worcester, MA: Clark University Press.

Turing, A. M. (1950). Computing machinery and intelligence. *Mind, 59,* 433–460.

Watson, J. B. (1913). Psychology as the behaviorist views it. *Psychological Review, 23,* 89–116.

CHAPTER 3

FROM BEHAVIORISM TO SITUATED COGNITION

An Examination of Learning and Instruction in the Second Half of the 20th Century through the Research and Writing of Richard C. Anderson

James M. Royer
University of Massachusetts, Amherst

The premise that underlies this chapter is that the history of a scientific discipline can be illuminated through an examination of the research and writing of a single prominent practitioner of that discipline. My goal is to try to understand the changes that have occurred in the learning and instruction area of educational psychology in the past 45 years. I am particularly interested in changes in theory, but I am also interested in changes in the way educational psychologists view their discipline and in changes in the methodology of research.

The individual who provides the material for my examination is Richard C. Anderson. Anderson certainly fits the description of a prominent practitioner of his discipline. In addition to being a prolific scholar, a partial listing of the honors he has received include: Director of a major research

The Impact of the Cognitive Revolution on Educational Psychology, pages 41–85

center, President of the American Educational Research Association, a two-time recipient of AERA's Palmer O. Johnson award for outstanding research, two awards for outstanding research from the International Reading Association, election to the International Reading Association Hall of Fame, appointment as a Fellow at Stanford University's Center for Advanced Study in the Behavioral Sciences, and recipient of APA's Edward L. Thorndike Award for distinguished career-long contributions to the psychology of education.

Anderson began his professional career believing that educational psychologists should be educational engineers who develop fault-proof instruction. He currently ascribes to a sociocognitive perspective that maintains that instruction takes place in a complex, multifaceted milieu. The educational scientist's role in the sociocognitive perspective is to try and understand the dynamics of that milieu, and to perhaps guide those dynamics along a more productive path. The starting belief system and the current perspective are vastly different and the principal goal of this chapter is to describe how and why those changes occurred.

The reader should be aware of the methodology that has produced the chapter. I have read or skimmed almost every one of the more than 150 separate publications Richard Anderson has produced. I read those writings in chronological order, beginning with the first article that Anderson published and ending with his most recent article. As I read I became increasingly aware that what he wrote made little sense without an understanding of the social-scientific context for the writing. To borrow a term from my literary colleagues, I needed to deconstruct a writing in order to judge its contribution to a constantly changing scientific world. I will share my deconstruction efforts with the reader as I go along. On occasion, I will also tell the reader what I think Anderson was thinking when he wrote something. I did this without consulting him about what he was thinking.

My primary justification for not consulting Anderson is that I think the written record probably provides a clearer view of thought processes than does reconstructed memories. Moreover, an introspective analysis of my own writings convinces me that my recollections are totally suspect. On more than one occasion upon rereading something I wrote I have asked myself, "What could you possibly have been thinking when you wrote that!" I suspect Anderson himself has had the same sense when he rereads some of the things he has written. When the chapter was finished I did send it to Anderson, and though I am sure he does not agree with everything that I have written, he does agree that my general methodology does no harm to truth.

I focus on a number of things as I describe Anderson's writings. First, I want the reader to understand his evolving theoretical belief system. Some of Anderson's former students (this one included) used to describe him as the Will Rogers of the Educational Psychology community—he never met a theory he didn't like. That turns out to be untrue. At the very most he seems to have believed in three identifiable theories and I attempt to describe those theories in sufficient detail so as to provide a context for interpreting what Anderson was doing. I also attempt to describe the research methodologies that Anderson used. This was a wonderful discovery for me. Anderson is a creative and innovative research methodologist, a fact that I was not aware of until I read his entire corpus of research. I also try and convey a sense of Anderson's research style. I have long believed that good psychologists develop a research style that is as distinctive and unique as a Miles Davis solo or a seemingly endless Faulkner sentence. Anderson has such a style and I could easily identify him as the author of any five sequential articles he produced.

Another point that needs to be mentioned is that Anderson is the star of this show and I write this chapter as if he alone was responsible for his printed work. This does not do justice to the over 100 collaborators who also have their name on his articles. I'm sure that Anderson would readily admit that much of the success he has achieved is due to the fact that he has had the opportunity to work with a large number of bright and energetic people during his academic career. I also apologize in advance for not mentioning every one of Anderson's publications. My purpose is to trace the development of themes and to describe landmark events. In so doing, I'm sure I will neglect to mention someone's favorite article, and perhaps even fail to mention an article that Anderson himself believes to be particularly important.

ANDERSON'S PRETHEORETICAL BEGINNINGS

Anderson received his doctoral degree from Harvard in 1960. Psychology in the 1960s was dominated by behavioral theory, which I will discuss in more detail shortly. There is little evidence, however, in Anderson's early writing that he subscribed to any theoretical position. In fact, as is the case with many newly minted researchers, Anderson seemed to be searching for a research issue that was worthy of his talents. His first published article was a review of authoritarian/democratic teaching studies (Anderson, 1959). He addressed two issues in the review. First, is there any evidence that one or the other of these teaching styles produces more learning? Second, does the authoritarian–democratic continuum provide an effective conceptualization of leadership? After a careful review of the evidence, he responded to both questions in the negative. Interestingly, he also interjected a very modern-sounding issue into his analysis of the research. He noted that a number of the studies reported evidence of statistical significance, but there was no effort to estimate practical significance. He called for evidence of both kinds in future research. This position has only recently been formally adopted by the American Psychological Association.

Anderson's second publication (Anderson, 1961a) was a book review that would not be noteworthy except for one remarkable coincidence. The book, written by an author with the interesting name of Minnie Louise Johnson Abercrombie, was concerned with a fundamental problem of science education; namely, the process of making judgments or inferences about data. I have not been able to find a copy of the book, but apparently Abercrombie described a theory of decision making and an instructional technique that purportedly enhanced students' abilities to make good judgments about data. Anderson panned the evidence regarding the instructional technique, but here is what he had to say about the theory:

> The author, who demonstrates an intimate acquaintance with the thought processes of students as well as the literature on perception and cognition, theorizes that mental organizations which she calls "schemata" are the basis for perception and thought. Schemata make "re-cognition" of events and objects possible. (p. 102)

The concept of schemata did not appear again in Anderson's writing until 1977, whereupon it became a central theoretical construct for him for the next 20 years. In all of his later writings that mentioned schemata as a construct, he never mentioned Abercrombie's book. He gave Piaget and Bartlett credit for the idea, and their writings did indeed precede Abercrombie's book. However, based on the description Anderson provides above, it would seem that Abercrombie's schemata construct was very close to Anderson's subsequent notion of a schemata. In contrast, Piaget and Bartlett used the term, but the fit between their ideas and the more modern usage of schemata was always a stretch.

After leaving Harvard, Anderson took a position as Assistant Superintendent of Schools in East Brunswick, New Jersey, and served a brief stint as a Lecturer at Rutgers University before moving on to the University of Illinois in 1963. Anderson's articles during the pretheoretical phase of his career seem to represent the activities of an ambitious and energetic researcher and writer who is not quite sure what he should be doing research on. One article made the case for nongraded homogeneous grouping, another argued that elementary teachers should be specialized (just as their junior high school and high school colleagues are), and a third described school record-keeping using electronic data-processing equipment. His empirical research examined a prediction of the McClelland and Atkinson achievement motivation theory using eighth-grade students as participants (rather than college students), and an effort to train creative behavior in sixth-grade students.

Anderson did publish an article though during this phase that I think provides great insight into the programmatic research agenda that he was soon to undertake (Anderson, 1961b). In the article Anderson argued that what education needed was "educational engineers" who could translate research findings into practical educational activities. He argued "reviews of research rarely say anything of importance to school personnel about educational practice, or indeed say anything of importance about educational practice to psychologists or sociologists" (p. 377). The solution to this problem was to train educational engineers who could bridge the gap between research and educational practice.

I think Anderson saw himself as assuming this role and it influenced his research agenda for many years to come. Given what I believe to be the centrality of the educational engineer idea to understanding the early part of Anderson's career, I want to trace out some of the implications of the analogy. An engineer is typically someone who translates basic science into usable products. The product that an engineer designs has a particular function and it accomplishes that function in a specified way. An electronic can opener is designed to open cans and it functions properly if it is used as the engineer designed it to be used. If it isn't used as the engineer designed it to be used, it doesn't open the can. I believe that Anderson had a similar conceptualization of what an educational engineer was supposed to do. That is, the engineer was supposed to design educational experiences that accomplished an educational goal. The goals were accomplished if the experiences were implemented in accordance with design principles. One problem with the formulation is that people (teachers) cannot always be counted on to use products as they were designed to be used. The solution to this problem was to either eliminate teachers altogether, or to design educational activities so tightly that they became "teacher proof." As we will see shortly, Anderson and his collaborators worked on both approaches.

If one is to be an educational engineer one must first master the basic science that provides the educational design principles. Accordingly, I now turn to an examination of the basic science that provided the design principles for educational engineers during the 1960s.

PSYCHOLOGICAL THEORY AND RESEARCH IN THE 1960s

Skinnerian Behaviorism

The dominant theory in the 1960s was behaviorism and it came in two flavors. One was Skinnerian behaviorism and I attempt to provide a simple view of the essential belief structure of that theory in Figure 3.1 below. The essence of Skinnerian behaviorism was the belief that the science of psychology identified functional relationships between stimulus events and response events. If a carefully defined stimulus condition could be shown to reliably elicit a particular response, then the behavior was "explained." There was no need at all to speculate about the possible mechanisms in the learner that mediated the relationship between the stimulus and the response. My Skinnerian friends have convinced me that there is a legitimate sense in which identifying functional relationships is indeed a type of scientific explanation. My sense has always been, however, that psychological theories should do something *besides* predicting the occurrence of behavior under specified conditions. Theories should lead to an *understanding* of behavior, and I find Skinnerian psychology vacuous with respect to understanding how and why behavior occurs.

I believed that Anderson had a similar sense. Despite the fact that Skinner is his intellectual grandparent (his Ph.D. advisor was John Carroll, who was a Skinner student), and that he undoubtedly received a goodly dose of Skinnerian psychology as a Harvard student while Skinner was on the faculty there, I find no evidence in Anderson's writings that he ever believed in Skinner's brand of behaviorism.

Associative Learning Theory

There were a number of variations on the second flavor of behavioral theory. The essence of the second type of theory, as represented in Figure 3.2,

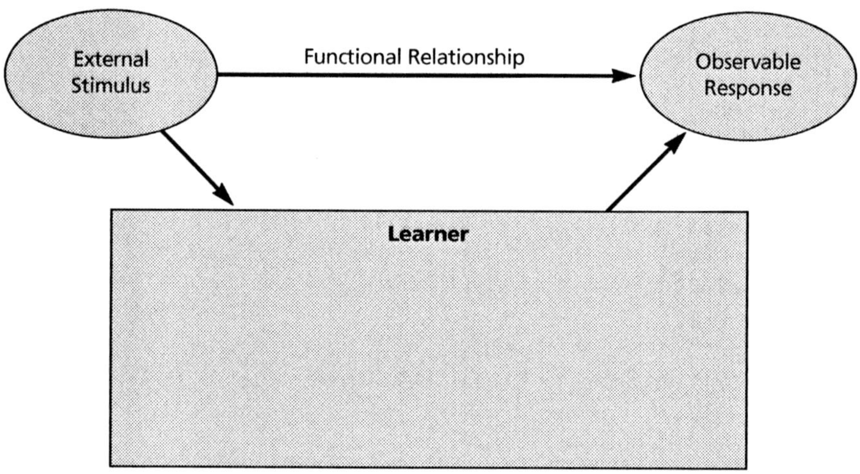

Figure 3.1. Skinnerian behaviorism.

was the supposition that chained stimulus–response (S–R) connections within the learner mediated between the overt stimulus and the observable response. The variation of the theory that Anderson seemed to believe in was often called interference theory (for reasons to be explained shortly), and again a little background might be in order to understand what it was about.

The content of 1960s associative learning theory was predetermined by Hermann Ebbinghaus, who had the bright idea that the way to study complex verbal learning was to begin with basics—that is, to begin with the learning of material that was uncontaminated by previous learning. In the late 1880s Ebbinghaus invented nonsense syllables (e.g., VQZ) as a means of accomplishing this purpose. Anderson restated Ebbinghaus's rationale for this choice years later in one of his own articles:

> The geneticist experiments with fruit flies and bread mold because these organisms are easier to study than say, oak trees, elephants or people. For similar reasons lists of nonsense syllables and discrete words have long comprised the chief experimental tasks of psychologists interested in memory. (Anderson & Myrow, 1971, p. 81)

The problem was that Ebbinghaus's method never progressed much beyond nonsense materials and during the 1960s the literature was filled with complicated studies involving the learning of lists of material that had little or no inherent meaning. The reasons for the label "interference theory" had to do with a fascination with the relationship between the acquisition and forgetting of successively learned lists. It was easy to show that forgetting of well-learned lists of nonsense material could be induced in a matter of minutes by having learners master other lists. In short, the learning of the second list of material interfered with the retention of the first list. The general consensus in the period was that interference experiments identified the mechanism responsible for forgetting in general. Moreover, it was believed that the identification of the events that seemed to facilitate the acquisition of nonsense materials could confidently be generalized to the learning of meaningful academic material. Thus, interference theory was believed to be concerned with the discovery of the basic science principles that could then be used by the educational engineer to design effective instruction.

The essence of interference theory as illustrated in Figure 3.2 was the idea that the learner created an internal representation of a stimulus–response event that had an isomorphic relationship with the observable external events. By isomorphic I mean that the internal copy contained important details of the external event, and it contained only those details. Learners themselves contributed nothing to the internal rep-

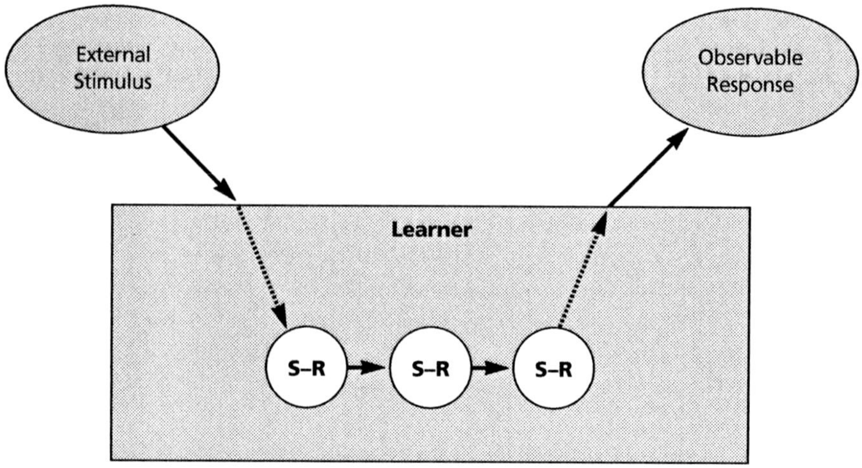

Figure 3.2. Associative learning theory.

resentation. Learning involved laying down the internal representation, recall involved reactivating the representation in the presence of the same or similar stimulus event, and forgetting involved breaking down the internal representation so that it could no longer be located and recalled. Learning of successive lists created forgetting because learning a new response to the same stimulus broke the bonds between the stimulus and the originally learned response. Given this framework, one could explain the acquisition of very complex learning by positing that it was achieved by chaining together many simpler S–R connections in a manner that allowed the accomplishment of a complex behavior.

The evidence that Anderson believed in this type of behavioral theory at an early point in his career comes from two 1965 publications. The first (Anderson, 1965) was a journal article that reported an effort to teach first-grade children an advanced problem-solving skill. In the article Anderson cited such well-known associative theorists as Charles Cofer, Howard and Tracy Kendler, and Irving Maltzman. Anderson used learning theory principles to design his instructional interventions, and he reported positive results for the training effort. He described these positive results as being an exercise in "human engineering" and were a "..triumph of the perspective that repertoires of behavior could be modified in accordance with the principles of learning." Anderson also took a swipe at fledgling cognitive psychologists who might perceive his triumph as involving the development of a problem-solving strategy (a "vacuous psychological construct" according to Anderson):

> Unlike other psychologists with a similar [behavioral] bias, the present author does not object to such concepts as strategy because they are theoretical, unobservable, or contain "surplus" meaning. Rather, the objection is based on the judgment that this kind of theory has proven unfruitful, lending itself more to ex post facto rationalization than explanation, prediction and control. (p. 284)

The second publication was a 1965 book that Anderson coedited with David Ausubel. Anderson is quite proud of the fact that the book (titled *Readings in the Psychology of Cognition*) was the first cognitive psychology book to appear as the cognitive revolution was looming on the horizon. However, I suspect that he would also admit that there was little cognitive psychology in the book and the cognitive content that was there was largely Ausubel's doing. Anderson called himself a "neo-behaviorist" in the introduction to the book. Later, in an introduction to a section of the book concerned with thinking, he defined in greater detail what it meant to be a behaviorist. He wrote that, "Above all the behaviorist believes that there is essentially one set of psychological laws that applies to all kinds of phenomena, to animals, and to humans, to children and to adults" (p. 535). He also described the behavioral perspective on problem solving by noting that, "Verbal behavior that gives direction to problem solving can profitably be conceived as response systems or perhaps habit families capable of being influenced by the external stimulating conditions (the problem) and by reinforcement." Consciousness was described as a "metaphorical use of language that muddied theoretical waters."

To summarize this section, the important parts of Anderson's early belief system were the ideas that important learning was solely dependent on the nature of the stimulus event the learner experienced, the response the learner made in the presence of that event, and the consequences the learner experienced after making the response. These ideas provided an educational engineer with all of the design principles needed to produce educational products.

THE EDUCATIONAL ENGINEER IN ACTION

Beginning around 1965 and lasting until the early 1970s, Anderson pursued the educational engineering analogy with vengeance. The early focus of his attention was concept learning that, interestingly enough, involved computer-controlled experiments using the Socrates system (Anderson, 1966; Anderson & Guthrie, 1966). These studies used artificial concepts created by making one attribute of complex stimuli relevant to a learning task. So, for example, objects might vary in shape (e.g., circles or squares), color (blue or green), number, and so on. These studies tested predictions derived from theories of concept learning and they examined issues such as whether repeated presentation of a relevant concept dimension produced superior learning compared to a presentation sequence involving both relevant and irrelevant attributes.

In 1967 Anderson published an *Annual Review of Psychology* article where he reviewed the research that an educational engineer should master (my description, not his). He began the review by indicating that, "It is now widely believed that behavioral science must be engineered into teaching methods and materials." The review was then divided into the following sections: (1) the technology of instruction, which included topics such as behavioral objectives, task analysis, lesson development, and tryout and revision cycles; (2) stimulus factors, which included errorless discrimination learning, and prompting, fading, and vanishing techniques that enhanced errorless learning; (3) response factors, which included topics like overt constructed responding, and techniques for controlling mathemagenic behaviors; (4) feedback factors, which examined the relative merits of types of reinforcers and aversive control; and finally, (5) the organization and sequencing of instruction, which considered issues like whether examples should be followed by rules, or vice versa, and whether practice and review should be spaced or massed. As a final note on this topic, Anderson expanded the ideas contained in his *Annual Review of Psychology* article into a textbook-length exposition (Anderson & Faust, 1973). In my mind, that book is the finest presentation of behaviorally oriented instructional design that I have seen.

Shortly after the review article was published, Anderson and his colleagues embarked on a series of five studies that examined programmed instruction as the vehicle for delivering engineered instruction. Programmed instruction was the perfect means of delivering the educational engineer's product because it was even better than teacher-proof; it eliminated the teacher all together. One issue that was examined in these studies was how to write program frames (an instructional segment in a program) so that errors were kept at a minimum while at the same time maintaining high levels of learning. The notion of low errors came from Skinner's research indicating that learners (both animal and human) would learn errors, in addition to correct responses, if allowed to make them. This lead to the notion that low error rates was a good thing, but one way of producing low error rates led to a bad thing called "copy frames." Copy frames were frames where the learner could copy the answer to the question posed by the frame without examining all of the material in the frame. Anderson's studies showed that learners would copy without reading if given the opportunity, thereby reducing what they learned from the program. This tendency by learners to circumvent instructional intent was pervasive enough that it was termed the "law of least effort." That is, learners will tend to engage in the minimal effort necessary to complete an instructional task. Awareness of the law of least effort added to the design principles the educational engineer needed to be concerned with, namely, procedures for assuring that learners did exactly what the engineer wanted them to do as they interacted with the instruction. Other research that extended into the 1970s examined feedback principles. Some of these studies involved complicated experiments using the PLATO computer system, which was created in part by cannibalizing the earlier Socrates system. This research lead to the discovery of the fact that retention of information was often enhanced if feedback about correct responses was delayed for a period of time.

It was also during this period that Anderson moved beyond doing research on the design principles of effective instruction and actually began to develop practical instruction. Anderson was involved in the writing of two biology texts (one a programmed text) in the late 1960s (Anderson, Faust, Guthrie, & Drantz, 1969; Liebherr, Anderson, et al., 1970).

There is evidence that Anderson began to have doubts about instructional products designed by educational engineers as he neared the 1970s. For instance, he published one study involving high school and college students that compared learning from a programmed text with learning from summaries of the text (Roderick & Anderson, 1968). It turned out that the programmed text was the better teaching vehicle for the high school students, but not for college students where there was little difference between the modes of instructional presentation. Even worse, the programmed instruction took the college students four times longer to complete than did the text summaries. In another article describing the wisdom of conducting field experiments that evaluated instructional effectiveness, Anderson wrote, "It is, in my judgment, a fair appraisal to say that right now, today, we cannot confidently predict the effectiveness of a lesson given a description of its philosophy, the style, the methods and the procedures of instruction" (Anderson, 1969). Anderson did not give up on the idea of the educational psychologist as an educational engineer for some years to come, but his confidence that the approach would produce failproof education seemed to be waning.

Before beginning the next section it is worthwhile to consider the enduring legacy of associative learning theory. Behaviorism has had a lasting impact on psychology, but the large majority of that impact comes from the Skinnerian style of behaviorism. Skinner was right. A good deal of the learning of nearly all organisms can be explained (in a Skinnerian sense) by a common set of principles. The associative theorists, however, wanted to go beyond Skinner's empty organism and to explain complex behavior in terms of internal events. They failed miserably in this effort, largely because of the belief that internal mental representations were solely determined by external events. At the beginning of the 1970s it became obvious that this was an untenable assumption and 30 years of research soon disappeared into the black hole of irrelevant data. I believe that this judgment is applicable to Anderson's research of the period. His studies on concept learning and programmed instruction, with the possible exception of the feedback research, add little to our understanding of human learning.

There was, however, a tremendous methodological benefit to be derived from the associative learning era. Researchers examining human learning during that period became superb experimentalists. They learned how to design experiments that contained controls that ruled out competing explanations, and they learned how to analyze data so as to identify results that could be replicated. If one looks at the current crop of well-known cognitive researchers, one typically finds that they learned their trade from advisors who were skilled practitioners of associative learning research. I think this is true of Anderson and his students as well. During the 1960s he learned how to design and conduct carefully controlled experiments. That learning proved of great benefit to both Anderson and his students as they entered the cognitive era.

THE BEGINNINGS OF THE TRANSITION FROM BEHAVIORAL TO COGNITIVE PSYCHOLOGIST

I was a member of Anderson's lab during the period when Anderson began his transformation from a behaviorist to a cognitivist and my recollection of one of the major events that changed the direction of his thinking was the simple demonstration that if learners are asked to learn a list of words that could be categorized, they tend to recall the words in category order. For instance, if a randomly presented 20-word list contains 5 animal names, 5 vegetable names, 5 furniture names, and 5 tool names, over a series of trials the words increasingly will be recalled in category order. This seems obvious now, but at the time the free-recall experiments (as they were called) shook the foundations of a belief system. The problem for the associative belief system was that internal representations from which recall is derived are supposedly based on the stimulus event that was experienced. The categorized free-recall experiments demonstrated that learners could, and would, change the representation of the event they experienced.

The demonstration that learners could willfully shape their own learning immediately became a research issue in Anderson's lab, which will be discussed in greater detail shortly. However, for several years to come there was a concerted rearguard effort that tried to incorporate the willful learner research into the associative theory framework. For instance, a series of studies (including the author's PhD dissertation) addressed the question of whether interference theory could explain the forgetting of successively learned free-recall lists. The answer was yes. Anderson and his students showed that if two successively learned lists contained the same categories, list 1 words would be selectively knocked out of memory by list 2 words from the same category. Alternatively, if list 1 and list 2 words were from different categories, there was far less memory loss.

The demonstration that interference theory explained forgetting of free-recall lists encouraged expansion of this line of research into more meaningful materials. Here Anderson received some help from David Ausubel, who had departed Illinois several years earlier. An examination of Anderson's research indicates that he always worked best when he had a competitor to joust with. Ausubel, who served as the adversary for the line of research to be described next, had argued that meaningful learning is fundamentally different from list learning and that interference theory had little to say about the acquisition and retention of material such as connected discourse. He even published a couple of studies that supported this position, or so he argued. Unfortunately for Ausubel, he was not a good experimentalist whereas Anderson was a brilliant one. In a series of carefully designed and controlled experiments that involved sentences as well as connected prose, Anderson was able to show that the interference effects that had been found with nonsense materials could also affect the learning and retention of prose materials.

It is apparent, however, that Anderson's certitude about associative learning theory as the theoretical bedrock for educational engineering was growing fragile. At the end of a study that successfully demonstrated interference effects with prose materials, Anderson wrote the following:

> Observers of the classroom may wonder how frequently in the "real world" forgetting analogous to retroactive inhibition actually occurs. How often does it happen that the preconditions to retroactive inhibition—similar stimuli paired with different responses—coincidently appear in ordinary classroom activity? We seldom teach students different answers to the same question. If retroactive inhibition is generated in prose only when the materials are so closely similar, we must question the efficacy of the interference model as an inclusive explanation of forgetting in the classroom. The atomistic approach that was required to make good the analogy between paired-associate retroactive inhibition and prose retroactive inhibition seems at once necessary and potentially misleading. It seems probable that students do forget without the presence of closely similar confusing material. Although proponents of interference theory believe that nonspecific interference accounts for a good deal of forgetting, there was little or no evidence for nonspecific interference in the study reported in this paper. The present

> experiment did reaffirm that retroactive inhibition can occur with realistic prose materials. But it is yet to be shown whether interference theory explains all of the phenomena of forgetting or only some of them. (Myrow & Anderson, 1972, p. 308)

The New Conceptualization of the Educational Engineer

It was also during this period in the early 1970s that Anderson began to reveal his new conceptualization of what an educational engineer should be doing. If one refers back to the previous discussion of the *Annual Review of Psychology* article (Anderson, 1967), one notes that the educational engineer was involved in controlling the stimulus events a learner experienced, in shaping the nature of the responses learners emitted, and in arranging appropriate consequences to the response. As the 1970s began, Anderson began talking about the additional necessity of controlling the cognitive activity of the learner. This perspective is very clear in his 1970 *Review of Educational Research* article concerned with controlling "student-mediating processes." In the article Anderson wrote:

> The general thesis of this paper is that the activities the student engages in when confronted with instructional tasks are of crucial importance in determining what he will learn. (p. 349)

> In this view, the chief problem for educational engineering is to discover how to alter the characteristics of instructional tasks so as to force students to do all of the processing required for learning. (p. 350)

One way that students could be "forced" to process information so that it would be learned was to ask them questions either prior to or after they had read something. Questions presented *after* a learner read a text passage had an interesting and unpredictable consequence. Learners receiving postquestions not only learned the material the questions were directed toward, they also learned ancillary material better. In short, questions presented after reading seemed to induce deeper processing of the text as a whole. This was a wonderful finding for the cognitive educational engineer and Anderson summarized this research and the educational implications for the research in 1975 (Anderson & Biddle, 1975).

THE COGNITIVE ERA

Prior to beginning a discussion of what I believe to be Anderson's most important and enduring line of research, it may be worthwhile to pause and consider yet another variation of the theoretical model that seemed to be driving Anderson's research. The demonstration that learners could influence the nature of the thing that they learned suggests the model illustrated in Figure 3.3.

As Figure 3.3 indicates, cognitive theory does not assume that the nature of the thing that has been learned is solely dependent on external events. Instead, learners now make contributions to their own learning. As one examines this model, several obvious questions come to mind. First, what is the nature of the thing that a learner contributes and how does that thing influence the representation of an experienced event? And second, what is the nature of the constructed representation? What is it made of? Once acquired, how is the constructed representation reactivated? These questions, and additional ones that followed, dominated the psychology of human learning for the remainder of the century and were very evident in Anderson's research.

Knowledge Representation Research

Anderson's first attempt to answer some of the questions about knowledge construction and representation was a study that examined the nature of the representation that is formed from learning meaningful sentences (Anderson, 1971). More specifically, the question was whether the learner remembered the sound of sentences, the orthography of sentences, or the meaning of sentences. The result of this cleverly designed first experiment showed that sentence learning produced a representation that preserved the meaning of the

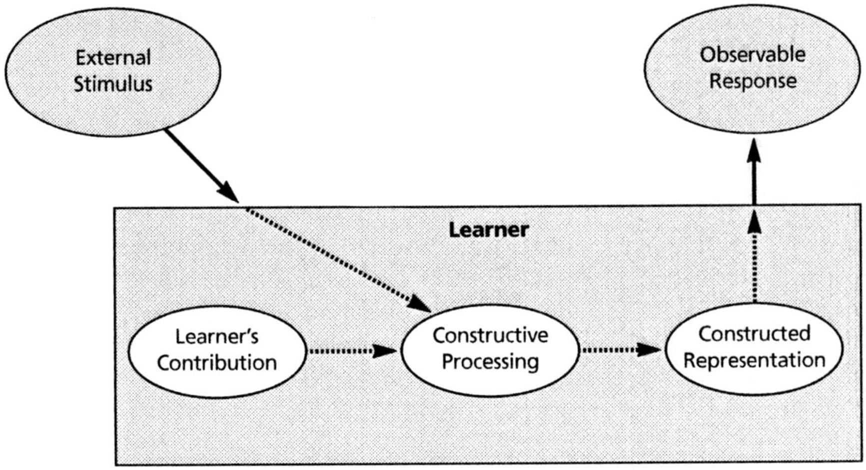

Figure 3.3. Cognitive learning theory.

sentence but little of the sentence's phonological and orthographic properties. Anderson was not the first to show that learners preserved the meaning of sentences rather than their form, but he was far more diligent than previous researchers in tracing the implications of this finding.

The next study examined what it means to "process" a sentence. This study showed that participants who filled an obvious blank while learning a sentence (e.g., Bill hit the with the hammer.) remember the sentence better than participants who simply read the sentence aloud.

The demonstration that learning sentences somehow created a semantic representation that preserved meaning, but not form, and that the nature of the representation that was formed could be manipulated through instructions to the learner, seemed to lead Anderson into a deeper reading of linguistic theory and the next few studies he completed on memory representations seemed to reflect that reading. However, ever the educational engineer, he paused in the early 1970s to consider at least one educational implication of the new cognitive view. Anderson knew that learners who processed information at a deeper level than mere surface reading would form a meaning preserving memory representation of that information. He recognized that that fact had implications for educational assessment. Then, as now, there was considerable debate about how to determine if students truly "comprehended" the instruction they received. Anderson's insight was that one way to do this was to see if they could recognize test items that had the same meaning as the instruction they experienced, but a different form. In short, one way to test for comprehension was to see if examinees could recognize paraphrases of instructional events they had experienced (Anderson, 1972a). Tracing the implications of this insight makes up a substantial part of my own research career that deals with measuring reading and listening comprehension (Royer, in press; Royer, Carlo, Dufrense, & Mestre, 1996).

Anderson's first two articles on sentence meaning had danced around a formal discussion of semantics. He now plunged whole-heartedly into the topic and soon found himself debating some of the leading linguists and cognitive psychologists in the world. One study, for example, pitted a view of semantics developed by Katz and Fodor against Quillian's semantic network theory. The Katz and Fodor view was that the meaning of sentences could be represented by a series of hierarchically arranged "semantic markers." Quillian, in contrast, viewed semantics as being encoded in a network consisting of semantic nodes and connections between nodes. In Quillinan's view, the meaning of a particular sentence was computed by evaluating the level of activity across the active nodes in the network (as an aside, this type of theory evolved into connectionism). Anderson's (1972) study favored the semantic marker interpretation, but it may also have triggered deeper thinking on his part about whether that type of theory had broad generality. For example, in a closely following study, he asked the question of how people seem to acquire the meaning of new terms and was able to show that meaning acquisition could be facilitated by embedding the term in a sentence context that provided an example of the meaning of the term (Anderson & Kulhavy, 1972).

The demonstration that embedding terms in context facilitates acquisition of the meaning of the terms leads to an interesting speculation. Perhaps our knowledge about what things mean is typically encoded in the form of the particular event that illustrates the meaning of the term. For example, perhaps hearing the term "dog" results in the activation of an image of a particular dog that serves as a token for the meaning of the general term. Anderson found support for this position in one study (Anderson & McGaw, 1973), and soon followed that study with another, demonstrating that the meaning of abstract terms could be facilitated by embedding the terms in sentences that made them more concrete and easier to image (Anderson, 1974).

The idea that general terms evoked a particular instantiation for the term leads to two further interesting ideas, one of which Anderson pursued immediately, and one of which occupied him several years down the line. The idea that he pursued immediately was the notion that if a particular term evoked a particular instantiation, perhaps it was the case that instantiations changed depending on the context in which the term was experienced. For example, what representation for the general term "ate" is evoked when one hears the following sentences: (1) The baby ate the steak. (2) The dog ate the steak. (3) The banker ate the steak. Anderson and Andrew Ortony (1975) argued that each of the sentences results in a different instantiation of the term "ate." I don't know enough about formal semantics to know this for sure, but my sense is that semanticists hated this idea. It makes formal descriptions of meaning very difficult because it allows the possibility that each speaker of a language can have an idiosyncratic representation for terms. From this perspective, the only reason that we can converse with understanding is that our shared experiences are sufficiently common so as to result in roughly similar instantiations for the terms we use in conversation.

Anderson continued to accumulate evidence for this perspective for several years. For example, in one study he gave formal voice to the supposition that "a word does not have a meaning, but rather has a family of potential meanings" (Anderson et al., 1976), and in a second study to follow he presented a formal model of how context could activate a particular instantiation among the word families (Half, Ortony, & Anderson, 1976). Anderson's research on word meanings continued to evolve and ultimately morphed into research on vocabulary acquisition. I will return to the discussion of his word meaning research shortly, but first I want to discuss the second strand of research mentioned above.

At the beginning of this section I mentioned the idea that learners influence the nature of knowledge representations through a constructive process that involves the external stimulus and the learner's prior knowledge. This idea raises a question about both the nature of the representations they form and about the nature of the knowledge the learner possesses that contributes to forming the representation of the learned event. Anderson addressed the representation question in his research on word meanings that I just discussed, but soon began to address the knowledge question in his schemata and prior knowledge research.

The Schema Research

The first time Anderson used the term "schemata" (other than the previously mentioned Abercrombie review) was in a book chapter that provided his summation of a small conference held in San Diego (Anderson, 1977). The notion of a schema was the driving theoretical construct at the conference (I was there) and Anderson indicated that he believed the construct would have considerable impact on "schooling" and on educational research. Anderson used the term "schooling" in the title of his article and I highlighted it in the previous sentence because I think it reflects a change in Anderson's thinking about what researchers should be doing. Previously, he had always used "education" or "training" when talking about the educational application of research findings. Both education and training imply a process whereby knowledge is passed from the instructing agent (teacher or machine) to the student. Schooling, in contrast, suggests a more active learning environment where teacher, student, and learning environment could all make a contribution to learning outcomes. I think Anderson was now thinking in terms of a more active learning environment. One thing Anderson attempted to do in his conference summation was define what a schema was and how it functioned. Anderson's description was very Piagetian. For instance, he talked about perception and comprehension as being analogous to Piagetian schemata that change as a function of assimilative and accommodative processes. His romance with Piagetian concepts seemed to change soon after, however, and his subsequent discussions of schema departed considerably from this original Piagetian-flavored view.

If one believes that the memory representation for an event is shaped, in part, by an individual learner's schemata, and if one believes further that a learner's schemata is shaped by idiosyncratic experiences, then it follows that two individuals experiencing the same event could very well have different representations of that event. Prior to discussing how Anderson pursued this idea, I want to note how radical a departure this idea is from the beliefs that directed Anderson's research in the 1960s. Then, the idea was that representations for experienced events were solely dependent on the stimulus properties of the event. Every individual experiencing an event would have the same memorial copy of that event, save for failure to encode some aspect of the event. The idea that two individuals experiencing the same event would have different representations of that event would have been absolute heresy in Anderson's lab in the 1960s.

The idea that event representation can be idiosyncratic leads to a variety of perspectives that form a continuum of belief about knowledge. One extreme position is that representations are so variable across individuals that discussions of reality are problematic. This perspective is balanced by another that says that experiences within a community of learners are sufficiently common so as to allow the expectation that members of a community will tend to represent an event in the same or similar way. Anderson's research on how prior knowledge and schemata influenced knowledge representation seemed to be based on this later alternative. It should be noted, however, that the first extreme is alive and well in today's intellectual discourse.

One of the first studies that pursued the theme about variable representations involved presenting a passage that could be interpreted as being about card playing or playing in a musical group to two groups of participants (Anderson, Reynolds, Schallert, & Goetz, 1977). One group had considerable experience with playing in musical groups and one did not. The results of the study indicated that the musical group did indeed tend to remember the passage as being about playing music, while the other group remembered it as being about playing cards.

The simple demonstration that learners acquired different things from reading the same passage, and that what they learned was predictable based on an analysis of the learning community they came from, suggested many questions that had both basic and applied significance. On the basic side was the question of what a schema was. The notion of schema that Anderson was developing suggested that it was a "data structure" with slots to be filled in for frequently experienced events. This data structure developed upon repeatedly encountering an event that had the same structural properties but different instantiations. One of the favorite examples of the era, for instance, was the restaurant schema where all restaurants share some features, but also have differences. The idea was that mention of a restaurant activated the general restaurant schema, which automatically filled in many details about what happened, and then the specifics of the experienced event would fill in some of the particular details (schema slots) such as what was eaten, whether one was formally seated or found one's own seat, and so on. This notion dovetailed nicely with the learning community idea in that learners from different learning communities could be expected to form different types of schemata. Thus, the schema possessed by a musician would be different than one possessed by an athlete.

On the applied side were questions about the role that schemata played in educationally important events. Schemata presumably enhance the acquisition of information. That is, if it is the case that an educational event activates a schema then acquisition of the details of that event should be easier to acquire and remember than in the situation where a schema is not activated. Anderson and his colleagues thought that it might be the case that schemata relevant to formal academic learning might be more common in the majority community than in the minority community, thereby giving majority students a learning advantage over minority students. Another possibility they pursued was that majority and minority students might have different schemata, and therefore learn different things from the same event. Since academic assessment procedures were likely to be insensitive to the minority perspective, it might erroneously appear that the minority student had acquired little from the event relative to the majority student.

These ideas were examined in a series of beautiful and important studies. One study showed that Asian Indians and U.S. college students acquired very different things from reading the same description of a wedding (Steffensen, Joag-dev, & Anderson, 1979). Two other studies examined differences in what white and Afro-American children learned from reading the same passages (Reynolds, Taylor, Steffensen, Shirey, & Anderson, 1982; Taylor-Delain, Pearson, & Anderson, 1985). Both of these studies involved presenting white and Afro-American children with passages or other materials describing the ritual insults that Afro-American children often engage in (e.g., "Yo momma so ugly she walks backward down the street so she don't scare people to death."). Anderson and his colleagues showed that white and minority students learned different things from the passages and that the attributes that predicted performance in the two groups were very different. The implications of these findings for the design of both instructional materials and assessment procedures were discussed at length in the studies.

Another series of schema studies that Anderson and his colleagues completed was concerned with the question of whether the schema that an individual used to process an event could be manipulated. Anderson knew that individuals from different learning communities had different schemata that would cause them to learn different things, but the question in the new series of studies was whether the same individual possessed multiple schemata that could be switched in and out depending on the situation. The reader may notice the similarity between this question and Anderson's earlier research on word meanings where he argued that learners use context to select the sense of meaning of a word from among a family of meanings. The multiple schemata research seemed to be based on a similar idea. Individuals possess multiple schemata that are potentially applicable in interpreting an event. The particular schema they use is determined by the context in which the event is experienced. As we will see later in this chapter, the complete unpacking of this idea motivated Anderson's transition from adherence to the principles of cognitive learning theory embodied in Figure 3.3, to the principles of social cognition theory that provides the theoretical framework for his most recent research.

The studies that Anderson and his colleagues completed that examined the idea of multiple individual schemata asked participants to change their view of what a passage was about at different times during or after exposure to the passage. In the first study participants were asked to read a passage about two boys visiting a house either from the perspective of a burglar or a home-buyer (Pichert & Anderson, 1977). The results indicated that participants reading the passage from the burglar perspective tended to recall things that might be stolen, while participants reading from the home-buyer perspective tended to recall information about the quality of the house. Anderson and his colleagues went to considerable length to show that this result was not associated with "response editing." Even when strongly encouraged to do so, participants tended not to recall a good deal of information relevant to the other perspective.

The study just mentioned had readers read the passage from a particular perspective and then attempt to recall all of the information they could. The results showed that recall could be predicted by the perspective that was operative during reading. This result leaves open the question of what might happen if readers read with one perspective, recalled everything they could, and then were asked, "By the way, the passage you just read can be thought of from another perspective. Why don't you think of that perspective and try to recall the passage again?" Using these directions Anderson and his colleagues (Anderson & Pichert, 1978) were able to show that information that was unrecallable when one perspective was active suddenly became recallable when the learner was asked to activate the second perspective.

The results of the burglar/home-buyer studies made sense when viewed from the framework of schema theory, but there was still a question about how schemata influenced performance. The slot filling analogy mentioned earlier suggests one answer. Readers with an active burglar schema fill slots in the schema with information about prospective items to steal. Since the burglar schema doesn't contain slots for home quality information, that information tends not to be stored. Therefore, readers tend to store burglar-relevant information with greater likelihood than home-buyer information. However, the study showing recall of previously unrecallable information as a function of shifting perspectives suggested that perspectives could influence retrieval as well as storage. Perhaps it is the case that a schema provides a retrieval plan for information. This possibility would suggest that information relevant to both the burglar and home-buyer schema gets stored, but only information relevant to the active schema is found and recalled. Anderson and his colleagues tested these alternatives in a study where perspectives were assigned prior to reading, immediately after reading, or long after reading (Anderson, Pichert, & Shirey, 1983). The results showed that schemata could influence both storage and retrieval, though the effects on retrieval seemed to be more substantial than those on storage.

While Anderson continued to mention this style of schema theory (a different style emerges later) for years to come, his empirical interest in the topic was clearly waning. He did, however, make a summative statement on the topic in his AERA Presidential address published in 1984. In this address Anderson provided a critical commentary on two versions of schema theory. The first, which he called the strong version, suggests that schemata consist of general principles that are abstracted from repeated experience. As experience increases, schemas grow in power to the point where when one encounters an experience that fits a schema, the schema then generates much of the particulars of the experience. Anderson was not in favor of this version and cited as evidence the fuzziness of many of the connections between schemas and particulars. For instance, he made note of his earlier research where the connection between an eating implement (e.g., spoon) and a food (soup, pabulum) could not be made by simple activation of an eating-with-spoon schema. The particular food that was instantiated upon encountering an eating schema was dependent upon the individual doing the eating (banker, baby). Anderson could not see how it would be possible to reconcile the computational properties of strong schema with the contextual dependencies illustrated in empirical research. He seemed to favor instead a weaker version of schema theory that had properties similar to prototype theories. Weak schemas, unlike their stronger cousins that fill in particulars of experience automatically, would activate a family of possible instances to instantiate and the contextual milieu would then provide the information that would allow selection of the most probable candidate. Taking the example mentioned above, in the weaker view, activation of the eating-with-spoon schema would nominate both soup and pabulum and the context (banker or baby) would result in selection of the most appropriate candidate. For the reader interested in this sort of thing, this is precisely the type of mechanism that

Kintsch (1988) proposed a number of years later in his very influential constructive integration theory of comprehension.

THE READING RESEARCH YEARS

The discussion of Anderson's research and thinking in the previous section is slightly out of historical step because of the desire to pursue Anderson's research on schema effects to their end, or at least to the end of the strong and weak versions of the theory. As it turns out, Anderson's very recent research revives the concept of schema, though in very different form than it appeared in the 1970s and 1980s. This research will be discussed near the end of the chapter.

I want to now step back in time to the mid-1970s when the University of Illinois was awarded a grant to establish the Center for the Study of Reading and Anderson became the first director of the Center. It may come as a surprise to those reading about the research that Anderson completed before the mid-1970s to know that he did not consider what he was doing during this era as reading research. Anderson, and those of us involved in similar research at this time, thought of ourselves as "verbal learning researchers" and later as "prose learning researchers," but not as reading researchers. I'm not sure why we didn't think of ourselves as reading researchers but I suspect it had something to do with an imagined pecking hierarchy between "pure educational researchers" and "applied educational researchers." The pure educational researchers were the experimentalists like us who were fully draped in the trappings of science, while the applied educational researchers were those mucking around in schools trying to teach kids to read, write, and do math. Anderson's anointment as director of the Reading Center changed that attitude and he and his Center colleagues were soon involved in a program of research that ultimately moved them forever out of ivory tower research, using college sophomores as the exclusive participants.

Early Reading Theory

Anderson's first formal foray into reading theory was an article contrasting bottom-up, top-down, and interactive theories of reading and how the three different perspectives might account for reading process such as word identification, syntactic and semantic analysis, and differences between listening and reading (Adams, Anderson, & Durkin, 1978). The dominant view at the Reading Center at the time (I was there as a visiting researcher) was that reading was an interactive process involving continuous hypothesis confirmation. The idea was the reader used knowledge sources such as word knowledge, knowledge of syntax and semantics, and world knowledge to formulate hypotheses about the information being pulled off of the printed page. Those hypotheses were then confirmed through a process of comparison of the printed orthography with the hypothesized input. Schema theory fit nicely within this perspective since it could be imagined that all of the knowledge sources contributing to this process consisted of schemata, and much of the research conducted in the early years of the reading center was strongly influenced by schema theory principles.

The Vocabulary Studies

Anderson's first empirical studies that could truly be called reading studies reexamined an issue that had long interested him—word meaning. But this time interest in word meanings was examined in the context of how vocabulary was acquired and how it influenced reading comprehension. As a package, Anderson's work on vocabulary is another of his absolutely beautiful bodies of research. The way this vocabulary research systematically develops and advances can serve as a model for young researchers of how programmatic research should be conducted.

Anderson's vocabulary research started with the question of whether difficult and unusual vocabulary would influence reading comprehension. On the face of it, this was an "of course" question, but Anderson and his coauthor pointed out conflicting evidence that made the answer less certain. On the one hand, there were studies showing that performance on vocabulary tests predicted performance on reading comprehension tests. However, this relationship leads to the prediction that one way to enhance reading comprehension performance would be to improve a reader's vocabulary, and Anderson and his coauthor cited research showing that training vocabulary had little impact on reading comprehension. Given this uncertain state of affairs, Anderson began to systematically untangle the relationship between vocabulary knowledge and reading comprehension. The first study (Freebody & Anderson, 1983a) involved measuring comprehension after having readers read a grade-appropriate text passage under three conditions. In the first condition they read the unaltered passage, in the second condition they read a passage where a rare and unusual word was substituted for every third substance word, and in the third condition a rare word was substituted for every sixth substance word. The results showed that there was little comprehension deficit in the every-sixth-word passage, but there was a significant deficit in the every-third-word passage.

The results of the rare word study were interesting in that they suggested that somehow readers were able to overcome the problem of difficult vocabulary in situations where the density of the difficult words was not very high. The next question Anderson pursued was how was this possible? One obvious answer that had been suggested by other researchers was that readers would use compensatory mechanisms to figure out text meaning, even when the text contained words they did not know. For example, a reader might use sentence context and problem-solving strategies to figure out the meaning of a sentence containing a difficult word. Anderson's next study empirically examined this possibility (Freebody & Anderson, 1983b). The study involved manipulating text coherence, vocabulary difficulty, and topic familiarity. The idea was that a low coherence text with easy vocabulary might yield the same comprehension results as a high coherence text with hard vocabulary. Likewise, a high topic-familiarity passage with hard vocabulary might be understood equally as well as a low-familiarity passage with easy vocabulary. The results of the study provided little support for the compensation idea. There was no evidence that readers could use an easy attribute of a passage to compensate for a difficult one.

The results of these first two studies showed that tracing the relationship between vocabulary knowledge and reading comprehension was not going to be an easy enterprise. Anderson's studies, in addition to previously described research showing a general lack of impact of vocabulary instruction, also raised the issue of how vocabulary is acquired. One possibility is that we acquire vocabulary through direct instruction. Another is that we acquire vocabulary "on the fly" by inferring the meaning of vocabulary words from surrounding context. These questions were pursued in the next series of studies Anderson completed.

One direction Anderson took in pursuing these thoughts was to examine the kinds of instructional experiences that seemed to facilitate vocabulary development and good reading skills. In an extremely ambitious and methodologically complex study conducted with elementary school students in reading groups, he examined the interaction of four classes of instructional variables: aspects of teaching methods, characteristics of instructional groups, attributes of instructional materials, and the personal characteristics of children making up the groups (Anderson, Mason, & Shirey, 1984). The central question in this study was whether reading competence developed most rapidly under conditions that emphasized accurate oral reading (e.g., correct word identification and pronunciation) or under conditions that emphasized extracting the meaning of what was being read. The results of the study indicated that students progressed best under conditions that emphasized the meaning of what was being read. From this finding, Anderson extracted what he referred to as the *law of meaningful processing*: "First, learning is facilitated when information is elaborated in such a way that all of the information is integrated into a coherent representation. Second, learning is facilitated when the reader is actually involved in attempting to generate a coherent representation."

A second direction that Anderson's research took was to examine vocabulary growth and informal (noninstructed) conditions that produce vocabulary growth. For example, in one study Nagy and Anderson (1984) calculated that there were 88,500 words in printed school English and went on to suggest it would be impossible to acquire that many words through direct instruction. Nagy and Anderson then provided an informed estimate of how many words middle school students might read during a year. The estimates ranged from 100,000 words a year for less-able students to an astounding 50 million words a year for the voracious middle school reader.

If it is impossible for readers to learn the vocabulary they have mastered from direct instruction, the obvious alternative source for their knowledge is learning from context. But interestingly enough, Anderson and his collaborators could find no empirical evidence showing that readers could acquire vocabulary from context. Given this, they proceeded to examine whether vocabulary could be acquired from context, and if so, what were the properties of context that facilitated that acquisition. The first published study (Nagy, Herman, & Anderson, 1985) showed that it was possible to acquire vocabulary from context and a second (Herman, Anderson, Pearson, & Nagy, 1987) examined features of text that lead to the strongest vocabulary growth. The text features manipulated in this second study included aspects of the macrostructure of the text, features associated with logical and temporal relations within the text, and features associated with concepts and the relations between concepts. Anderson and his coauthors were able to show that incidental learning of vocabulary was best in a text where there was lots of emphasis on concepts and the relations between concepts. This study was followed by another, much larger study that involved over 350 students enrolled in grades 3, 5, and 7. This study had students read either expository or narrative texts and then after 6 days their knowledge of difficult words contained in the passage was tested. The results of this study were similar to the previous one in that conceptually rich texts produced the best vocabulary acquisition. The study also showed that it was possible for a student to acquire a new vocabulary word after a single reading of that word in context, though the likelihood of that happening was mediated by a number of word-level and text-level variables.

In the years to come Anderson's interest in word meanings and vocabulary ceased for a time to be his central research interest, though it is clear that interest in the topic never disappeared entirely. For instance, he published an article in 1989 that examined the question of whether the speed of recognition of a word is related to frequency of related words (Nagy, Anderson, Schommer, Scott, & Stallman, 1989). The answer to the question was yes, and this lent support to a particular class of semantic representation theories. Anderson also published two theoretical/review articles in this later period that summarized much of his research and thinking on word meanings (Anderson, 1990; Anderson & Nagy, 1991). The first of the two articles is particularly interesting in that the reader gets to see Anderson in full battle with some pretty heavyweight psycholinguist types. Anderson's interest in word acquisition also resurfaced in the late 1990s in the guise of learning to read Chinese. This research will not be discussed in this article, though the reader should be aware that Anderson has now completed and published a substantial number of articles on learning to read Chinese.

All in all, I think that Anderson's work on word meanings represents his best and most enduring body of work. He began with some empirical work that captured his attention, read a lot of the literature on the philosophy and psychology of language, completed some more beautiful empirical work, and then moved on to examine the educational implications of the topic. This work alone could stand as the life's work of a very influential scientist.

Shifting the Focus from the Reader to the Setting

In 1985 Anderson and his colleagues published a monograph, entitled *Becoming a Nation of Readers*, that foreshadowed his next major research interest (Anderson, Hiebert, Scott, & Wilkinson, 1985). The authors devoted the first chapter of the book to describing reading as a psychological process, and then spent the remaining six chapters discussing the situational conditions that facilitated the development of good reading skills. There was a chapter that related the development of reading skills to language acquisition in general and to the concept of emerging literacy, and another chapter that talked about how reading beyond the elementary years had to be viewed within the larger context of learning from reading. The heart of the book involved three chapters that were concerned with the teaching of reading. One chapter reviewed the evidence that the quality of teaching played a critical role in the acquisition of reading skill; another examined the impact of different teaching methods and different ways of assessing reading competence, and the third discussed teacher education and the professional development of teachers. The topic of these chapters is not particularly noteworthy in that the content would be expected in a book that was directed to an audience outside of the research community. What does seem to be noteworthy, however, is Anderson's response as a researcher to the perspective that things going on outside of the learner are critically important in understanding reading performance.

I think the point of shifting focus may be worth elaborating on. Anderson's research during the first 25 years of his career had focused on the individual learner and he used experiments to examine the learner. He operated as the typical laboratory researcher operates in that he attempted to control the stimulus events learners experienced and then examined their behavior in response to those events. The educational payoff of this approach is that the properties of stimulus events that facilitate performance can then be built into all instructional materials. Notice that the concept of a stimulus event has a narrow bandwidth. Stimulus events as considered by the experimentalist almost always refers to learning materials provided to the student. Activities occurring outside the context of the learning materials are sources of noise that should be minimized to the extent possible. The most extreme form of this view was the "teacher-proof" instruction position that Anderson had adopted early in his career. If you want to reduce the teacher-related noise in an educational situation, you either script teachers' behavior so that they all do the same thing, or you eliminate the teacher altogether.

The discussion in the *Nation of Readers* book focused on the sources of noise rather than the learning materials. In a shift of interest that seems surprising given his previous history, Anderson soon became fascinated with these issues as topics for research and moved from doing well-controlled experiments to engaging in studies of reading behavior in real-world settings. It was almost as if he convinced himself that the real bang for the buck in influencing reading performance was not going to be found in the manipulation of reading materials. Rather, it was going to be found in manipulating aspects of the macroenvironment that surrounded the process of learning to read. This led to an even more surprising outcome—he soon found himself mucking around in classrooms teaching kids to read.

One of the earliest studies in this new phase examined the relationship between out-of-school activities and reading performance (Anderson, Wilson, & Fielding, 1988). The study involved having students write down what they did every day when not in school. The study found that time spent reading books out of school was a good predictor of reading achievement, and that watching television had a negative effect on reading achievement. Another study published at about the same time looked at the role of errors in learning to read (Anderson, Wilkinson, Mason, Shirey, & Wilson, 1988). One school of thought was that errors were bad and the best reading development would occur if the difficulty level of books was carefully calibrated so as to minimize word identification errors. Anderson and his colleagues were able to show that this was not the case and that errors were often beneficial to skill development, particularly in situations where the student engaged in meaningful processing in an effort to resolve an error situation.

An Interest in Teacher Training

Anderson also began to publish articles addressing teacher training. One of the first articles in this genre involved an examination of how reading is taught in Japan (Mason, Anderson, Omura, Uchida, & Imai, 1989). This article was soon followed by three additional articles that were concerned with teacher education in the United States. In the first of these articles (Anderson, Armbruster, & Roe, 1990) he noted that the quality of teaching that a child receives is a major determinant of their progress in reading, writing, and language arts, and went on to suggest that a teacher's typical preservice training experience fell far short of the ideal. Several specific techniques such as reciprocal teaching and reading recovery were mentioned, and he suggested that a good way to teach teachers to use these techniques was by showing them videotapes that modeled the techniques. Never one to let a good idea lie fallow, Anderson and his colleagues soon produced a very influential six-part video training series that did exactly what he had called for (Anderson, Au, & Jacobs, 1991).

About this time Anderson was trained as a reading recovery teacher and he was spending part of his day working with individual children who were at risk for reading problems. This experience was reflected in his next teacher training article (Anderson & Armbruster, 1990) where he and his colleague provided nine "rules of thumb" for teaching reading that were grounded both in the reading research literature and in direct classroom experience. Anderson's actual experience as a reading teacher was invaluable in lending examples and credibility to the maxims contained in the article. The final teacher training article in the series evaluated an actual teacher training sequence (Armbruster, Anderson, & Mall, 1991). Anderson and his colleagues developed a training sequence that began with teachers writing lesson plans and then teaching the lesson that was videotaped. This was followed by having the perspective teacher view and critique the tape, view the tape again with the cooperating teacher followed by another round of suggestions for improvement, and a final revision and submission of both tapes and notes as a classroom assignment.

Anderson as a Commentator on Reading Research and as a Mediator and Participant in the Reading Wars

On more than one occasion during the 1990s Anderson offered his view of the state of reading research. One noteworthy example is his 1993 article describing the future of reading research. In the article he describes a growing gulf between scientific studies of reading and discussions about how reading and writing are acquired and how they are taught. The scientists, he suggested, see little value in the writings of the people interested in teaching children to read and write, and the educators teaching children to read and write haven't the foggiest idea of what the scientists are talking about.

Anderson (1993) noted that there were several reasons for this gulf. One was the tendency for educators to engage in "interpretative scholarship" where the goal of the scholarly activity was to describe the meaning of instructional events from the perspective of the various participants in the events. This approach bears little resemblance to the traditional scientific method and has tended to be discounted by the scientific community. On the other side, Anderson pointed out that the activities of the scientific reading community have become increasingly opaque to anyone other than direct participants. Some of the research (such as connectionist modeling) does not even involve human participants, and other research (such as fMRI brain studies) requires a highly esoteric knowledge base to even understand descriptions of the research. This research says nothing to the teacher who wants to do a better job in the classroom.

Anderson also commented on the reading wars in the 1993 article. Even informed lay people are aware that the professional reading community was engaged in a fierce debate during the 1990s. One side of the debate (often called the whole languages side) argued that children should be taught to read by immersing them in literature. The analogy that is often used by this camp is that we naturally learn to speak by being immersed in an environment that is filled with spoken language. Since reading, according to this view, is also a natural linguistic skill, learning to read can be achieved by immersion in literary events that emphasize that writing is conveying meaning through print, and reading is extracting that meaning. The opposite side (often called the skills, phonics, or code side) argued that direct instruction in decoding is often necessary to assure skilled reading.

Anderson made an effort to find an agreed-upon middle ground between various positions in the reading community in a conference that was held in 1996 and reported in a 1998 book (Osborn & Lehr, 1998). In his introduction to the book (Anderson, 1998) Anderson noted that the goal of the conference was to answer four questions:

- What are the essential components of reading and writing instruction in American schools?
- What is the agreed-upon common knowledge base for the teaching and learning of reading and writing?

- How do we prepare teachers and manage schools so that all students achieve acceptable standards of performance in reading and writing?
- What are the strategies to use to change educational policy and to communicate with teachers, administrators, and the general public?

He noted that the hope was that the conference would lead reasonable people to seek an end to the seemingly endless conflict between the whole language and the skills camps. This hope was to go unfulfilled. Anderson summed up the divisiveness present in the conference in this way:

> So the conference proved to be a microcosm of the entire field, revealing a fair amount of conflict and some confusion. That the conference was unable to fulfill the hopes that we had for it prompts this warning: A field divided into warring camps is not a profession. An agreed-upon body of knowledge is typical of and I will go further to say, essential to a true profession. Although an enormous amount of writing and research has focused on various aspects of reading, it is sad to have to acknowledge that this work has not coalesced into an agreed-upon body of knowledge about the content or form of reading instruction, particularly beginning reading instruction. (p. 3)

Anderson also became involved in another controversy regarding reading research and instructional policy. He and several colleagues (Taylor, Anderson, Au, & Raphael, 2000) called for reading researchers to exhibit caution and moderation in the claims they make about their research findings. This call for moderation was elicited by a famous study that claimed to show that the reading performance of children from economically disadvantaged backgrounds could be greatly enhanced by instructional experiences that placed an emphasis on the explicit teaching of decoding skills. Anderson and his colleagues argued that the authors of this study had overstated their evidence, and that the media and policymakers had then used these overstatements to argue for a simple solution to what Anderson and his coauthors felt was a very complicated problem.

The article described in the above paragraph is interesting for reasons other than the argument it makes about the policy implications of reading research. It is also interesting for what it suggests about Anderson's belief system. The article contains a figure that purports to show the factors that influence a student's literacy achievement. These factors include influences from family and community, from the school, from the classroom, from the curriculum, and finally from factors that are centered in the individual such as exposure to language conventions, comprehension processes, and ownership of literacy competence. The striking feature of the figure is that it bears almost no resemblance to Figure 3.3 presented earlier in this chapter. The stimulus and response events mentioned in that figure have been subsumed into much larger entities such as "school," "teaching," and "curriculum." Events occurring within the individual are no longer described in terms that attempt to define knowledge, but instead are described in socially laden terms such as "language conventions," "ownership," and "literacy aspects." This 2000 article does not reflect the beginnings of a paradigm shift in Anderson's thinking, but instead reflects mature thinking that had had begun to incubate at least 10 years earlier.

THE SHIFT TO A SITUATED COGNITION PERSPECTIVE

At this point another pause is in order to catch up on Anderson's changing theoretical perspective. The cognitive theoretical perspective illustrated in Figure 3.3 indicates that the nature of the representation stored in the learner's head is constructed from an interaction between the incoming stimulus event and the learner's prior knowledge. Ample research had accumulated showing that the general form of this characterization was correct. That is, it was clear that what learners have in their head was different than the thing that was experienced and it was clear that the nature of this difference is associated with prior experience. Moreover, Anderson's research and the research of others suggested that the concept of a schema provided a good way of thinking about how this prior experience was organized and structured. Beyond this, things were pretty fuzzy in the Figure 3.3 characterization, and efforts to resolve some of this fuzziness lead to shifting the focus from individual learners to learners embedded within a social-cognitive milieu.

Anderson was one of a group of very talented and bright researchers who moved out of the laboratory and into the classroom during the late 1980s and early 1990s. The shift from the university laboratory to the classroom as a context for educational research was accompanied by an increasing awareness that an understanding of academic learning had to move well beyond considering what learners experienced as "stimulus events." Specifically, the research focus shifted to considering learners-within-groups, and even to considering the possibility that there were group schemata that played a role in the learning of group members. This new framework is described in Figure 3.4.

There are a number of important differences between the perspective illustrated in Figure 3.4 and the one illustrated in Figure 3.3. In Figure 3.3 the source of input in the model is external stimulus events. The implicit assumption in the model is that the contents of mind (schemata and memory representations) are ultimately determined by events that are unique to the individual. Educational theorists operating from a cognitive perspective as represented in Figure 3.3 might concede that events other than a stimulus event experienced by an individual learner influence the nature

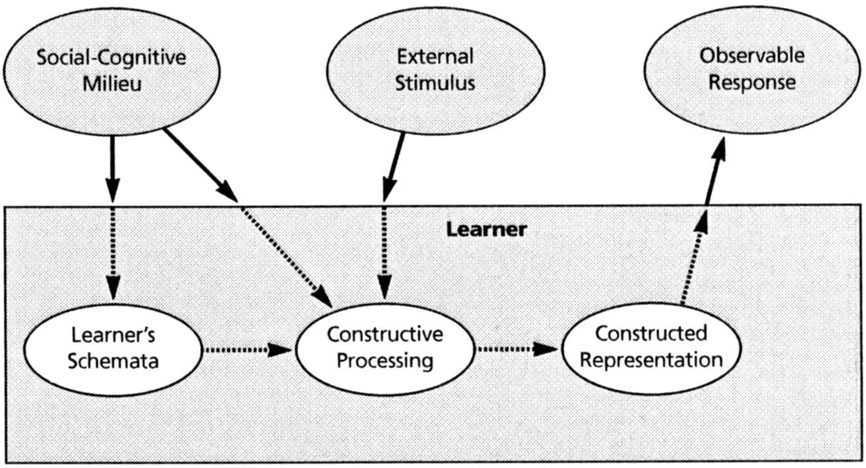

Figure 3.4. Situated cognition learning theory.

of what gets represented in mind, but those things are outside of the purview of the cognitive scientist. Rather, they are grist for the mill of social psychologists, curriculum developers, and teacher trainers.

Figure 3.4 makes the stuff outside of the stimulus event a very large part of what a cognitive scientist is interested in. The situated cognition perspective says that the social-cognitive milieu learners are embedded in influences their formation of schemata, and this in turn directly impacts on their success in benefiting from educational experience. Moreover, Figure 3.4 suggests that the social-cognitive milieu the learner is embedded in contributes to learning in a manner that is qualitatively different that the contribution associated with "external stimulus events." To be specific, learning derived from individual activities like reading, thinking, studying, and so on is different from learning derived from group interaction.

This change in perspective is very important, as will be illustrated by Anderson's next series of empirical studies. To foreshadow the research to be discussed, the idea that the social-cognitive milieu plays a direct role in influencing the formation of knowledge structures expands the educational researcher's realm of interest to the classroom context, to teacher activities, and most importantly for Anderson's new line of research, to the peer group that the student is part of.

The Microanalysis Studies

The fact that Anderson's research interests had shifted focus was clearly demonstrated in a series of very innovative studies that involved a very fine-grained analysis of reading groups. The first study (Anderson, Wilkinson, & Mason, 1991) involved having teachers teach two kinds of reading lessons, one that focused on story meaning and the other that focused on word analysis and accurate reading. These lessons were video taped and the innovative aspect of the research effort was that the performance of the students and the group as a whole was captured on tape and then synchronized with a detailed text analysis of the reading materials. This allowed the analysis of an intertwining of four aspects of the instructional experience: the continuous behavior of the student, the continuous behavior of the teacher, the continuous behavior of the entire reading group, and the changing properties of the text the group was interacting with. Anderson suggested that this type of a study had the virtue of allowing moment-by-moment understanding of what was going on in the reading group. The study provided many interesting results with two of the most interesting being the finding that an emphasis on understanding in a reading lesson resulted in better performance than an emphasis on accurate reading, and the finding that the collective performance of the group as a whole was a better predictor of individual performance than was a measure of individual student ability.

The next microanalysis study attempted to identify events occurring in a reading lesson that would facilitate or inhibit attention to the lesson (Iami, Anderson, Wilkinson, & Yi, 1992). Over 100 grade 2 and grade 3 students were videotaped while participating in a reading group. Anderson's shift in focus was clearly evident in his statement of the starting assumption for the study where he wrote "attention arises as much from the social logic of groups as from the inner logic of individuals." The tapes of the lessons were analyzed to provide a moment-by-moment measure of attention for each group member during each lesson. Again, there were a number of findings of interest but among the more interesting was the finding that the attention of group members declined when one member of the group made an error.

A third microanalysis study had the ambitious goal of predicting and explaining behavior while students were engaged in an oral reading episode (Chinn, Waggoner, Anderson, Schommer, & Wilkinson, 1993). Again the entire reading group was videotaped and analysis of the videotapes was synced with a flowing analysis of the text being read. The main purpose of the study was to test the hypothesis that a consideration of group behavior would add something to understanding individual reading performance. The study found support for this hypothesis and the authors concluded that the results were consistent with a sociocognitive theory that says that thought and actions are situated in the sense that they depend in a fundamental way on the immediate context.

The next study using the microanalysis methodology asked the question of what were the relative merits of silent versus oral reading when both occurred in the context of reading group instruction (Wilkinson, & Anderson, 1995). The general sequence of the reading lesson was that students would either orally or silently read a text segment and this would be followed by a group discussion of the segment. Again, lessons were taped and tape analysis was synced with an analysis of the text that was being read. The results of the study showed that children reading silently were more attentive to the story and they remembered more of the story than they did during oral reading. This positive finding was accompanied by a negative finding, which showed that the silent reading was slower paced than the oral reading. The slower pace was associated with the fact that faster readers would have to wait for slower readers to finish before they could engage in a discussion of the story. The article concludes with several recommendations on how best to use both silent and oral reading when conducting reading lessons.

The microanalysis studies are the most complex and ambitious studies that Anderson has completed to this point in his research career. Having completed a number of text analyses on my own, and having watched a colleague who does television research score moment-by-moment videotapes, I have some sense of the amount of effort that went into completing the studies. Aside from the data collection and data scoring effort, Anderson and his colleagues also had to figure out how to analyze the massive amounts of data they had collected and how to decide which stream of independent variable data (text variation, teacher behavior, group behavior) was influencing the ongoing behavior of individual students. This involved the construction of complex multivariate procedures that allowed statistical analysis of the data. The studies stand as a monument to a wonderfully skilled and innovative educational scientist.

There is another important way that the microanalysis studies depart from Anderson's previous research efforts. Anderson's previous research was clearly in the quantitative research tradition. He designed studies that collected numbers that could then be plugged into inferential statistical formulas. The microanalysis studies did collect data that could be analyzed in this tradition, but it was clear in the later studies that the process of transforming the data into indices that were readily quantifiable was, to some degree, forced. The fact of the matter was that the data that was being collected was more qualitative than quantitative. One indication of Anderson's talent as a researcher is that he didn't bat an eye at this sea change, he simply learned new ways of dealing with the data stream that he was collecting.

The Argumentation Studies

The next topic that Anderson began to examine was argumentation in children. It is not obvious why an interest in this topic emerged, but a good guess would be that children often presented arguments that were captured in the microanalysis studies and Anderson's inquisitive mind soon began to explore their nature and properties. An initial study (Anderson, Chinn, Chang, Waggoner, & Yi, 1997) examined the quality of naturally occurring arguments that children make. Anderson and his colleagues began their analysis by posing a theory of argument that could be used to evaluate argument quality. This theory consisted of a mix of informal deductivist theory and Gricean speech act theory. After examining the transcripts of 20 discussions conducted by approximately 80 fourth-grade children, Anderson and his coauthors concluded that most of the arguments that children make do not meet the strong standards imposed by formal deductive theory, but when examined from the perspective of the mixed theory they devised, the arguments were generally logical.

Having completed an initial study on the occurrence and quality of classroom argument, Anderson and his colleagues turned their attention to the instructional implications of the topic. The starting premise for the first instruction study was that classroom discussions can be used to promote the development of argument and that logical argument, in turn, has potential for increasing motivation and developing reasoning ability (Chinn & Anderson, 1998). The data reported in the study was gathered in the context of fourth-grade children engaging in "collaborative reasoning." The children in the study took positions on important issues, presented reasons and evidence for their positions, challenged each others' reasons and evidence, and changed their mind when they felt that the evidence warranted it. The principle aim of the study was to analyze the structure of group argumentation with the ultimate aim of allowing educators to improve the quality of group discussion and to plan discussions that would effectively promote the development of reasoning ability. The structure of group argumentation was analyzed using two techniques: argument networks and causal networks. Argument networks represent interactive argumentation as a series of premises and conclusions in an intricate web of arguments and subarguments.

Using the argument network technique, Chinn and Anderson (1998) noted the benefits and insights obtained from the approach, but also identified several significant shortcomings of the procedure. Specifically, they found that the argument network approach failed to capture the temporal and causal relationships among ideas expressed during discussions and it obscured the fact that students were often considering alternative consequences of the same event. These shortcomings led to the development of causal networks that represent sequences of events that are causally related. The authors then traced the instructionally useful insights to be gleaned from this second method of analyzing the structure of group argumentation.

Having conducted an initial exploratory analysis of student arguments, followed by a study that developed procedures for capturing the nature and properties of arguments, Anderson and his colleagues turned to a much more fundamental question: Where do arguments come from? More exactly, what role does the social milieu play in the acquisition of argument form? This is a question that has a great lineage, with Vygotsky and Piaget posing quite different answers to the question. Piaget proposed that interactions between children created sociocognitive conflict within an individual child that subsequently promoted cognitive growth. In essence, social interaction served as a catalyst that spurred the child to do the individual cognitive work that resulted in growth. Vygotsky, in contrast, proposed that the elements of growth were collective and first emerged in the group before they could be owned by the individual child. In short, cognitive growth emerged in the group as a whole and only then became part of the individual's cognitive repertoire. Despite the fact that the parameters of this debate have been well known for 30 years, there is little evidence from empirical studies that supports one view over the other. That is, there was little evidence until Anderson and his colleagues conducted the "snowball" study.

Anderson and his colleagues (Anderson et al., 2001) proposed that the arguments they had been studying in earlier research snowballed in discussion groups. That is, an argument would be absent during the initial phases of a discussion, suddenly appear, and then be quickly adopted for use by the group as a whole. The reader will note that this seems to be exactly what Vygotsky had proposed, but failed to support with empirical evidence.

Anderson and his colleagues had to confront a number of conceptual and methodological difficulties in order to evaluate the snowball hypothesis. The first was how to recognize argument forms. They proposed that argument forms could be conceptualized as schemata that contain invariant forms and placeholder variables. For example, one argument form that Anderson and his colleagues identified was, "I think [POSITION], because [REASON]." The capitalized part of the schema proposition is a variable that can take a variety of forms within the schematic structure. As a methodological aside, the process of recognizing comparable argument forms in thousands of lines of taped transcripts is a major undertaking that was greatly facilitated by the availability of qualitative analysis software that allowed the identification of schema forms.

Having developed a procedure for differentiating one argument from another, Anderson and his colleagues developed a means for determining the extent to which an argument was a social phenomenon. This entailed analyzing the probability that an argument form would occur in group members as a function of how many times it had occurred before. If it was a group occurrence, an argument should be absent from group discourse until it appeared once, whereupon the frequency with which it appears should rapidly increase with increasing usage. This would provide evidence that the transmission of the argument form is, in fact, a socially mediated event.

The evidence supported the snowball hypothesis. Anderson and his colleagues identified 13 different argument schemas and sorted through 48 discussion transcripts (involving about 15,000 lines of text) in order to calculate the probability an argument form would occur dependent on how many times it had occurred before. The large majority of the forms they examined were quickly adopted and used by group members after they initially appeared in group discourse, thereby providing evidence of the strong role groups play in individual learning.

Consider the implications of this finding for the schematic presented in Figure 3.4. The snowball study says that argument structure schemas can be almost instantaneously transmitted from individual to individual if the schema is encountered in a group context. The concept of the easy schema transmitted in the snowball study is in striking contrast to the schema acquired in Anderson's earlier schema studies. There, schemas were acquired only after many encounters with a particular experience such as going to a restaurant.

Sorting out the differences between schemata acquired via group experience and those acquired via repeated individual experience will undoubtedly be a major research topic in the years to come. Also of interest will be the identification of the characteristics of individuals who introduce original schemata to group discussion, and whether or not there are some individuals who are more successful at having their original contributions accepted than others. This writer has the feeling that Anderson will provide answers to these questions before the average researcher identifies them as issues.

CONCLUSIONS

This chapter has focused on a description and analysis of the research activities of one of the preeminent educational researchers of the last half of the 20th century. I believe that the description not only chronicles changes in the beliefs and activities of one researcher, but also describes major changes that have occurred in the entire field of learning and instruction. In the sections below I summarize the changes that occurred in the field as illustrated by Anderson's research.

Changes in Theory

Most educational psychologists interested in learning and instruction during the early 1960s ascribed to some version of behavioral theory described in an early section of this chapter. Instructional goals were couched in the language of behavioral objectives that specified the observable performances learners would be able to accomplish if learning was successful. Instruction consisted of specified activities that learners were to experience, and if a particular version of instruction did not achieve specified goals, it was revised until it did. All of this activity was governed by a theory that suggested that all important learning outcomes could be achieved with the right mix of stimulus environments, consequences occurring at the time of learning, and appropriate attention to measuring observable student responses. Teachers in this perspective were deliverers of engineered instruction and learners were passive recipients of the events they were experiencing.

This view changed radically with the advent of the cognitive revolution. Initially, this change was focused solely on the learner. That is, the learner was no longer viewed as a passive recipient of experience, but rather, was viewed as an active change agent that altered the nature of what was acquired from experience. However, there was little change in the view that the important aspects of learning could be identified by focusing on the experience of individual learners. There is a sense in which there was little change from behavioral to cognitive theory in this regard. That is, learning and instruction theorists ascribing to behavioral theory believed that manipulations of a learner's stimulus environment could produce desired outcomes. Likewise, the cognitive theorists also believed that manipulating the instructional environment of individual learners would produce positive change. The view of what to manipulate in that environment changed, and the view of the consequences of experiencing the environment changed, but the perspective that the experience of individual learners was of primary importance did not.

This certainty that educational goals could be accomplished by focusing on the experience of individual learners crumbled with the introduction of theory and research showing that it was necessary to broaden the definition of the instructional environment to include the social milieu in which learning was occurring. The important aspects of learning from this perspective included group processes as well as individual experience.

To summarize, I would suggest that Anderson's shifting theoretical perspectives and the shifting theoretical perspectives of mainstream learning and instruction research were similar in nature. Anderson began with the view that learners copied instructional experiences and that they behaved in a manner predictable from the principles of behavioral learning theory. This perspective was replaced with the view that learners changed the nature of what they experienced, but nonetheless, the important level of analysis for producing positive learning outcomes was still the individual learner. The current view is that learning can no longer be understood by focusing on individual experience. Instead, the frame of analysis must broaden to include the social-cultural milieu the learner comes from.

Changes in Beliefs about What Educational Researchers Should Do

As noted earlier in the chapter, Anderson believed early in his career that educational psychologists interested in learning and instruction should be educational engineers. They should use the scientific principles of psychology to produce educational products that could then be delivered to schools where they could be used to benefit learning. This meant that instructional researchers were interposed between the basic science producers and educators, the consumers of the educational products. I think the educational engineer view was one that was held by many learning and instruction researchers during the period. I certainly held the view when I began my own career during the early 1970s.

The problem with this view of the role of the learning and instruction specialist was the separation from both the basic science and the field of educational practice. The separation from educational practice was particularly problematic. I am sure that many psychologists during this era felt a sense of frustration brought on by the fact that educators seemed to have little interest in the marvelous instructional methods they were describing in their research. I certainly felt this frustration, and it was only after many years that it became obvious to me that my agenda was not the agenda of educators, and that if I did want to have a positive impact on instructional practice I needed to enter into partnerships with educators, instead of serving as an outsider who provided them with information they had little interest in.

I believe that Anderson made this shift in role perception during the late 1970s and much of his research during the 1980s and 1990s was firmly grounded in the concerns of educators and in his own experience in working with educators, and actually filling the role of a reading teacher. I think that many of the more prominent learning and instruction researchers from the 1980s on made a similar shift.

Changes in Research Methodology

The manner in which research was conducted and analyzed was very different when Anderson started his career than it is now. In the 1960s and 1970s researchers did experiments and they analyzed their quantitative data using inferential statistical procedures, usually analysis of variance. The participants in the research effort were most often college students, though an occasional study was conducted with primary or secondary students. There was typically one dependent variable in a study.

As research increasingly moved into schools, it was no longer possible to conduct true experiments because experimental niceties such as random assignment of participants to treatments was simply not possible. In addition, it became common practice to collect more than one dependent variable. These changes forced a change in the way research was conducted and analyzed. Increasingly, studies depended on naturally occurring variation in the environment as a means of manipulating independent variables, and multivariate regression procedures became as common as analysis of variance. As a personal aside, I took a multivariate statistics course as a graduate student in the late 1960s, but quickly forgot everything I learned because I never used it. Ten years later I had to painfully relearn everything I had forgotten.

Research methodologies and analysis procedures continued to change and grow in complexity during the 1990s, as illustrated by Anderson's research. His microanalysis and argumentation studies are highly complex, involving the simultaneous collection of a variety of quantitative and qualitative measures, collected in ongoing real time. The analysis and interpretation of the data from the studies is equally complex, and is possible only because of the development of techniques (such as software that analyzes qualitative data) that were not even dreamed of during the 1960s.

I think the changes in theory, role view, and research methodology has also produced changes in the way in which our science influences educational practice. I recently reread a 1964 book titled *Theories of Learning and Instruction*, published by the National Society for the Study of Education (Hilgard & Richey, 1964). The book summarized most of what was known at the time about how psychological theory and research could be used to influence educational practice. I saw little or nothing in the book that would be useful today, and in fact, I don't think it was very useful at the time the book was published. In contrast, I have also recently read a National Research Council Monograph, titled *How People Learn* (Bransford,

Brown, & Cocking, 1999). I think the contents of the recent book provide a wonderful summary of the impact our science is currently having on educational practice. Much of that impact has to do with changes in theory, beliefs, and research practices.

ACKNOWLEDGMENTS

I would like to thank Tom Andre, Don Cunningham, Ray Kulhavy, and John Guthrie for their comments on an earlier draft of this chapter. Requests for information regarding the chapter should be sent to: James M. Royer, Department of Psychology, University of Massachusetts, Amherst, MA 01003, or via email to: royer@psych.umass.edu.

REFERENCES

Adams, M., Anderson, R. C., & Durkin, D. (1978). Beginning reading: Theory and practice. *Language Arts, 55,* 19–25.

Anderson, R. C. (1959). Learning in discussions: A resume of the authoritarian-democratic studies. *Harvard Educational Review, 29,* 201–215.

Anderson, R. C. (1961a). Review of Abercrombe, *Anatomy of Judgment. Harvard Educational Review, 31,* 102.

Anderson, R. C. (1961b) The role of the educational engineer. *Journal of Educational Sociology, 34,* 377–381.

Anderson, R. C. (1965). Can first graders learn and advanced problem solving skill? *Journal of Educational Psychology, 56,* 283–294.

Anderson, R. C. (1966). Sequence constraints and concept identification. *Psychological Reports, 19,* 1295–1302.

Anderson, R. C. (1967). Educational psychology. *Annual Review of Psychology, 18,* 129–164.

Anderson, R. C. (1969). The comparative field experiment: An illustration from shigh school biology. *Proceedings of the 1968 Invitational Conference on Testing Problems,* pp. 3–30. Princeton, NJ: Educational Testing Service.

Anderson, R. C. (1970). Control of student mediating processes during verbal learning and instruction. *Review of Educational Research, 40,* 349–369.

Anderson, R. C. (1971). Encoding processes in the storage and retrieval of sentences. *Journal of Experimental Psychology, 91,* 338–340.

Anderson, R. C. (1972a). How to construct achievement tests to assess comprehension. *Review of Educational Research, 42,* 145–170.

Anderson, R. C. (1972b). Semantic organization and the retrieval of information from sentences. *Journal of Verbal Learning and Verbal Behavior, 11,* 794–800.

Anderson, R. C. (1974). Concretization and sentence learning. *Journal of Educational Psychology, 66,* 179–183.

Anderson, R. C. (1977). The notion of schemata and the educational interprise. In R. C. Anderson, R. J. Spiro, & W. E. Montague (Eds.), *Schooling and the acquisition of knowledge* (pp. 415–431). Hillsdale, NJ: Erlbaum.

Anderson, R. C. (1984). Some reflections on the acquisition of knowledge. *Educational Researcher, 13*(9), 5–10.

Anderson, R. C. (1990). Inferences about word meanings. In A. Graesser & G. Bower (Eds.), *Psychology of learning and motivation: Influences and text comprehension.* New York: Academic Press.

Anderson, R. C. (1993) The future of reading research. In A. Sweet & J. Anderson (Eds.), *Reading Research into the Year 2000* (pp. 17–36). Hillsdale, NJ: Erlbaum.

Anderson, R. C. (1998). Reflections on literacy education. In J. Osborn & F. Lehr (Eds.), *Literacy for all* (pp. 1–10). New York: Guilford Press.

Anderson, R. C., & Armbruster, B. B. (1990). Some maxims for learning and instruction. *Teachers College Record, 91,* 396–408.

Anderson, R. C., Armbruster, B. B., & Roe, M. (1990). Improving the education of reading teachers. *Daedalus, 119,* 187–209.

Anderson, R. C., Au, K., & Jacobs, A. S. (1991) *Teaching reading: Strategies from successful classrooms* [Videotape]. Champaign, IL: Center for the Study of Reading.

Anderson, R. C., & Ausubel, D. P. (Eds.). *Readings in the psychology of cognition.* New York: Holt Rinehart & Winston.

Anderson, R. C., & Biddle, W. B. (1975). On asking people questions about what they are reading. In G. Bower (Ed.), *Psychology of learning and motivation* (Vol. 9, pp. 89–132). New York: Academic Press.

Anderson, R. C., Chinn, C., Chang, J., Waggoner, M., & Yi, H. (1997) On the logical integrity of children's arguments. *Cognition and Instruction, 15(2),* 135–167.

Anderson, R. C., & Faust, G. W. (1973). *Educational psychology: The science of instruction and learning.* New York: Dodd-Mead.

Anderson, R. C., Faust, G. W., Guthrie, J. T., & Drantz, V. (1969). *Population genetics: A self-instructional program.* Morristown, NJ: Silver Burdett.

Anderson, R. C., & Guthrie, J. T. (1966). *Journal of Experimental Psychology, 72,* 501–504.

Anderson, R. C., Hiebert, E. H., Scott, J. A., & Wilkinson, I. G. (1985). *Becoming a nation of readers.* Center for the Study of Reading, Champaign, IL.

Anderson, R. C., & Kulhavy, R. W. (1972). Learning concepts from definitions. *American Educational Research Journal, 9,* 385–390.

Anderson, R. C., Mason, J., & Shirey, L. (1984). The reading group: An experimental analysis of a labyrinth. *Reading Research Quarterly, 20*(1), 6–38.

Anderson, R. C., & McGaw, B. (1973). On the representation of the meanings of general terms. *Journal of Experimental Psychology, 101,* 301–306.

Anderson, R. C., & Myrow, D. L. (1971). Retroactive inhibition of meaningful discourse. *Journal of Educational Psychology, 62,* 315–321.

Anderson, R. C., & Nagy, W. E. (1991). Word meanings. In P. D. Pearson (Ed.), *Handbook of reading research* (Vol. 2, pp. 690–724). New York: Longman.

Anderson, R. C., Nguyen-Jahiel, K., McNurlen, B., Archodidou, A., Kim, S.-Y., Reznitskaya, A., Tillmanns, M., & Gilbert, L. (2001) The snowball phenomenon: Spread of ways of talking and ways of thinking across groups of children. *Cognition and Instruction,* 19, 1–46.

Anderson, R. C., & Ortony, A. (1975). On putting apples into bottles—A problem of polysemy. *Cognitive Psychology, 7*, 167–180.

Anderson, R. C., & Pichert, J. W. (1978). Recall of previously unrecallable informantion following a shift in perspective. *Journal of Verbal Learning and Verbal Behavior, 17*, 1–12.

Anderson, R. C., Pichert, J. W., Goetz, E. T., Schallert, D. L., Stevens, K. V., & Trollip, S. R. (1976). Instantiation of general terms. *Journal of Verbal Learning and Verbal Behavior, 15*, 667–669.

Anderson, R. C., Pichert, J. W., & Shirey, L. (1983). Effect's of the reader's schema at different points in time. *Journal of Educational Psychology, 75*, 271–279.

Anderson, R. C., Reynolds, R. E., Schallert, D. L., & Goetz, E. T. (1977). Frameworks for comprehending discourse. *American Educational Research Journal, 14*, 367–382.

Anderson, R. C., Wilkinson, I. A. G., & Mason, J. M. (1991). A microanalysis of the small-group guided reading lesson: Effects of an emphasis on global story meaning. *Reading Research Quarterly, 26*(4), 417–441.

Anderson, R. C., Wilson, P., & Fielding, L. (1988). Growth in reading and how children spend their time out of school. *Reading Research Quarterly, 23*(3), 285–303.

Anderson, R. C., Wilkinson, I., Mason, J. M., Shirey, L., & Wilson. P. (1988). Do errors on classroom tasks slow growth in reading? *Elementary School Journal, 88*, 267–280.

Armbruster, B. B, Anderson, R. C., & Mall, V. C. (1991) Preparing teachers of literacy. *Educational Leadership, 49*(3), 21–24.

Bransford, J. B., Brown, A. L., & Cocking, R. R. (1999). *How people learn: Brain, mind, experience and school.* Washington, DC: National Academy Press.

Chinn, C.A., & Anderson, R. C. (1998). The structure of discussions intended to promote reasoning. *Teachers College Record, 100*, 315–368.

Chinn, C., Waggoner, M., Anderson, R. C., Schommer, M., & Wilkinson, I. A. G. (1993) Situated actions during reading lessons: A microanalysis of oral reading error episodes. *American Educational Research Journal, 30*(2), 361–392.

Freebody, P., & Anderson, R. C. (1983a). Effects on text comprehension of differing proportions and locations of difficult vocabulary. *Journal of Reading Behavior, 15*(3), 19–39.

Freebody, P., & Anderson, R. C. (1983b). Effects of vocabulary difficulty, text cohesion, and schema availability on reading comprehension. *Reading Research Quarterly, 18*, 277–294.

Half, H. M., Ortony, A., & Anderson, R. C. (1976). A context sensitive representation of word meaning. *Memory and Cognition, 4*, 378–383.

Hilgard, E. R., & Richey, H. G. (Eds.). (1964). *Theories of learning and instruction. The Sixty-Third Yearbook of the National Society for the Study of Education.* Chicago: University of Chicago Press.

Imai, M., Anderson, R. C., Wilkinson, I., & Yi, H (1992). Properties of attention during reading lessons. *Journal of Educational Psychology. 84*, 160–173.

Kintsch, W. (1988). The role of knowledge in discourse comprehension: A construction-integration model. *Psychological Review, 95*, 163–182.

Liebherr, H., & Anderson, R. C., et al. (1970). *Biological science: Patterns and processes* (2nd ed.). New York: Holt, Rinehart & Winston.

Mason, J. M., Anderson, R. C., Omura, A., Uchida, N., & Imai, M. (1989). Learning to read in Japan. *Journal of Curriculum Studies*, 21, 389–440.

Myrow, D. L., & Anderson, R. C. (1972). Retroactive inhibition of prose as a function of the type of test. *Journal of Educational Psychology, 63,* 303–308.

Nagy, W. E., & Anderson, R. C. (1984). How many words are there in printed school English? *Reading Research Quarterly, 14,* 304–330.

Nagy, W. E., Herman, P., & Anderson, R. C. (1985). Learning words from context. *Reading Research Quarterly,* 20(2), 233-253.

Nagy, W. E., Anderson, R. C., Schommer, M., Scott, J. A., & Stallman, A. (1989). Morphological families in the internal lexicon. *Reading Research Quarterly, 24*(3), 262–282.

Osborn, J., & Lehr, F. (Eds.). (1998). *Literacy for all.* New York: Guilford Press.

Pichert, J. W., & Anderson, R. C. (1977). Taking different perspectives on a story. *Journal of Educational Psychology, 69,* 309–315.

Roderick, M. C., & Anderson, R. C. (1968). A programed introduction to psychology versus a textbook-style summary of the same lesson. *Journal of Educational Psychology, 59,* 381–387.

Reynolds, R. E., Taylor, M. A., Steffensen, M. S., Shirey, L. L., & Anderson, R. C. (1982). Cultural schemata and reading comprehension. *Reading Research Quarterly, 17,* 353–366.

Royer, J. M. (in press). Uses for the sentence verification technique for measuring language comprehension. *Progress in Education.*

Royer, J. M., Carlo, M. S., Dufrense, R., & Mestre, J. (1996). The assessment of levels of domain expertise while reading. *Cognition and Instruction, 14,* 373–408.

Steffensen, M. S., Joag-dev, C., & Anderson, R. C. (1979). A cross-cultural perspective on reading comprehension. *Reading Research Quarterly, 15,* 10–29.

Taylor, B., Anderson, R. C., Au, K., & Raphael, T. (2000) Discretion in the translation of reading research to policy: A case from beginning reading. *Educational Researcher, 29*(6), 16–26.

Taylor-Delain, M., Pearson, P. D., & Anderson, R. C. (1985). Reading comprehension and creativity in black language use: You stand to gain by playing the sounding game. *American Educational Research Journal, 22,* 155–173.

Wilkinson, I. A. G., & Anderson, R. C. (1995) Sociocognitive processes in guided silent reading: A microanalysis of small-group lessons *Reading Research Quarterly, 30,* 710–740.

CHAPTER 4

MAY YOU TEACH IN INTERESTING TIMES![1]

Donald J. Cunningham
Indiana University

ABSTRACT

The cognitive revolution has revolutionized more than theory, research and methodology in psychology. It has also had a major impact upon the content and pedagogy of introductory classes—particularly those classes focusing on learning, cognition, and instruction. In this brief paper I will recount my experiences with these classes, first as a student and then as a teacher. I make no pretense that my experience is typical but when you've been doing something for this long, you just have to talk about it now and again.

MY LIFE AS A STUDENT

I distinctly remember the class "Psychology of Learning" that I took in the Spring semester of 1964 from Robert Roselius at Wittenberg University (at 8:00 AM!). The class followed a rather traditional read–lecture–test format and the textbook, James Deese's (1958) *The Psychology of Learning*, was solid, if uninspiring. The usual topics of the time were covered: origins in

The Impact of the Cognitive Revolution on Educational Psychology, pages 87–103

the British Associationist philosophers, Pavlov, Ebbinghaus, Skinner, Thorndike, Hull, and the like. Beyond a few derogatory comments about Tolman's purposive behaviorism and a brief excursion into Gestalt psychology, cognition was hardly mentioned. Gardner (1985) reminds us that there was already considerable activity within the cognitive revolution[2] by the time I took this class, but little or none of it had worked its way into the textbook or Roseluis's lecture notes!

My next encounters with classes in learning and cognition were at the University of Illinois where I was pursuing my doctorate in a program called the "Psychology of Classroom Learning and Instruction." I had originally applied to a number of graduate schools in the field of industrial psychology—I hadn't been all that impressed by the field of learning psychology during the aforementioned class. I was admitted to several schools, Michigan State and Tennessee, if I remember correctly, and also was turned down at several. Out of the blue, I received a letter from one Richard C. Anderson asking if I would be interested in applying for this new doctoral program and offering me an assistantship far better than any place else had offered. It seems that Professor Anderson, newly arrived at the University of Illinois and working to build up this doctoral program, had been sifting through the rejection files of other graduate programs in psychology at the University of Illinois and had come across my file. Well, I'd never heard of such a program and neither had my professors at Wittenberg, but I decided to accept his offer.

During my first semester, I took a class with Anderson on learning and instruction as well as a class from the arch Hullian G. Robert Grice. Of the latter I will say little except to report that the rat finally did learn the discrimination I was required to teach and I vowed never again to consort with rodents. Anderson's class was a real eye opener. It was a curious mixture of hard core behaviorism and learning in school contexts. One of our textbooks was a computer-mediated presentation of Holland and Skinner's (1961) *The Analysis of Behavior* that Anderson had incorporated into a CAI system called SOCRATES (actually a glorified slide projector). Although the delivery left something to be desired, this book may yet be the best single instructional resource for learning principles of classical and operant conditioning. The other textbook was Robert M. W. Travers's (1963) *Essentials of Learning: An Overview for Students of Education.* This class and these readings were my first exposure to learning theories that focused on instructional settings and I remember the class as somewhat of an epiphany in my professional career. For the first time I knew what I wanted to be when I grew up: a researcher in the psychology of learning applied to education.

I began my career in educational psychology as a behaviorist, a staunch advocate of B. F. Skinner and the programmed instruction movement, just like my mentor, Dick Anderson. I became deeply enmeshed within and committed to the scientific study of learning and to a systems approach to the design and development of instruction. I believed that with the proper application of rational canons of science and utilizing an evolving arsenal of methodological tools, we educational psychologists would be able to generate lawful descriptions of the teaching–learning process and, eventually, prescriptions for educational practice.

I continued to take every class I could find on learning and related topics and began to discover worlds of research and theory that were as yet undreamed of in our doctoral program. Jack Adams taught a course on human learning that introduced me to the rapidly emerging information-processing approach to the study of cognition. His book, *Human Memory* (1967), had not yet been published and we were offered dittoed copies of the chapters soon after he had finished typing them and encouraged to comment and critique. Instead of waiting for the content to make its way into the textbook, we were assisting in the process as it happened! Memory was rapidly emerging as a pervasive topic and it became not only permissible but mandatory to theorize about the processes underlying learning. While much of this work maintained its behaviorist origins (e.g., Adams adamantly promoted an interference theory of forgetting in both long-term and short-term memory), we strongly felt that we were moving far beyond the notion of learning as a simple change of behavior as a result of a stimulus. Other classes such as mathematical learning theory from Bill Batchelder and psycholinguistics from Charlie Osgood made it seem as if the methodological tools and domains of application were unlimited.

But the class and book that had the biggest impact on me was a doctoral seminar I took with Bill Montigue where we systematically worked our way through Ulrich Neisser's *Cognitive Psychology*[3] (1967) and taught each other as much as we could about the topics. His definition of cognition was widely cited: "....cognition refers to all the processes by which the sensory input is transformed, reduced, elaborated, stored, recovered and used" (Neisser, 1967, p. 4). Although the research described in the book was focused on very basic cognitive processes of sensation, perception, and simple memory, Neisser also included a provocative and visionary chapter titled "A Cognitive Approach to Memory and Thought." Here he speculated about a reconstructive view of memory, the nature of cognitive structures, the role of the executive in managing cognitive processes, and so forth, themes that would reemerge in his more accessible book, *Cognition and Reality* (Neisser, 1976).

All of my exciting new learning was taking place against a backdrop of our doctoral program on classroom learning and instruction that was still very much entrenched within a behaviorist foundation. The first research I remember Dick Anderson (1965) talking about at our weekly seminar was his attempt to debunk the work of cognitivists such as Bruner and Piaget on concept learning and problem solving. He argued that careful analysis of the task and systematic instructional strategies could produce high-level learning outcomes even in students who were cognitively immature in terms of Piaget's model. Anderson had several lines of research underway in this same vein, often in collaboration with his graduate students. I and a number of my fellow graduate students were busy completing master's theses and doctoral dissertations designed to support a more behavioral, direct instruction model. My master's theses, for example, was a very basic paired associate task designed to show that differences in the total time available for learning number–nonsense syllable pairs had confounded the results of prior research on whether prior prompts or feedback played important roles in learning (Cunningham & Anderson, 1968). Compared to the heady stuff we were learning about in courses described above, this was pretty mundane!

And yet by the end of my doctoral program there were signs that the program, and Anderson along with it, were moving toward fully participating in the cognitive revolution. David Ausubel, a radical cognitivist, was also on the faculty at that time and he and Anderson had actually coedited *Readings in the Psychology of Cognition* (Anderson & Ausubel, 1965). Interestingly, however, none of the readings included in the book was authored by Anderson, while Ausubel had three. The two instructional texts (Anderson & Faust, 1973; Anderson, Faust, Roderick, Cunningham, & Andre, 1969) he produced during this period, while still very much embedded in the behaviorist tradition, nevertheless introduced more cognitively oriented content (e.g., a chapter on memory stuck in awkwardly as the last chapter of Anderson & Faust, 1973). Anderson's transition from behaviorist to cognitivist is well documented by Royer (Chapter 3, in this volume) and will not be repeated here. I recount this history to note that we, who were about to become professors of educational psychology in universities throughout the country (and world, for that matter) and teach our own classes in learning and cognition, could certainly "borrow" content and pedagogical strategies from a master of the behavioral approach to learning and instruction, but we were on our own with respect to the explosion of research and theory in cognition. Interesting times indeed!

MY LIFE AS A PROFESSOR

I arrived at Indiana University School of Education in August 1969, not quite yet a PhD, but ready to change the world. One of my primary quests was to teach the beginning learning and cognition class. In those days, as now, this course is offered by the Department of Educational Psychology as a graduate, three-credit-hour class to all comers. There are no prerequisites, so for some students the class could be a repeat of an undergraduate class as part of their psychology major while for others it might be their first course ever in psychology. It is a required class in many of the master's and doctoral programs in Indiana's School of Education, further increasing the diversity of the students.

Fortunately for me, the regular instructor for the class had recently been appointed as an administrator and was looking to decrease his teaching load. My assignment was to be a temporary one-semester appointment, but this has lasted for 35 years and counting. I would guess that well over 2,500 students have passed through this class and spent a little time with me thinking about EDUC P540 *Theories of Learning and Cognition in Education,* the official title.

Does anyone ever forget their first class as a teacher? While I had taught as an associate instructor at the University of Illinois and was concurrently teaching other classes, this is the teaching experience that I will always remember as my first. I actually had very little warning that I was going to teach and therefore decided to follow closely the syllabus and readings of the previous instructor. He had adopted seven(!) books, including primary sources such as Ausubel's (1963) *The Psychology of Meaningful Verbal Learning,* Gagne's (1965) *The Conditions of Learning,* and Bruner, Oliver, and Greenfield's (1967) *Studies in Cognitive Growth,* as well as some challenging secondary sources such as Flavell's (1963) *The Developmental Psychology of Jean Piaget* and Reynolds's (1968) *A Primer of Operant Conditioning.* My first priority was *reading* all of these books, some for the first time! Fortunately, the students were as overwhelmed as I and did not read ahead of me.

As is true of most new assistant professors, I had received absolutely no training in college pedagogy and therefore taught as I had been taught. In my original course with Anderson, he had used unannounced quizzes to motivate reading, a kind of variable ratio schedule of reinforcement. Coming to class you never knew if this was to be the day for a quiz. Most of the students disliked the procedure but admitted that it did make it more likely that they would read the material. I continued this practice for more than 20 years, dropping it only when my theoretical views shifted (of which more later). I also adopted Anderson's final examination procedure. He distributed a pool of 12–15 questions a few weeks ahead of the examination and announced that the final would consist of three of those questions, arbitrarily chosen. He also encouraged students to collaborate in studying for the examination and preparing model answers. Here again his behaviorist leanings came to the foreground. He announced the objectives for the final examination and carefully defined the behaviors required to meet those objectives, a kind of early version of the mastery learning approach. Getting the students to prepare answers to 12 questions while only having to read their answers to three of them was a wonderful example of managing his own contingencies of teaching! By encouraging students to collaborate, he may even have been showing early symptoms of becoming a constructivist, although I think he would disavow the label.

P540 followed this same basic pattern for several years: reading a mixture of primary and secondary sources, class sessions that were primarily lectures by me summarizing and supplementing the readings, and assessments where students mostly reported back what they had read or heard. The content shifted a bit over these years as the cognitive information-processing approach to cognitive psychology in education began to compete with and, in some cases, achieve dominance over the Piaget, Bruner, and Ausubel versions. I had students read the Atkinson and Shiffrin (1968) paper on human memory and "models of memory" became a standard unit in the class for many years. Donald Norman's (1967) *Memory and Attention: An Introduction to Human Information Processing* was one of the required books at this time.

The organization of the class thus primarily reflected my lingering behaviorist leanings even as I introduced more and more cognitive psychology. It was very teacher centered in that I was the primary (exclusive?) determiner of the outcomes of the class; that is, the students were to master the essentials of the various theories of learning and cognition that were presented in the lectures and readings. I treated the students as if they were all preparing to be experts in the field of learning psychology. Once they learned the principles, they could then apply them to their various fields of study. That was none of my business! The cognitive revolution changed all that.

In accord with Neisser's (1967) speculations, the field of learning began to look very carefully at cognitive processes underlying learning and the factors that influenced those processes. After an initial flurry of studies about the structure of memory, the focus shifted to levels of processing (e.g., Craik and Lockhart's [1972] paper was widely cited) and the consequences of different types and depths of processing. Alan Paivio's (1971) work on the role of imagery was enormously popular during this period. Theoretical and empirical papers on meta-cognition began to appear with increased frequency (e.g., Flavell, 1970). Books on mnemonics and other strategies for remembering information found their way on to bestseller lists (e.g., Lorayne & Lucas, 1974) as well as scientific publications lists (Luria, 1968). CRM Productions produced a highly entertaining film called *Human Information Processing* (Norman, 1971), starring a very young-looking Donald Norman, that illustrated many memory phenomena through the reenactment of a cocktail party. It was a big hit with my class!

Another emergent strand in the cognitive literature was the influence of prior knowledge on learning. While the concept of schema had been around for many years (e.g., Bartlett, 1932), it began to play an increasingly important role for cognitive psychologists such as David Rumelhart (1975), Ulrich Neisser (1976), and (wait for it!) Richard C. Anderson (Pichert & Anderson, 1977). Roger Shank (Shank & Abelson, 1977) and others began investigating script knowledge defined as stereotypical memory for events like going to a restaurant or attending a birthday party. More formal models of knowledge structures such as HAM (Anderson & Bower, 1973) and ACT (Anderson, 1976) also appeared with increasing frequency.

All this work changed not only the content of P540 but also how it was taught. As well, during this period, I (e.g., Cunningham, 1972) was carrying out a modest program of research on "prose learning"; that is, learning by reading. What, I wanted to know, are the features of prose materials that lead to efficient and effective learning outcomes? What strategies should we teach students in order to optimize comprehension? I thought deeply about the ways in which I could make the P540 class more relevant to my students. Was it enough for them to become knowledgeable about the various theories of learning or did I want to be able to link these theories to their respective fields of application? As the literature of cognitive psychology shifted to the study of learning of meaningful tasks in more realistic and relevant contexts (e.g., Bransford & Franks, 1971; Bransford & Johnson, 1972), I tried to incorporate that work into my classroom. I began to include much more demonstration and discussion. Instead of lectures about the principles of behaviorism or information-processing cognitive psychology, many classes were devoted to exercises where the students either experienced the principles based upon some activity that I devised or worked in groups to develop applications of their own. To this day, I get letters from students who have vivid memories of one or more of these class sessions. One of my favorites is what I sometimes refer to as my "Fuck" class. While we are studying classical conditioning my students usually roll their eyes, assuming that salivating dogs have little relevance to education. At a propitious moment I will walk to the board and write the word "fuck" in large letters and wait for the reaction that usually comes—nervous laughter, smirks, wide eyes, and so on. The discussion that follows centers on how words such as racial/ethnic invectives and "dirty" words come to have emotional connotations and the power to shock, offend, and wound. Classical conditioning provides a useful model of how such emotional meaning might have developed and how it might be minimized. Exercises on mnemonic techniques and problem solving have also proved very popular over the years. Bransford and Stein's (1984) *The Ideal Problem Solver* was a great source of exercises.

I moved away from primary source readings and began to select texts that were more focused on application and integration. Most of the comprehensive textbooks on learning at the time (e.g., Hilgard & Bower, 1966) were more appropriate for a course in a psychology department than educational psychology, so I experimented for a number of years to find a useful mix of more focused books. Neisser's (1976) *Cognition and Reality* filled a real gap for a number of years but it was not particularly strong (in fact quite hostile toward!) basic information-processing cognitive theory. Bransford's (1979) *Human Cognition* replaced Neisser for a few years. Barry Wadsworth's (1971) *Piaget's Theory of Cognitive Development* was popular with my students. Gagne and Briggs's (1974) *Principles of Instructional Design* was accorded almost reverential status by students in our Department of Instructional Systems Technology. Paul Chance (1979) put out a lovely little book on behavioral theories (including Bandura's early work on observational learning) titled *Learning and Behavior* that I used for a number of years. I never eliminated behaviorism from my class as I know others have done because it is still a foundational theory in a number of areas within my School of Education (e.g., special education, counseling psychology) and because I still find it to be a useful tool for analyzing aspects of learning and cognition. Besides, it was where I started!

I maintained the practice of unannounced quizzes and a comprehensive final examination since I found that students who had read the material prior to coming to class were much better prepared to discuss and apply it. While these techniques had their origin in behavioral theories, I was quite impressed by the cognitive processes they engendered. My quizzes changed from simple knowledge questions (e.g., "List and briefly describe Piaget's stages") to questions that required the students to integrate and reconcile perspectives (I got the idea from Ausubel [1963]!). Now students might have to tell me how Gagne's levels of intellectual skills were different from *and* similar to Piaget's stages. The problem was that as the questions become more complex and pitched at higher-level cognitive strategies, they became quite onerous and time-consuming for the student to complete in the 20 minutes I allocated for the quiz. Both the quizzes and the final examination generated a lot of student anxiety and I was actively looking for alternatives.

At the end of the 1970s and into the early 1980s a few comprehensive textbooks that really addressed applications to education began to appear and I used most of them at one time or another, supplementing any areas that they omitted. Rich Mayer's (1981) *The Promise of Cognitive Psychology* was one of the best but it was organized by content areas (e.g., teaching reading, mathematics) and I had my course organized by theorist. The book that best fit my course organization was Margaret Gredler's (1986) *Learning and Instruction*, with separate chapters on most of the theoretical perspectives I covered in P540 along with explicit links to applications of the theories. The one weakness of the text, I thought, was the brevity of the treatment of cognitive information processing. Glover, Ronning, and Bruning's (1990) *Cognitive Psychology for Teachers* filled that gap and I often adopted the two books together when the latter became available. By the 1990s, then, I had mostly moved away from primary source readings by the theorists themselves through a series of more focused secondary texts to a comprehensive textbook. Today there are numerous comprehensive textbooks, including the latest editions by Gredler (2001), Mayer (2003), Schunk (2004), Ormrod (2004), and my current favorite, Driscoll (2000). As is true in most fields, textbook content tends to lag 5 to 10 years behind developments in the field, so there will always be a need to supplement with focused readings.

In the late 1980s and into the 1990s, the cognitive revolution spawned theories that began to challenge and extend some of the foundational assumptions of cognitive theory: basic notions of knowledge and what it means to know (e.g., constructivism as in the work of Ernst von Glasersfeld (1984), the importance of context (e.g., situated cognition as in the work of Brown, Collins, & Duguid, 1989), the role of experts in apprentice models of instruction (e.g., Collins, Brown, & Newman, 1989), the social nature of learning and the importance of learning communities (e.g., Lave & Wenger, 1991), and the importance of culture and history in cognition (e.g., Cole, 1996), to mention just a few. Connectionist models of cognition (e.g., Rumelhart & McClelland, 1986) set the stage for distributed models of cognition, such as those discussed in Salomon (1993) where cognition is moved out of the head and into the connections with artifacts in the world and other people (see also Cobb, 1994).

It was during this period that my own theoretical interests began to stray away from the information-processing view of cognition. My excursion in prose learning took an unexpected turn. Prose, connected discourse, and text learning were under investigation by a growing number of psychologists of the day. Under the dominant metaphor, reading became text processing, an information-processing activity coordinated with many other information-processing activities (pattern recognition, feature analysis, recoding, semantic analysis, storage, retrieval, etc.). My move away from this metaphor began when I wondered exactly what I (and others, as well) meant by prose, text, discourse, and so on. To cut a very long story short, this inquiry led me through a maze of readings in discourse theory, literary criticism, hermeneutics, genre, intertextuality, poetics, and inevitably to semiotics. I remember attending a meeting of the Semiotic Society of America in Bloomington in 1979 and being dazzled by the magnificent array of theoretical conceptions on display. The notion of text takes on quite a different meaning when applied to such diverse phenomena as a ballet, a bar room conversation, the Empire State Building, the seating arrangement at the United Nations, clothing, and so forth. At the time, the majority of research on text learning relied upon perhaps two dozen texts (old timers will remember Circle Island and a chapter from Rachel Carson's *The Sea Around Us*). Yet psychologists continued to make sweeping generalizations about text processing and reading as if text were a fixed effect in an ANOVA model. What an impoverished conception of text we had, and consequently what an impoverished view of cognition! I was appalled at my own insularity as a psychologist from the research and writing of so many legendary figures as C. S. Peirce, Roland Barthes, Claude Levi-Strauss, Gregory Bateson, Thomas Sebeok, Umberto Eco, and many more too numerous to mention. I grew resentful of the fact that my training had been so narrow and biased toward a particular worldview of scholarship and not a little depressed about how insignificant I had come to regard my own previous work and that of educational psychologists in general. In the 1980s I began publishing semiotic models of cognition and have continued to do so but since hardly anyone else considers semiotics as part of the "cognitive revolution," I will resist the temptation to elaborate here.[4]

I continued to teach the standard theories in the P540 class but my newfound theoretical perspectives began to intrude, first in how I taught the class and later as I introduced more and more material from "beyond the fringe," as I sometimes characterized things. First, I began to talk about theories as tools. Here is a brief excerpt from a recent syllabus:

> The course is designed to provide you with the basic distinctions and concepts necessary to apply various theories of learning, cognition, and cognitive development to educational settings, broadly defined. These theories are tools that parents, educators, trainers, counselors, and many others can use to make their endeavors more productive and useful. This course will introduce and illustrate the proper use of these tools in providing insights into defining and solving problems. The emphasis will always be on the use of these theories to solve realistic and relevant problems drawn either from your own personal experience or from cases we will study. It is only when you have experienced applying these theories to actual problems that you can readily see their strengths and weaknesses.
>
> I have used the word "tool" several times already and it is an important one to understand in the context of this course. I claim that theories are tools much like hammers and screwdrivers are tools. A hammer is a useful and effective tool if your task is to drive a nail into a board. You can try to use a screwdriver to drive a nail, but I suspect you will fail and wind up with cuts and bruises. Likewise, if you try to drive a screw with a hammer, the wood will split and the bond is unlikely to hold—but a screwdriver will accomplish the task very well. And despite the best efforts of our most clever toolmakers, there is no such thing as a tool that meets all of our needs.
>
> The analogy I am drawing is that a theory like Skinner's operant conditioning is a useful and appropriate tool for certain problems like toilet training young children but not for other problems. Similarly, Piaget's theory is espe-

> cially helpful when considering the appropriateness of certain mathematics tasks for a grade level, but not very helpful for many other problems. During this course we will review a number of theories and theoretical concepts because no one theory is applicable to all of the problems one is likely to encounter.
>
> But theories also carry with them a worldview, a conception of what it means to be human: what it means to learn something, to teach something, to know something, to be a person. It is important that we examine these worldviews so that we can better judge the appropriateness of using a particular theory. To extend the analogy, to a 2-year-old child with a hammer in her hands, everything in the world needs a good pounding. If we limit ourselves to one or a few theories, or if we fail to see the kinds of assumptions that theory makes about the world, we will commit a similar error. If we limit ourselves to operant conditioning, for example, we run the risk of regarding all of our behavior (from simple forms like disliking broccoli to complex forms like understanding quantum mechanics) as strengthened or weakened according to external consequences—rewards and punishments. One of my major aims for the course is that you carry away with you a variety of tools and a sense of when they are and are not best applied.

Those who have kept up with the cognitive revolution will certainly recognize the influence of Lakoff and Johnson's (1980) *Metaphors We Live By* and Vygotsky's (1978) *Mind in Society* in this use of the concept of theories as tools. I began designing class activities that encouraged the students to identify the "metaphors" that various theories embodied so that they could better compare the suitability of the theory for application in their fields of study. In one such exercise, I have the students consider the following set of assumptions about teaching and learning:

SOME DEBATABLE ASSUMPTIONS

1. Knowledge is external to a learner and can be represented in some format akin to a task analysis or semantic map. Our job as educators is therefore to package this knowledge in some optimal format.
2. Instruction is primarily a process of efficient communication where knowledge in this optimum format is efficiently and effectively transferred to the pupil.
3. The effectiveness of the teaching/learning process is best measured by assessing the success of this communication process, whether the knowledge as pre-specified has been acquired by the pupil.
4. The context in which learning should take place is that which minimizes disruptions to the communication process—for example,

where students can concentrate without distraction. The context per se is irrelevant to the knowledge being acquired.

5. At its heart, learning is an abstractive process. Concrete experience is very useful in illustrating the abstractions and algorithms that make up mature thought, but it is abstractions that free one from the immediacy of experience and allow generalized thought. These abstractions are most conveniently represented in linguistic and/or mathematical codes. In the long run, these abstractions are capable of being simulated on a computer.
6. Tools such as computers and calculators can assist the knowledge communication process, taking over such aspects that are inconvenient for or not well suited to the human teacher.
7. Learning is fundamentally an individual act and is best conceptualized from an individual perspective. Group learning may sometimes be an effective pedagogical strategy, but nothing more. In instruction we are dealing with individual selves, not collectives.
8. Knowledge of efficient strategies for storing, retrieving and processing information can help the learner optimize the communication process.

And, of course, the BIG one:

9. TEACHING CAUSES LEARNING.

Each of these "debatable" assumptions is embodied in one or more of the theories of learning, cognition, and cognitive development that we study in P540, and it is instructive for the students to speculate what a theory that held different or opposite assumptions would look like. Constructivists would certainly find assumption 1 objectionable, situated cognition theorists stress the importance of context in learning (assumption 4), sociocultural theorists would reject the centrality of the individual learner (assumption 7), and so on. The focus of our class discussions and activities switched to these meta-theoretical factors in determining the assumptions underlying the various theories we reviewed. Since every theory is in essence a worldview, a set of glasses through the world is observed, what is revealed by the theory, what aspects of the world are brought to the forefront? Likewise, what things are hidden, obscured, or disrupted? What is gained by using a particular theory to analyze a situation and what is lost?

In accord with this approach to understanding theory, I stopped using the unannounced quizzes and substituted an assignment I call a "reflexive" paper. One of these is due three times throughout the semester. The word reflexive means directed or turned back upon itself or self-referential. Umberto Eco put the matter this way:

> To speak about "speaking," to signify signification, or to communicate about communication cannot but influence the universe of speaking, signifying and communicating. (1976, p. 29)

This notion of reflexivity is related to, but of a different order than reflection as described by Schon (1983). I see reflexivity more as a refection on our reflections, thinking about our thinking process, knowing how we know. When I ask my students to write reflexively about theories, they initially see their task as one of reflection: Can I identify elements of Bandura's observational learning in this particular situation with which I am familiar? But I am asking them to take their analysis a step further. What worldview does Bandura's theory bring to the situation? What things are highlighted when looking thorough the lens of the theory? What things are ignored or overlooked? What is gained by looking at the situation in this way? What is lost? What new insights did the analysis provide? What potential avenues of insight might be overlooked unless an alternative perspective is brought to bear? For example, one student wrote a series of reflexive papers documenting her ongoing struggle with bulimia as viewed through the lenses of various theories. By the end of the semester we agreed that she had a much better perspective on her situation than might have been provided by simple reflection on isolated theories.

In a related vein, I have adopted a co-teaching model for the class, where groups of students are responsible for preparing a portion of the classroom activities for each of the theorists. Over the years I have been doing this I have had a few disasters but more often than not I wind up learning something new about the theory as the students bring in the perspectives of their various fields of application and expertise. I have retained the final examination procedure since I am reluctant to let go of the notion that there are some basic distinctions about the field of learning and cognition as it is applied to instruction that everyone should be able to use.

TODAY AND BEYOND

One feature of the "interesting times" we live in is that they are constantly changing. As I write, Indiana University is in the midst of creating a new department and degrees (masters and doctorate as well as an undergraduate "specialization") in Learning Science. Here is our description of the field:

> Learning Sciences is an interdisciplinary field that draws on multiple theoretical perspectives and research paradigms from the human sciences. Researchers in the Learning Sciences attempt to understand the nature and conditions of learning, cognition, development, and related areas of human performance. They investigate cognition in its material, social, and cultural context

> but with the broad goal of conducting both laboratory-based and naturalistic investigations. The intent of this research is to develop evidence-based claims about how people learn that have practical, pedagogical, and theoretical implications. Given this focus on theory, practice, and learning in context, learning sciences research frequently involves carrying out design, implementation, and policy work intended to improve the learning experience of all learners.

We are busy designing new courses and redesigning existing courses in accord with this description. For example, I just completed teaching a learning science class for undergraduate preservice teacher education students where we spent most of the semester using learning science concepts to analyze video classrooms—a kind of virtual field experience. Could the students use learning science concepts (scaffolding, transfer, multiple intelligence, etc.) to interpret the teacher and student behaviors in the video? I also had them compare their analyses with other students to make them aware that different analysts often see different things. Their final task is to design a lesson in which they explicitly include content and strategies that, based upon the learning science research literature, ensure that their students will have a successful learning experience. Our eventual goal for this class is that our students will incorporate these strategies into their actual teaching. It will be exciting over the next few years to see how successful this approach to teaching undergraduates will be.

Here then is one person's journey in teaching a class on theories of learning and cognition during the "cognitive revolution. "Over the years I gradually changed from a behaviorist-oriented instructor concerned with presenting the theories to my students and motivating them to learn the material, to an instructor concerned with helping students apply the theories by providing concrete examples of appropriate uses of theory, to a coteacher who is interested in helping students learn how to "think" with theories, using theories as tools to analyze situations and solve problems of importance to them. This is *my* cognitive revolution!

NOTES

1. The title of this chapter is an allusion to the phrase "May you live in interesting times," reputed to be an ancient Chinese curse. There is good reason to doubt its origins, however—see http://hawk.fab2.albany.edu/sidebar/sidebar.htm. Nevertheless, the last 35 years have proved to be both challenging and exhilarating for university professors who take their teaching seriously.
2. For an alternative history of the cognitive revolution, see Bruner (1990, pp. 1–32).
3. This is one of the few books that I have retained from my graduate school days.
4. Those interested in this saga could consult Cunningham (1992, 1998).

REFERENCES

Adams, J. (1967). *Human memory.* New York: McGraw Hill.

Anderson, J. (1976). *Language, memory and thought.* Hillsdale, NJ: Erlbaum.

Anderson, J., & Bower, G. (1973). *Human associative memory.* Washington, DC: Winston.

Anderson, R. C. (1965). Can first graders learn and advanced problem solving skill? *Journal of Educational Psychology, 56,* 283–294.

Anderson, R. C., & Ausubel, D. P. (Eds.). (1965). *Readings in the psychology of cognition.* New York: Holt, Rinehart & Winston.

Anderson, R. C., & Faust, G. W. (1973) *Educational psychology: The science of instruction and learning.* New York: Dodd-Mead.

Anderson, R. C., Faust, G. W., Roderick, M., Cunningham, D. J., & Andre, T. (Eds.). (1969). *Current research on instruction.* Englewood Cliffs, NJ: Prentice Hall

Atkinson, R., & Shiffrin, R. (1968). Human memory: A proposed system and its control processes. In K. W. Spence & J. T. Spence (Eds.), *The psychology of learning and motivation: Advances in research and theory* (Vol. 2, pp. 89–195). San Diego, CA: Academic Press.

Ausubel, D. P. (1963). *The psychology of meaningful learning.* New York: Grune & Stratton.

Bartlett, F. (1932). *Remembering.* Cambridge, UK: Cambridge University Press.

Bransford, J. (1979). *Human cognition: Learning, understanding and remembering.* Belmont, CA: Wadsworth.

Bransford, J., & Franks, J. (1971). The abstraction of linguistic ideas. *Cognitive Psychology, 2,* 331–380.

Bransford, J., & Johnson, M. (1972). Contextual prerequisites for understanding: Some investigations of comprehension and recall. *Journal of Verbal Learning and Verbal Behavior, 11,* 717–726.

Bransford, J., & Stein, B. (1984). *The ideal problem solver.* New York: Freeman.

Brown, J. S., Collins, A., & Duguid, P. (1989). Situated cognition and the culture of learning. *Educational Researcher, 18,* 32–42.

Bruner, J. (1990). *Acts of meaning.* Cambridge, MA: Harvard University Press.

Bruner, J., Oliver, R., & Greenfield, P (Eds). (1967). *Studies in cognitive growth.* New York: Wiley.

Chance, P. (1979). *Learning and behavior.* Belmont, CA: Wadsworth.

Cobb, P. (1994). Where is the mind? Constructivist and sociocultural perspectives on mathematical development. *Educational Researcher, 23,* 13–20.

Cole, M. (1996). *Cultural psychology: A once and future discipline.* Cambridge, MA: Harvard University Press.

Collins, A., Brown, J., & Newman, S. (1989). Cognitive apprenticeship: Teaching the crafts of reading, writing, and mathematics. In Lauren B. Resnick (Ed.) *Knowing, learning, and instruction: Essays in honor of Robert Glaser.* Hillsdale, NJ: Erlbaum.

Craik, F., & Lockhart, R. (1972). Levels of processing: A framework for memory research. *Journal of Verbal Learning and Verbal Behavior, 11,* 671–684.

Cunningham, D. J. (1972). The retention of connected discourse: A review. *Review of Educational Research, 42,* 47–71.

Cunningham, D. J. (1992). Beyond educational psychology: Steps toward an educational semiotic. *Educational Psychology Review, 4,* 165–194.

Cunningham, D. J. (1998). Cognition as semiosis: The role of inference. *Theory and Psychology, 8,* 827–840.

Cunningham, D. J., & Anderson, R. (1968). Effects of practice time within prompting and confirmation presentation procedures on paired associate learning. *Journal of Verbal Learning and Verbal Behavior, 7,* 613–616.

Deese, J. (1958). *The psychology of learning.* New York: McGraw Hill.

Eco, U. (1976). *A theory of semiotics.* Bloomington: Indiana University Press.

Flavell, J. (1963). *The developmental psychology of Jean Piaget.* Princeton, NJ: Van Nostrand.

Flavell, J. (1970). Developmental studies of mediated memory. In H. W. Reese & L. P. Lipsett (Eds.), *Advances in child development and behavior* (Vol. 5). New York: Academic Proess.

Gagne, R. (1965). *The conditions of learning.* New York: Holt, Rinehart & Winston.

Gagne, R. & Briggs, L. (1974). *Principles of instructional design.* New York: Holt, Rinehart & Winston.

Gardner, H. (1985). *The mind's new science.* New York: Basic Books.

Glover, J., Ronning, R., & Bruning, R. (1990). *Cognitive psychology for teachers.* New York: Mcmillan.

Gredler, M. (1986). *Learning and instruction: Theory into practice.* New York: Macmillan.

Gredler, M. (2001). *Learning and instruction: Theory into practice* (4th ed.). Upper Saddle River, NJ: Merrill/Prentice Hall.

Hilgard, E., & Bower, G. (1966). *Theories of learning* (3rd ed.). New York: Appleton Century Crofts.

Holland, J. G., & Skinner, B. F. (1961). *The analysis of behavior.* New York: McGraw Hill.

Lakoff, G., & Johnson, M. (1980). *Metaphors we live by.* Chicago: University of Chicago Press.

Lave, J., & Wenger, E. (1991). *Situated learning.* New York: Cambridge University Press.

Lorayne, H., & Lucas, J. (1974). *The memory book.* New York: Ballantine.

Luria, A. (1968). *The mind of a mnemonist.* New York Basic Books.

Mayer, R. (1981). *The promise of cognitive psychology.* New York: Freeman.

Mayer, R. (2003). *Learning and instruction.* Upper Saddle River, NJ: Merrill/Prentice Hall.

Neisser, U. (1967). *Cognitive psychology.* New York: Appelton, Century, Crofts.

Neisser, U. (1976). *Cognition and reality.* San Francisco: Freeman.

Norman, D. (1967). *Memory and attention: An introduction to human information processing.* New York: Wiley.

Norman, D. A. (1971). Technical director and "anchor man." *Human Information Processing* [Film]. Beverly Hills, CA: CRM Productions.

Ormrod, J. (2004). *Human learning* (4th ed.). Upper Saddle River, NJ: Merrill/Prentice Hall.

Paivio, A. (1971). *Imagery and verbal processes.* New York: Holt, Reinhart & Winston.

Pichert, J., & Anderson, R. C. (1977). Taking different perspectives on a story. *Journal of Educational Psychology, 69,* 309–315.

Reynolds, G. (1968). *A primer of operant conditioning.* Glenview, IL: Scott, Foresman.

Rumelhart, D. (1975). Notes on a schema for stories. In D. G. Bobrow & A. M. Collins (Eds.), *Representation and understanding: Studies in cognitive science.* New York: Academic Press.

Rumelhart, D., & McClelland, J. (1986). *Parallel distributed processing.* Cambridge, MA: MIT Press.

Schon, D. (1983). *The reflective practitioner: How professionals think in action.* New York: Basic Books.

Schunk, D. (2004). *Learning theories: An educational perspective.* Upper Saddle River, NJ: Merrill/Prentice Hall.

Shank, R., & Abelson, R. (1977). *Scripts, plans, goals and understanding.* Hillsdale, NJ: Erlbaum.

Travers, R. M. W. (1963). *Essentials of learning: An overview for students of education.* New York: Macmillan.

von Glasersfeld, E. (1984). An introduction to radical constructivism. In P. Watzlawick (Ed.), *The invented reality* (pp. 17–40). New York: Norton.

Vygotsky, L. (1978). *Mind in society.* Cambridge, MA: Harvard University Press.

Wadsworth, B. (1971). *Piaget's theory of cognitive development: An introduction for students of psychology and education.* New York: Mckay.

CHAPTER 5

CONCEPTUAL UNDERSTANDING VERSUS COMPUTATIONAL SKILL

How Cognitive Science Helps Resolve the Great Debate of Mathematics Education

Richard E. Mayer
University of California, Santa Barbara

ABSTRACT

This chapter explores the relation between cognitive science and mathematics education in historical context. The chapter starts by describing the ongoing debate on how and what to teach in mathematics classrooms—that is, whether to take a basic skills approach or a conceptual understanding approach. Then, the chapter briefly reviews the historical roots of the debate, three approaches to defining the objectives of mathematics education, three views of learning in mathematics education, four enduring contributions of cognitive science to mathematics education, and how the merger of cognitive science and mathematics education helps resolve the debate.

The Impact of the Cognitive Revolution on Educational Psychology, pages 105–118

THE GREAT DEBATE OF MATHEMATICS EDUCATION

In the field of mathematics education, the issue of what and how to teach represents a bitterly contested and long-running debate, paralleling what Chall (2002) calls the "great debate" on teaching children to read. As shown in Table 5.1, mathematics education's great debate concerns both how to teach (e.g., by guided methods or discovery methods) and what to teach (e.g., computational skill or conceptual understanding). On one side of what has been called the "math wars" (Klien, 2002) are those who favor direct instruction of computation skill (represented in the upper left corner in what I call the "skills quadrant"), and on the other side are those who favor discovery learning of conceptual understanding (represented in the lower right corner in what I call the "understanding quadrant").

Table 5.1. Locating the Great Debate in Mathematics Education

What to Teach	*How to Teach*	
	Guided methods	Discovery methods
Basic skills	**Skills quadrant**	
Conceptual understanding		**Understanding quadrant**

When framed in terms of what to teach, the debate concerns what Bosse (1995, p. 180) describes as the "collision between skills instruction and understanding." In adding the dimension of how to teach, teaching of skills is thought to be accomplished through guided methods—such as direct instruction or seeing worked-out examples—whereas teaching of understanding is thought to be accomplished through discovery methods—such as group discussions or open-ended problem-solving activities. In short, the debate concerns the relative place of computational skill (taught with guided methods) and conceptual understanding (taught with discovery methods).

Viewed from some perspectives, the great debate in mathematics education offers a puzzling choice between helping students get the right answer versus helping students understand what they are doing. An important contribution of cognitive science is to offer a research-based reconciliation based on cognitive task analysis and the architecture of the human information-processing system. First, the focus is on authentic tasks (such as solving an important problem) rather than on isolated skills. In cognitive task analysis, the goal is to determine the knowledge and cognitive processes required to accomplish a specific task (such as solving a word problem). In general, a cognitive task analysis of an authentic mathematical task reveals that students need both procedural knowledge (emphasized in the skills approach) and conceptual knowledge (emphasized in the understanding approach), as well as other kinds of knowledge.

Second, by specifying the to-be-learned knowledge in detail, the learning goal is clarified. Given the limited working memory capacity of the human cognitive system, discovery methods may fail when attention is applied to irrelevant parts of the task. Thus some amount of guidance may be needed to help focus students' attention and cognitive processing. Another way to reduce cognitive load is for students to automatize their procedural knowledge—such as computational procedures—and to build a repertoire of problem cases or types that can be reconciled with new problems. A limitation of discovery methods (emphasized in the understanding approach) is that students might not discover the required knowledge, whereas a limitation of guided methods (emphasized in the skills approach) is that students may become proficient in only one kind of knowledge. It follows that a balance of direct instruction and guided discovery can be used to foster the knowledge needed for the to-be-learned task.

HISTORICAL OVERVIEW: THE GREAT DEBATE BEFORE AND AFTER THE COGNITIVE REVOLUTION

The basic struggle in mathematics education has long been to determine the role played by computational skill and conceptual understanding. The battle lines can be seen in two early yearbooks from national societies concerned with mathematics education—the 1930 Yearbook of the National Society for the Study of Education (NSSE), entitled *Report of the Society's Committee on Arithmetic: Part 1, Some Aspects of Modern Thought on Arithmetic; Part II, Research in Arithmetic*, and the 1935 Yearbook of the National Council of Teachers of Mathematics (NCTM), entitled *The Teaching of Arithmetic*. For example, Knight's (1930, p. 5) chapter in the1930 NSSE Yearbook described arithmetic learning in terms of building stimulus–response (S-R) connections and emphasized drill as the preferred method of instruction: "Theoretically, the main psychological basis is a behaviorist one viewing skills and habits as fabrics of connections." In contrast, Wheeler's (1935, p. 247) chapter in the 1935 NCTM Yearbook proposed Gestalt theory as the basis for mathematics education: "The whole purpose of arithmetic is to discover number relationships and to be able to reason with numbers. It is not to learn the tables." Wheeler (1935, p. 243) urged teachers to "forget drills" and "concentrate on relations."

Yet, even as the great debate raged in the 1930s, thoughtful observers recognized the futility of framing the situation as an either/or choice. In a review of the great debate during the first half of the 20th century, Buswell

(1951b, p. 144) noted: "Much harm has been done to the organization and teaching of arithmetic by trying to force all learning situations to fit any one theory of learning." Buswell argued for combining a new focus on "understanding of number relations and of the number system" with the traditional focus on "computational efficiency, which was almost the sole criterion of arithmetical ability in the 1920s" (p. 154). While warning against "thinking of arithmetic mainly in terms of separate...operations" (p. 154), Buswell clearly noted that "there is, however, a place for practice and drill in arithmetic" (p. 147).

Even in the heyday of behaviorism in the 1920s, few educational psychologists ever presented the situation as an either/or choice. Although Thorndike is usually associated with the idea that learning involves drill and practice of basic facts, he clearly acknowledged the need for both basic arithmetic facts and accompanying reasoning strategies. In *New Methods in Mathematics* (1921, pp. 58–59), Thorndike stated: "Arithmetic consists, not of isolated, unrelated facts, but of parts of a total system, each part of which may help the knowledge of other parts, if it is learned properly.... Almost all arithmetic knowledge should be treated as an organized interrelated system." In *The Psychology of Arithmetic*, Thorndike (1922, pp. 73, 109) wrote: "[T]he psychologists of today do not wish to make learning arithmetic a mere matter of acquiring thousands of disconnected habits, nor to decrease by one jot the pupil's genuine comprehension of its general truths.... [W]e do not wish the pupil to be a repository of separated abilities.... Rather he is to be an effective organization of abilities, cooperating in useful ways to meet the quantitative problems life offers."

Sometimes the debate is cast as whether to learn arithmetic procedures in isolation or in the context of everyday quantitative reasoning situations (Nunes, Schliemann, & Carraher, 1993). However, more than 50 years ago mathematics educators had reached consensus on the importance of contextualizing mathematics instruction within actual everyday situations. In the Introduction to the 50th Yearbook of the National Society for the Study of Education, entitled *The Teaching of Arithmetic*, the consensus was summarized as follows (Buswell, 1951a, p. 3): "A...major point of emphasis by the committee is that, while the number system and number processes are the heart of arithmetic, they are important only as they function in the actual thinking of people. There is no excuse for arithmetic in a social vacuum.... There is no dichotomy requiring a choice between mathematical and social arithmetic. These are two aspects of the same thing." Wheat (1951, p. 32) made a similar point concerning the context of mathematical thinking: "In short, the pupil does not first learn his arithmetic and then apply it, as separate items of experience.... He does them all at once...as he moves along a single road of number-thinking." In the 1941 Yearbook of the NSSE, Brownell (1941)—a strong proponent of the role of understanding in mathematics learning—listed "computational skill" as an important component in mathematical learning along with "mathematical understanding" and "sensitivity to number in social situations and the habit of using number effectively in such situations." Overall, scholars have been working toward a reconciliation of the debate for generations.

THREE APPROACHES TO DEFINING EDUCATIONAL GOALS

Much of the great debate in mathematics education concerns how to define the teacher's goals. Table 5.2 lists three approaches to defining goals in mathematics education. First, from a mathematician's perspective, mathematics learning is defined in terms of subject content. At a gross level, subject content can be defined as broad topics such as arithmetic, algebra, geometry, calculus, and so on; at a finer level, a broad topic such as arithmetic can be broken into narrower topics including addition of whole numbers, subtraction of whole numbers, multiplication of whole numbers, division of whole numbers, addition of fractions, subtraction of fractions, multiplication of fractions, division of fractions, and so forth.

Table 5.2. Three Approaches to Defining Educational Goals

Approach	*Defines Goals in Terms of:*	*Examples of Goals*
Mathematician	Subject content	Learn algebra (broader topic), learn functions (narrower topic)
Educator	Learning activities	Complete workbooks, make change for purchases, discuss solutions to word problems
Psychologist	Learner's knowledge	Build automatic procedures for addition, develop differentiated categories of word problem types, form concept of a mental number line

Second, from an educator's perspective, mathematics learning is defined in terms of instructional activities. Examples include asking students to complete computational worksheets, having students discuss how they solve various problems, or having students work on a project such as running a store where they must give correct change for purchases. The educator's and mathematician's approaches sometimes can be combined, such as when a teacher says the class is learning about some topic using some activity (e.g., learning about addition and subtraction by having to make change for purchases).

Third, from a psychologist's perspective, mathematics learning is defined in terms of cognitive changes in the learner's knowledge. Educational psychologists seek to define learning outcomes in terms of the degree to which students form the specific knowledge and cognitive processes required for success on a class of problems. Examples include conceptual knowledge such as having a mental number line, conceptual knowledge such as knowledge of problem types, strategic knowledge such as knowledge of how to develop solution plans, and procedural knowledge such as how to carry out arithmetic computations. A major contribution of cognitive science has been to advance the application of cognitive task analysis in mathematics education—that is, to contribute to the usefulness of techniques for analyzing and describing the knowledge required for various mathematical tasks (Mayer, 1999).

THREE VIEWS OF WHAT CHANGES WHEN SOMEONE LEARNS

Learning is generally defined as a "relatively permanent change in a learner's knowledge or behavior that is a result of the learner's experience" (Mayer, 2001, p. 38). This definition consists of three parts: (1) learning is *relatively permanent*, (2) learning is a *result of experience*, and (3) learning is a *change in behavior or knowledge*. Although the first two components of the definition have been consistent for more than a century, the third component—concerning what is learned—has changed greatly.

Much of the great debate in mathematics education concerns how to define what students learn. Table 5.3 lists three conceptions of what is learned that have emerged over the past 100 years of research on learning (Mayer, 1992, 1996). The earliest conception—popularized by the behaviorist approach in the first half of the 20th century—is that learning involves the strengthening and weakening of stimulus–response associations. When a student is rewarded for giving a response to a question, the link between the question and the response automatically becomes stronger for the student. When a student is punished for giving a response to a question, the link between the question and the response automatically becomes weaker for the student. This view of learning—first proposed by Thorndike (1911/1965) as the law of effect—has direct implications for mathematics education. Concerning the issue of what to teach, the emphasis is on learning procedures (such as arithmetic computation) and basic facts (such as definitions). Concerning how to teach, the emphasis is on drill and practice.

The second conception—popularized by the information-processing approach in the middle part of the 20th century—is that learning involves

Table 5.3. What Changes When Someone Learns?

Approach	*What Changes*	*What to Teach*	*How to Teach*
Behaviorist	S–R Connections	Procedures and facts	Drill and practice with feedback
Information processing	Information	Components of tasks	Lectures and textbooks
Cognitive constructivist	Knowledge	Integrated components of tasks	Guided knowledge construction

placing new information into one's memory. If learning is a change in the information in a learner's memory, then the implications for mathematics education are clear. Concerning what to teach, the emphasis is on defining the information that is specifically needed to carry out a particular task (such as solving story problems). Concerning how to teach, the emphasis is on efficient methods for information delivery such as lectures and textbooks.

The third conception—which evolved from the information-processing approach in the late 20th century—is that learning involves building cognitive structures. A central concept is that what is learned depends on cognitive processing during learning. According to the cognitive constructivist view, knowledge is different than information. Knowledge is subjective—a construction that exists only in a person's mind—whereas information is objective—a commodity that exists outside of people's minds. The implications for mathematics education are that, concerning what to teach, the focus is on helping learners organize and integrate knowledge that is specifically needed to carry out a particular task, and concerning how to teach, the focus is on methods for guiding the learner's cognitive processes during learning such as through guided discussions and activities.

Although all three conceptions have some validity, a major contribution of cognitive science has been to advance a deeper understanding of the cognitive constructivist view of learning with its focus on guiding cognitive processing during learning. One important line of research and practice in mathematics education growing out of this approach involves the teaching of learning strategies, or what has been called cognitive strategy instruction (Pressley & Woloshyn, 1995).

FOUR CONTRIBUTIONS OF COGNITIVE SCIENCE TO THE GREAT DEBATE OF MATHEMATICS EDUCATION

This chapter is concerned with the contributions of cognitive science to the ongoing debate about what and how to teach mathematics. The cognitive revolution—which began in 1956 (Gardner, 1985) and spread quickly by the 1960s (Neisser, 1967) and 1970s (Lachman, Lachman, & Butterfield, 1979)—is concerned with understanding how the mind works. Research and theory in cognitive science offers some insight concerning what is learned and how learning occurs. The field of mathematics education is concerned with improving how people solve mathematics problems. Research and theory in mathematics education seeks to understand how instruction can promote mathematics learning. Thus, the interaction of cognitive science and mathematics education can produce advances in both fields—extending cognitive theory to mathematical domains and informing educational practice that is consistent with how people learn.

Cognitive Task Analysis

A major contribution of cognitive science is cognitive task analysis—a technique for describing the knowledge and cognitive processes required to perform a mathematical task (Anderson et al., 2001). In short, cognitive science seeks to describe what someone needs to know in order to perform authentic mathematical tasks such as solving an arithmetic word problem, determining the cost of a building project, or making a purchase decision.

For example, consider the task of solving a word problem such as the following (Hegarty, Mayer, & Green, 1992; Hegarty, Mayer, & Monk, 1995):

> At Lucky, butter costs 65 cents per stick.
> This is 2 cents less per stick than butter at Vons.
> If you need to buy 4 sticks of butter,
> how much will you pay at Vons?

In a previous analysis, I (Mayer, 2003) have shown how this problem requires four iterative steps: (1) *translating*, which involves converting each sentence into a mental representation (such as converting "a stick of butter costs 65 cents at Lucky" into a representation such as "LUCKY = .65"); (2) *integrating*, which involves building a mental model of the problem situation (such as categorizing the problem as a "total cost problem" of the form, "total cost = (LUCKY + .02) × 4"); (3) *planning*, which involves devising a plan to solve the problem (such as "first find the cost of butter at Vons by adding .02 to 65, and then find the total cost by multiplying the result by 4"); and (4) *executing*, which involves carrying out the plan (such as .65 + .02 = .67, .67 × 4 = 2.68).

A cognitive task analysis has important instructional implications because it can help pin-point difficulties that a student is having and can help the teacher establish clear instructional goals. For example, students tend to have the most difficulty with translating and planning (Mayer, 2003), whereas most instruction focuses on executing (Mayer, Sims, & Tajika, 1995).

Types of Knowledge

A related contribution of cognitive science is to distinguish among several qualitatively different kinds of knowledge. Table 5.4 lists six kinds of knowledge that learners may need for accomplishing various cognitive tasks: factual, conceptual, procedural, strategic, metastrategic, and attitudinal. Factual knowledge refers to basic facts about the world (such as knowing that "$" means dollars or that there are 60 minutes in an hour). Conceptual knowledge refers to knowledge of problem types (such as having the mental category for work problems) or organizing frameworks (such as the concept of a mental number line) or mental models (such as using bundles of sticks to represent borrowing in two-column subtraction). Procedural knowledge refers to knowledge of how to perform a sequence of operations (such as knowing how to calculate an answer for 7.2 × 5.4 = ___). Strategic knowledge refers to knowledge of how to plan a solution for a problem (such as breaking a problem into parts). Metastrategic knowledge refers to knowledge of how to monitor and coordinate one's knowledge (such as coordinating several procedures or determining that a solution plan needs to be repaired). Attitudinal knowledge refers to knowledge of one's abilities, interests, and preferences (such as believing that "I am a good in math" or "My performance depends on my effort").

The taxonomy in Table 5.4 is similar to that proposed by Anderson and colleagues (2001), which consists of factual, conceptual, procedural, and metacognitive knowledge. The major difference is that in Table 5.4, metacognitive knowledge is broken down into strategic, metastrategic, and attitudinal knowledge. The taxonomy in Table 5.4 is also similar to the five strands of proficiency proposed by Kilpatrick, Swafford, and Findell (2001): conceptual understanding (which corresponds to conceptual knowledge), procedural fluency (which corresponds to procedural knowledge), strategic competence (which refers to strategic knowledge), adaptive reasoning (which corresponds somewhat to metastrategic knowledge), and productive disposition (which refers to attitudinal knowledge).

In previous analyses, I (Mayer, 2003) identified five types of knowledge involved in mathematical problem solving: semantic and linguistic (which correspond somewhat to factual knowledge in Table 5.4), schematic (which corresponds to conceptual knowledge in Table 5.4), strategic (which corresponds to strategic and metastrategic knowledge in Table 5.4), and procedural (which corresponds to procedural knowledge in

Table 5.4. Some Kinds of Knowledge Involved in Mathematical Problem Solving

Kind	*Other Names*	*Definition*	*Example*
Factual	Facts	Knowledge of facts about the world	$ means dollars; there are 60 minutes in an hour
Conceptual	Concepts, schemas, models	Knowledge of problem types, organizing frameworks, and mental models	Category for work problems, concept of number line
Procedural	Procedures, processes, algorithms, skills	Knowledge of how to perform a sequence of operations	7.2 × 5.4 = 38.88
Strategic	Heuristics, methods, techniques	Knowledge of how to develop a solution plan	Breaking a problem into parts, such as multiplying 7.2 × 5.4
Metastrategic	Monitoring, executive control, coordinating, orchestrating	Knowledge of how to monitor and coordinate one's knowledge	Determining that a solution plan needs to be repaired
Attitudinal	Attitudes, beliefs, dispositions, attributions, interests	Knowledge of one's abilities, dispositions, and preferences	The belief that "I am good in math" or "My performance depends on my effort"

Table 5.4). Thus, overall, one contribution of cognitive science is to identify several different kinds of knowledge that may be involved in mathematical problem solving.

Psychologies of Subject Matter

A third major contribution of cognitive science involves the development of psychologies of subject matter (Bruer, 1993; Mayer, 1999, in press). Rather than trying to build general theories of how people learn, current cognitive science theories of learning focus on learning in specific subject domains (Bransford, Brown, & Cocking, 1999; Mayer, 1999). For example, research in the psychology of mathematics learning includes evidence that teaching aimed at fostering number sense in primary grade learners can greatly enhance their development of arithmetic competence in later grades (Griffin, Case, & Capodilupo, 1995). Thus, fine-grained research on the psychology of mathematics learning helps target the teaching of specific concepts and strategies.

Architecture of the Human Information-Processing System

Finally, a fourth major contribution of cognitive science is to provide a model of how learners build knowledge in their memories (Mayer, 1999; Sweller, 1999). For example, Figure 5.1 presents a basic information-processing model that highlights three memory stores (indicated by the boxes labeled "sensory memory," "working memory," and "long-term memory") and three cognitive processes for learning (indicated by the arrows labeled "selecting," "organizing," and "integrating"). When stimuli impinge on the ears, sensory memory can briefly hold incoming sounds before they fade away. When stimuli impinge on the eyes, sensory memory can briefly hold incoming images before they fade away. If the learner pays attention to portions of the incoming sounds and/or images (as indicated by the selecting arrow), they can be transferred to working memory for further processing. Working memory is highly limited so only a few pieces of information can be selected for active processing in working memory at any one time. In working memory, the learner may mentally organize the incoming material into a coherent mental representation—a process indicated by the organizing arrow in Figure 5.1. In long-term memory, the learner may possess relevant prior knowledge including schemas, which are activated and integrated with incoming material in working memory. The process of integrating incoming material with activated knowledge from long-term memory is indicated by the integrating arrow in Figure 5.1. For meaningful learning to occur, the learner must engage in the cognitive processes of selecting, organizing, and integrating, and store the newly constructed knowledge in long-term memory.

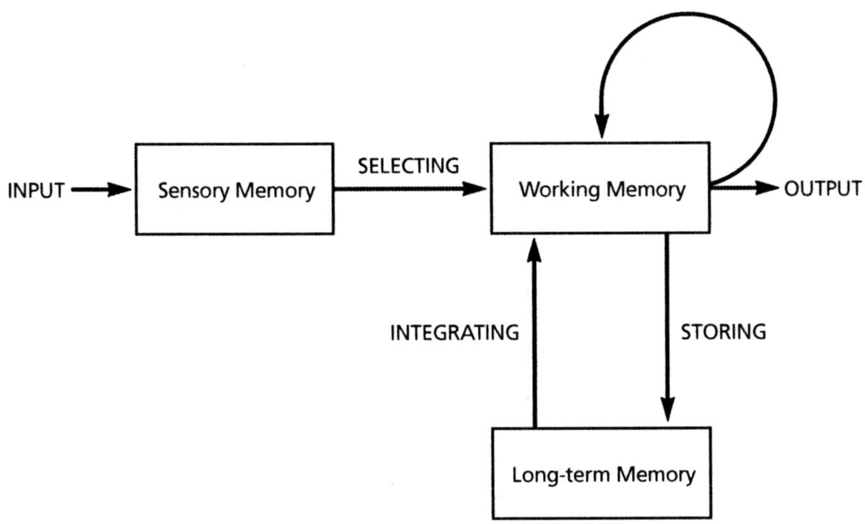

Figure 5.1. Basic information-processing model.

According to this view, one of the major challenges of instruction is to avoid overloading the learner's working memory. Sweller (1999) and his colleagues have developed cognitive load theory, which posits that instructional episodes can create demands for extraneous cognitive load—cognitive processing that does not produce meaningful learning—so instructional designers should attempt to minimize extraneous cognitive load in learning environments. For example, students may learn more from studying worked-out examples of mathematics problems than from actually solving them because the worked-out examples allow the learner to allocate more cognitive capacity to understanding how to the solve problem (Sweller, 1999).

RESOLVING THE GREAT DEBATE OF MATHEMATICS EDUCATION

Research and theory in cognitive science can contribute to resolving the great debate in mathematics along the same lines that thoughtful scholars have suggested for generations. The resolution is that students need a collection of specific kinds of knowledge to be able to solve mathematics problems—including both conceptual knowledge and procedural knowledge. However, a major advance offered by cognitive science is the detailed specification of the knowledge a learner needs to become proficient on a mathematical task.

Given the domain specificity of mathematics learning, the cognitive science approach involves clearly describing an authentic mathematical task. If the goal is to prepare learners to be able to perform the task, then the next step is to determine the knowledge that is needed to perform the task. Using cognitive task analysis, it is possible to create a step-by-step analysis of the cognitive processes involved in the task, as well as a description of the knowledge needed to support the steps. In most complex tasks, a full range of knowledge types will be involved—drawing from the six types listed in Table 5.4. In creating instructional experiences to help students build these skills, it is important to be sensitive to the issue of cognitive load—that is, to not overload the learner. At the same time, it is important to ensure that students learn to coordinate the knowledge within the context of the task rather than to simply learn how to carry out each part of the task in isolation. For example, Van Merrienboer (1997) suggests a part-task approach that involves practice on parts after students have developed an understanding of the whole task.

In many ways, the return of psychologies of subject matter have saved cognitive science from becoming bogged down in the study of contrived laboratory tasks (Neisser, 1976, 1982). By focusing on how people learn mathematics, for example, cognitive scientists are challenged to developed theories of how people learn in real contexts. So far a major contribution of this fruitful merger of cognitive science and mathematics education has been the analysis of specific knowledge needed for success in various types of mathematics tasks—including both conceptual and procedural knowledge as well as strategies and metastrategies for coordinating them. Thus a major step in reconciling the great debate of mathematics education is to recognize the role of knowledge in mathematical performance.

REFERENCES

Anderson, L. W., Krathwohl, D. R., Airasian, P. W., Cruikchank, K. A., Mayer, R. E., Pintrich, P. R., Raths, J., & Wittrock, M. C. (2001). *A taxonomy for learning, teaching, and assessing: A revision of Bloom's taxonomy of educational objectives.* New York: Longman.

Bosse, M. J. (1995). The NCTM standards in light of the new math movement: A warning! *Journal of Mathematical Behavior, 14,* 171–201.

Bransford, J. D., Brown, A. L., & Cocking, R. R. (1999). *How people learn.* Washington, DC: National Academy Press.

Brownell, W. A. (1941). The evaluation of learning in arithmetic. *Arithmetic in general education: Sixteenth yearbook of the National Council of Teachers of Mathematics* (pp. 231–232). New York: Teachers College Bureau of Publications.

Bruer, J. T. (1993). *Schools for thought: A science of learning in the classroom.* Cambridge, MA: MIT Press.

Buswell, G. T. (1951a). Introduction. In N. B. Nelson (Ed.), *The fiftieth yearbook of the National Society for the Study of Education: Part II, The teaching of arithmetic* (pp. 1–5). Chicago: University of Chicago Press.

Buswell, G. T. (1951b). The psychology of learning in relation to the teaching of arithmetic. In N. B. Nelson (Ed.), *The fiftieth yearbook of the National Society for the Study of Education: Part II, The teaching of arithmetic* (pp. 143–154). Chicago: University of Chicago Press.

Chall, J. S. (2002). *The academic achievement challenge.* New York: Guilford Press.

Gardner, H. (1985). *The mind's new science.* New York: Basic Books.

Griffin, S. A., Case, R., & Capodilupo. S. (1995). Teaching for understanding: The importance of central conceptual structures in the elementary school mathematics curriculum. In A. McKeogh, J. Lupart, & A. Marina (Eds.), *Teaching for transfer: Fostering generalization in learning.* Hillsdale, NJ: Erlbaum.

Hegarty, M., Mayer, R. E., & Green, C. (1992). Comprehension of arithmetic word problems: Evidence from students' eye fixations. *Journal of Educational Psychology, 84,* 76–84.

Hegarty, M., Mayer, R. E., & Monk, C. A. (1995). Comprehension of arithmetic word problems: A comparison of successful and unsuccessful problem solvers. *Journal of Educational Psychology, 87,* 18–32.

Kilpatrick, J., Swafford, J., & Findell, B. (Eds.). (2001). *Adding it up: Helping children learn mathematics.* Washington, DC: National Academy Press.

Klien, D. (2002). A brief history of American K–12 mathematics education in the 20th century. In J. M. Royer (Ed.), *Mathematical cognition* (pp. 175–225). Greenwich, CT: Information Age.

Knight, F. B. (1930). Methods of teaching. In F. B. Knight (Ed.), *Twenty-ninth Yearbook of the National Society for the Study of Education: Part I, Some aspects of modern thought on arithmetic; Part II, Research in arithmetic.* Chicago: University of Chicago Press.

Lachman, R., Lachman, J. L., & Butterfield, E. C. (1979). *Cognitive psychology and information processing.* Hillsdale, NJ: Erlbaum.

Mayer, R. E. (1992). Cognition and instruction: On their historic meeting within educational psychology. *Journal of Educational Psychology, 84,* 405–412.

Mayer, R. E. (1996). Learners as information processors: Legacies and limitations of educational psychology's second metaphor. *Educational Psychologist, 31,* 151–162.

Mayer, R. E. (1999). *The promise of educational psychology: Learning in the content areas.* Upper Saddle River, NJ: Prentice Hall.

Mayer, R. E. (2001). Changing conceptions of learning: A century of progress in the scientific study of education. In L. Corno (Ed.), *Education across a century: The Centennial volume. One hundredth yearbook of the National Society for the Study of Education* (pp. 34–75). Chicago: University of Chicago Press.

Mayer, R. E. (2003). Mathematical problem solving. In J. M. Royer (Ed.), *Mathematical cognition* (pp. 69–92). Greenwich, CT: Information Age.

Mayer, R. E. (in press). Teaching of subject matter. *Annual Review of Psychology.*

Mayer, R. E., Sims, V., & Tajika, H. (1995). A comparison of how textbooks teach mathematical problem solving in Japan and the United States. *American Educational Research Journal, 32,* 443–460.

Neisser, U. (1967). *Cognitive psychology.* New York: Appleton-Century-Crofts.

Neisser, U. (1976). *Cognition and reality.* New York: Freeman.

Neisser, U. (1982). *Memory observed.* New York: Freeman.

Nunes, T., Schliemann, A. D., & Carraher, D. W. (1993). *Street mathematics and school mathematics.* New York: Cambridge University Press.

Pressley, M., & Woloshyn, V. (1995). *Cognitive strategy instruction that really improves children's academic performance.* Cambridge, MA: Brookline Books.

Sweller, J. (1999). *Instructional design in technical areas.* Camberwell, Australia: ACER Press.

Thorndike, E. L. (1911/1965). *Animal intelligence.* New York: Hafner.

Thorndike, E. L. (1921). *New methods in arithmetic.* Chicago: Rand McNally.

Thorndike, E. L. (1922). *The psychology of arithmetic.* New York: Macmillan.

Van Merrienboer, J. J. G. (1997). *Training complex cognitive skills.* Englewood Cliffs, NJ: Educational Technology Publications.

Wheat, H. G. (1951). The nature and sequencing of learning activities in arithmetic. In N. B. Nelson (Ed.), *The fiftieth yearbook of the National Society for the Study of Education: Part II, The teaching of arithmetic* (pp. 22–52). Chicago: University of Chicago Press.

Wheeler, R. H. (1935). The new psychology of learning. In *Tenth yearbook of the National Council of Teachers of Mathematics: The teaching of arithmetic.* New York: Teachers College, Columbia University.

CHAPTER 6

THE IMPACT OF THE COGNITIVE REVOLUTION ON SCIENCE LEARNING AND TEACHING

Eugenia Etkina
Rutgers, The State University of New Jersey
Jose P. Mestre[1]
University of Massachusetts, Amherst
Angela O'Donnell
Rutgers, The State University of New Jersey

ABSTRACT

In this chapter we discuss ways in which the "cognitive revolution" has affected conceptualization of how people learn science, approaches to science instruction, and preparation of science teachers. Prior to the cognitive revolution, learning a complex process was conceived as demonstrating mastery through observable behaviors of all the subcomponents of the complex process. Within the cognitive perspective, learning a complex process is perceived as constructing knowledge, meaning, and sense-making by the learner. Hence, the shift has been from a view that learning is the acquisition of desired behaviors, to a view that learning is the construction of knowledge

The Impact of the Cognitive Revolution on Educational Psychology, pages 119–164

by the individual, construction that is mediated by the context of the learning, the social environment, and the prior knowledge of the learner. We begin with an overview from a cognitive perspective of several areas relevant to science teaching and learning, including the role of prior knowledge in learning, the nature of expertise, transfer of learning, metacognition, and assessment. We then consider instructional implications suggested by the science of learning and formulate nine instructional principles for successful science instruction. Four specific examples of science instruction based on the instructional principles are then presented. We conclude with an analysis of the current instructional practices at the elementary, secondary, and tertiary levels and discuss whether or not they are aligned with our best understanding of how people learn, and provide practical recommendations for teacher preparation and professional development that will allow teachers to translate the "cognitive revolution" into sound instructional practices.

INTRODUCTION

The "cognitive revolution" started about 40 years ago and signaled a major shift in the study of and view of learning. Prior to this (as well as concurrent with the emergence of the cognitive revolution) the behaviorist view of learning prevailed. In the behaviorist view learning consists of making connections between stimuli and responses, and evidence of learning is restricted to observable behaviors that are interpretable as "competence." The behaviorist approach to teaching a student a complex process (e.g., knowledge and procedures for solving a certain class of problems) consists of breaking up the process into component parts, teaching the student each component, then teaching the student how to string together the various components until ultimately, the desired behavior is obtained. Learning a complex process within the behaviorist view is achieved once the student exhibits behavior consistent with behaviorally defined performance goals.

Absent from the behaviorist view is an interest in the cognitive mechanisms used by the individual to learn the complex process. This would seem to be an important consideration since knowledge about the cognitive mechanisms underlying learning might provide insights into how to shape instruction to make learning more efficient. Recent cognitive research, in fact, suggests that a complex process cannot be completely learned by decomposing and teaching individuals subprocesses without regard to the context within which the complex process will be performed. Knowing how the subprocesses interact within the context of performing the entire process is as important as knowing how to perform the individual subprocesses. In short, knowing the individual sub-processes does not "add up to" knowing the entire complex process (Resnick & Resnick,

1992). Also absent from the behaviorist approach is an interest in whether or not the process learned made sense to the individual. This also would seem to be an important consideration because if the process learned conflicts with knowledge already possessed by the individual, then the individual either may not be able to accommodate in memory the process learned in any meaningful sense or will construct parallel, conflicting knowledge structures.

Hence, the focus of the behaviorist approach is the final manifestation of competent behavior by the learner, not whether the knowledge learned makes any sense to the learner, or whether the learner is able to use the acquired knowledge flexibly in novel contexts. This latter issue remains a problem for most approaches to learning. A behavioristic approach is quite successful at teaching many skills, for example, multiplication of two multi-digit numbers. Sequential mastery of progressively more complex procedures in the presence of feedback can be accomplished in limited domains. However, learning of procedures is quite different from developing conceptual understanding of a domain (Anderson, 1990).

In stark contrast, the term *learning* within cognitive science is synonymous with understanding, and the study of learning with understanding in cognitive science is often approached from a multidisciplinary perspective. Current views of learning include the idea that individuals construct knowledge (Center for Psychology in the Schools and Education, 2003). These views are not new but only recently were they widely accepted. For example, De Lisi and Golbeck (1999) note that between 1920 and 1980, Piaget developed a constructivist theory of cognitive functioning and development. Learners not only construct knowledge but the knowledge they already possess affects their ability to learn new knowledge. If new knowledge that we are trying to construct conflicts with previously constructed knowledge, the new knowledge will not make sense to us and may be constructed in a way that is not useful for flexible application (Anderson, 1987; Resnick, 1983, 1987; Schauble, 1990; von Glasersfeld, 1989, 1992). In contrast to behaviorism, prior knowledge and sense-making are very conspicuous in the constructivist view of learning.

Historically, we find that views about learning, instruction, and teacher education in the sciences share some of the views that are held in the present. For example, among the earliest modern references to an introductory general science course that embodied the scientific method of inquiry is one attributed to Thomas H. Huxley (Twiss, 1929) who offered a series of illustrated lectures to London children in 1869. Huxley's (1898) views about science education were quite similar to views consistent with research on learning today. He emphasized science as a process for learning about nature, not as learning facts about nature, and believed students should be prepared "to face scientific problems, not by knowing at once the conditions of every problem, or by being able at once to solve it; but by being familiar with the general current of scientific thought, and by being able to apply the methods of science in the proper way, when they have acquainted themselves with the conditions of the special problem" (p. 122). By arguing that students' own knowledge of science is what counted and that filling students with scientific information was counterproductive, it can be easily argued that Huxley's views are consistent with constructivist epistemology.

Similar views are expressed in some early 20th century textbooks about science teaching (Twiss, 1929). With goals such as "development in the methods of thinking and investigation that are common to all the sciences," the learning process described as the "accumulation of meanings, and the acquisition of motor skill and useful habits of all kinds, through first-hand experience with the phenomena of everyday life, and in connection with the satisfaction of intimate personal interests and needs," and teaching goals described as "to show the children the errors and absurdities of untested generalizations by leading them to work out the implications of their ideas" (pp. 44–45), we again see views consistent with constructivism. We even find statements, such as "the teacher should constantly be aware that a general statement means nothing to any one unless it serves to call up in his mind a large number of images and memories of particular concrete facts that are included under it, and that have been repeatedly associated with it and with one another" (p. 67), that reflect the importance of a hierarchy in knowledge organization, and the importance of context on knowledge application and transfer. This statement also reflects the importance of prior knowledge. Twiss also advocates continuous formative assessment in the science classroom by advocating that the science teacher continuously monitor and estimate her or his students' abilities against some ideal standard.

These views were not the only ones prominent at that time, as illustrated by several examples that are more consistent with behaviorist thought. For example, a book entitled *Teaching of Science* (Woodhull, 1918) suggests that a discovery-style learning approach is much too inefficient to implement, and that information-laden lectures from the teacher are the way to promote students' understanding of science subject matter. Despite what would appear to be a "transmissionist" view of learning, Woodhull advocates the need to start from what the student already knows, as implied by constructivism: "It would seem self-evident that the first thing one must do is to find out the exact mental equipment of his students—to find out what the foundation is before he begins to build upon it. But the schools are full of persons just out of college, teaching not the pupils, but their own self-respecting course in physics" (pp. 83–84). Beginning with a foundation of what the learner knows is not unique to constructivism—behavioral views also emphasize determining entering behavior.

Near the middle of the last century, perspectives on science teaching begin to show consistent trends. For example, Hunter (1934) advocates science instruction that is less prescriptive and more discovery-oriented. He argued against a purely lecture approach to science teaching, stating, "It is the method handed down from the earliest medieval times when an authoritative statement was considered the only way to get material since few people had access to the manuscripts then in use" (p. 165). Instead, Hunter proposes the increased use of laboratory work, demonstrations, Socratic questioning, and other forms of deductive/inductive approaches. He also saw a need for reliable assessment methods and proposed more standardized assessments for all students, and also advocated continuous formative assessment in the form of a short quiz to begin each class period.

A few years later in a yearbook of the National Society for the Study of Education (Henry, 1947), we begin to see other trends, as evidenced by the statements, "Overt 'doing' is an essential element in the child's learning" (p. 67) and "Evaluation through the interpretation of behavior is especially useful in attempting to determine individual progress in relation to such basic objectives as skills, interests, appreciations, and attitudes, and in relation to the functional understanding of facts, large conceptions, and basic principles of science" (p. 116). Unlike the view today that students come to classes with conceptions formed as a result of constructing their own private understanding from observations, misconceptions in this era were attributed to the use of terminology without initially acquiring firsthand experience: "It would appear in many cases that these misconceptions occur as the result of introducing children to the use of terms and symbols before they have had sufficient firsthand experience, which is the raw material of conceptual knowledge" (pp. 68–69). The book acknowledges a trend toward general science courses, from junior high school through college, to educate the many rather than the few. A well-rounded content preparation of prospective science teachers was proposed, with courses in biology, chemistry, and physics (with at least 6–12 credit hours in each, and a total of 60 credits in all sciences). Finally, it is suggested that some psychology coursework should be integrated into the preparation of prospective science teachers.

Although the views of learning expressed by the cognitive revolution had their antecedents in earlier decades, it was not until the mid-1950s that learning was widely viewed as an active and constructive process. The cognitive revolution that occurred in the 1960s may have been the inevitable outgrowth of ideas and trends that were in place in the earlier part of the century, but the emergence of that revolution depended on the development of more adequate tools for measuring cognitive performance and on the invention of the computer that was then used as an analog for the human mind. In the early part of the 20th century, there were few instruments available for measuring intellectual performance (O'Donnell & Levin, 2001). The measurement of learning was often confined to counting incidences of observable behavior that could be reliably measured. The onset of the cognitive revolution depended on the availability of more sophisticated measures of complex thinking and performance. In concert with the increased availability of better measurement tools, the development of the computer and its use as a model for human thinking provided a methodology for testing theories about complex mental performance. Simulating that performance in a computer environment allowed researchers to test theories about how experts solved problems and to provide some understanding of the architecture of human cognition. The cognitive revolution made it possible to go beyond observable behavior to make inferences about the mental activities of learners, to distinguish among kinds of learners, and to provide a methodology for testing hypotheses about the mental processes implicated in complex thinking. The work begun by the cognitive revolution in the 1960s has developed into a rich science of learning in the early part of the 21st century.

In this chapter we begin with an overview of research findings from the emerging science of learning (Bransford, Brown, & Cocking, 1999; Mestre & Cocking, 2000, 2002) as they pertain to learning science. We then extract from this body of research instructional principles for teaching science, and provide several examples of science instruction aimed at promoting learning with understanding that are based on the science of learning. Current teaching practices in K–16 classrooms as well as current practices for preservice and in-service teacher education are examined through the lens of the science of learning. We end with concluding remarks about the kind of learner envisioned by the adoption of principles emerging from the science of learning and the important role of teacher preparation and development in providing opportunities for this kind of learning.

INSIGHTS ABOUT THE LEARNER FROM THE SCIENCE OF LEARNING

Perhaps the most succinct characterization of the learner and his or her role in the learning process is provided by the 14 learner-centered psychological principles articulated by the American Psychological Association's (APA) *Center for Psychology in the Schools and Education* (CPSE, 2003). The articulation of each principle resulted from discussions of the research supporting these principles by members of a special task force of APA's Board of Educational Affairs. These principles reflect the current consensus of researchers in the area of human learning and development. Among the principles that are relevant to this chapter are the following:

1. The learning of complex subject matter is most effective when it is an intentional process of constructing meaning from information and experience.
2. The successful learner, over time and with support and instructional guidance, can create meaningful, coherent representations of knowledge.
3. The successful learner can link new information with existing knowledge in meaningful ways.
4. The successful learner can create and use a repertoire of thinking and reasoning strategies to achieve complex learning goals.
5. Higher-order strategies for selecting and monitoring mental operations facilitate creative and critical thinking.
6. Learning is influenced by environmental factors, including culture, technology, and instructional practices.

In the sections that follow, we elaborate on some of the implications of these principles as they apply to educational practice.

Constructivism and the Role of Prior Knowledge in Learning

Constructivism has important implications for learning and instruction. Constructing knowledge is a lifelong, effortful process requiring significant mental engagement from the learner. The learner's mind within constructivist epistemology is not a blank slate upon which new knowledge can be inscribed. According to von Glasersfeld, constructivism is a "vast and wooly area on contemporary psychology, epistemology, and education" (1997, p. 204). Phillips (1995) also notes that there are many types of constructivism. Moshman (1982) describes three types of constructivism, which suggest that knowledge construction is directed from outside, from within, and from a combination of internal and external directions. Externally directed knowledge construction occurs as students construct a representation of the outside world. Information-processing approaches to cognition allow for such construction as learners must always construct their own representation from "given" knowledge or experiences. Knowledge construction that is internally driven requires the learner to utilize existing knowledge to transform, organize, and reorganize existing knowledge. Piagetian theory describes this process as one of adaptation, occurring through mechanisms of assimilation, accommodation, and equilibration

(De Lisi & Golbeck, 1999). Social constructivism describes knowledge construction that occurs as a result of both internal and external directions and is the result of the reciprocal interaction between individual knowledge construction and the external direction of the outside world. Vygotsky's theory of cognitive development is a good example of this latter type of constructivism (Bruning, Schraw, & Ronning, 1999). In all three approaches to constructivism, knowledge previously constructed by the learner will affect how she or he interprets new knowledge that the teacher is attempting to impart.

A constructivist teacher focuses on probing the knowledge that learners have previously constructed in order to make appropriate instructional choices with respect to the content to be learned. The teacher needs to evaluate if sufficient prior knowledge is available and evaluate whether this knowledge conflicts with the knowledge being taught. If it does, the teacher should guide learners in reconstructing knowledge. The teacher encourages knowledge construction or reconstruction on the part of the learner that is compatible with current scientific thought so that learners will store the knowledge in memory in a form that is optimal for long-term recall or for application in problem-solving contexts (Anderson, 1987; Mestre, 1994; Resnick, 1983; Schauble, 1990; von Glasersfeld, 1989, 1992). To ignore learners' preknowledge makes it highly probable that the message intended by the teacher will not be the message received by the student. For example, when children who believe Earth is flat are told that it is round, they may understand this to mean that it is round like a pancake, with people standing on top of the pancake (Vosniadou & Brewer, 1992). When subsequently told that Earth is not round like a pancake, but rather round like a ball, children envision a ball with a pancake on top, upon which people could stand (after all, children reason, people would fall off if standing on the side of a ball!).

Another instructional implication of constructivism is that instructional strategies that facilitate the construction of knowledge should be favored over those that do not. Sometimes this statement is interpreted to mean that we should abandon all lecturing and adopt instructional strategies where students are actively engaged in their learning. Although the latter goal is certainly desirable, the former is an overreaction. Moshman's (1982) analysis suggests that active learning is still possible when explanations are provided, although clearly a more social constructivist approach will result in greater learner activity. Hence, instructional approaches where students are discussing science, doing science, teaching each other science, and offering problem solution strategies for evaluation by peers will facilitate the construction of science knowledge.

The Nature of Expertise: Organization and Application of Knowledge

Much of what is known about the acquisition of knowledge, its storage in memory, and its application to solving problems has come from studies of expertise and how it develops. Experts have extensive knowledge that they can apply flexibly to solve problems, and so cognitive scientists have focused on characterizing the development of expertise, as well as the organization, acquisition, retrieval, and application of experts' knowledge (see Bransford et al., 1999, Ch. 2). Among the salient findings is that experts' knowledge is highly organized (Chi & Glaser, 1981; Glaser, 1992; Larkin, 1979; Mestre, 1991). The organization is hierarchical, with the top of the hierarchy containing the major principles/concepts of the domain, and ancillary concepts, related facts, equations, and other details occupy the middle to lower levels of the knowledge pyramid. Because of the highly organized nature of their knowledge, experts are able to access their knowledge quickly and efficiently. Ericsson and Kintsch (1994) described experts as being able to retrieve large amounts of information during problem solving because of the retrieval cues that were stored in long-term working memory that allowed for swift access to long-term memory stores. Furthermore, procedures for applying the major principles and concepts are closely linked to the principles, and retrieved with relatively little cognitive effort when a major principle is accessed in memory. Automation of these processes makes it possible for experts to perform high-level cognitive tasks rapidly (Anderson, 1990). This allows experts to focus their cognitive efforts on analyzing and solving problems, rather than on searching for the appropriate "tools" in memory needed to solve the problems. Knowing more, by virtue of having an efficient organizational structure of the knowledge, means that it requires relatively little effort for experts to learn even more about their area of expertise since new knowledge is integrated into the knowledge structure with the appropriate links to make recall and retrieval relatively easy.

Experts also approach problem solving differently from novices (Chi, Feltovich, & Glaser, 1981). For example, when asked to categorize problems (without solving them) according to similarity of solution, experts categorize according to the major principles that can be applied to solve the problems (e.g., conservation of momentum), whereas novices categorize according to the superficial attributes of the problems (e.g., according the objects that appear on the problem statement, such as "pulleys" and "inclined planes"). This difference between attention to deep structural characteristics of problems rather than surface-level characteristics distinguishes experts from novices. When asked to state an approach they would use to solve specific problems, experts discuss the major principle they would apply, the justification for why the principle can be applied to the problem, and a procedure for applying the principle. In contrast, novices jump immediately to the quantitative aspects of the solution, discussing the equations they would apply to generate an answer. Compared to novices, experts spend more time planning for problem solving, monitoring their problem-solving efforts, and regulating their use of strategies.

Research on the nature of expertise indicates that expert performance involves the use of automated knowledge and procedures. One implication for instruction resulting from this is that the tacit knowledge that experts use to solve problems should be made explicit during instruction, and that students should actually practice applying this (no longer tacit) knowledge while solving problems. In addition, students will need to practice the kind of metacognitive activity in which experts engage. Simply telling students how major ideas apply to problems will not lead to the construction of knowledge. Students need to engage actively in applying and thinking about how the big ideas are relevant for solving problems so that they become internalized as used purposely as problem-solving tools. Possessing automated knowledge and procedures translates to being able to perform lower-level activities without much thought, thereby freeing up more working memory for focusing on more complex activities.

Transfer of Learning

We define transfer of learning (hereafter transfer) as the ability to apply knowledge or procedures learned in one context to new contexts. A distinction is commonly made between near and far transfer. The former consists of transfer from initial learning that is situated in a given setting to ones that are closely related. Far transfer refers both to the ability to use what was learned in one setting to a different one as well as the ability to solve novel problems that share a common structure with the knowledge initially acquired. We note that there is an emerging third way to talk about transfer, one that meets a criterion of generativity (Carey & Spelke, 1994; Gelman & Williams, 1998). The idea here is that learners can, on their own, come up with novel instances or solutions. A number of overviews of issues in transfer can be found in Chapter 3 of the National Research Council report *How People Learn* (Bransford et al., 1999), Brown and Campione (1998), and Brown, Bransford, Ferrara, and Campione (1983); for topics related to generativity, see Carey and Spelke (1994) and Gelman and Williams (1998). Barnett and Ceci (2002) provide a literature review of the salient research on transfer, a taxonomy that can help organize the field, and directions for future study.

A major but often tacit assumption in education is that the knowledge that students learn in school will transfer to situations and problems encountered outside of school. Yet, some classic studies of analogical transfer illustrate that transfer of relevant knowledge is not common from one situation to a second situation, where both situations are isomorphic (i.e., share the same structure) but differ in context (Gick & Holyoak, 1980; Hayes & Simon, 1977; Reed, Dempster, & Ettinger, 1985; Reed, Ernst, & Banerji, 1974). Only after receiving hints pointing out that two situations are isomorphic are students able to transfer relevant knowledge. More recently, Blanchette and Dunbar (2001) found that although students can spontaneously draw analogical inferences from one domain to another, they do not make enough inferences to support a fully-fledged transfer from one domain to another. If, as these studies suggest, the ability to apply knowledge flexibly is context-bound, then an important question for education is how to structure instruction to encourage transfer, short of the impossible task of covering in detail all the relevant contexts in which the knowledge being taught could be applied. What, then, is known about successful transfer and upon what does it depend?

Research suggests that several factors affect transfer. First, initial learning is necessary for transfer (Brown et al., 1983; Carey & Smith, 1993; Chi, 2000). Although this seems obvious, it is noteworthy that many failures to produce transfer have resulted from inadequate opportunities for students to learn effectively in the first place (e.g., see discussions by Brown, 1990; Klahr & Carver, 1988; Littlefield et al., 1988). Attention to initial learning is very important for transfer, especially when measures of transfer are used to evaluate the degree to which educational interventions are, or are not, "effective." Whereas rote learning does not tend to facilitate transfer, learning with understanding does (Bransford et al., 1983; Mandler & Orlich, 1993; see also literature review in Barnett & Ceci, 2002). Thus, attempts to learn too many topics too quickly may hamper transfer since the learner may simply be memorizing isolated facts with little opportunity to organize the learned material in any meaningful fashion or to link it to related knowledge. Although previous learning can enhance transfer, it can also obstruct it (Bransford et al., 1999; Gelman & Lucariello, 2002). For example, new learning may not proceed rapidly if knowledge that the learner possesses that is relevant to the new learning remains inactivated; on the other hand, when tasks share cognitive elements, transfer is facilitated (Glaser & Baxter, 1994; Singley & Anderson, 1989). This is true even for young children (Brown & Kane, 1988).

Context also plays a pivotal role in transfer. If the knowledge learned is too tightly bound to the context in which it was learned, transfer to superficially different contexts will be reduced significantly (Bjork & Richardson-Klavhen, 1989; Carraher, 1986; Eich, 1985; Lave, 1988; Mestre, 2002; Saxe,

1989). For example, students who learn to solve arithmetic progression problems can transfer the method they learned to solve similar physics problems involving velocity and distance, but students who learn to solve the physics problems first are unable to transfer the method to solve isomorphic arithmetic progression problems (Bassok & Holyoak, 1989). The transfer from physics to arithmetic was apparently blocked by embedding the physics equations within that specific context, which then precluded students from seeing their applicability to another context. These findings also suggest that because students had more general knowledge about arithmetic/algebra, those who learned to solve the problems within a mathematical context first were able to screen out the content-specific details of the problem-solving procedures, whereas those who learned to solve the physics problems first appear to attribute the underlying physics context as crucial to the application of the problem-solving procedures, and hence were unable to apply those procedures to a math context. Furthermore, the context within which quantities/variables in a problem are presented affects transfer, as a study by Bassok (1990) demonstrated; students exhibited spontaneous transfer from problems involving speed (meters per second) to problems involving price rate measured in dollars per minute, but not to problems involving salary rate measured in dollars per year. It appears that dollars per minute was interpreted by students as a continuous rate similar to meters per second, but dollars per year was interpreted more like a discrete quantity rather than a rate, and hence the lack of transfer.

In summary, research suggests that transfer is enhanced when the learner abstracts the deep principles underlying the knowledge being learned, and that abstraction is facilitated by opportunities to experience concepts and principles in multiple contexts. People's prior knowledge and experience in a domain affects their subsequent transfer, although sometimes the effect is initially negative because previously learned concepts and routines must be changed to deal with new settings (e.g., Barnett & Ceci, 2002; Bransford et al., 1999; Bransford & Schwartz, 1999; Hartnett & Gelman, 1998; Singley & Anderson, 1989). In educational settings, this is frequently referred to as the "implementation dip" or the "J curve effect" (e.g., Fullan, 2001).

Metacognition: Self-Reflecting about Learning

Research suggests that acquisition and transfer of knowledge can be improved, by the use of metacognitive strategies and metacognitive awareness (Brown, 1975, 1980; Flavell, 1973). Metacognition includes the awareness of self as a learner (e.g., "I am good at mathematics"), knowledge of strategies for successful learning (e.g., "Drawing diagrams can aid understanding"), and knowledge of when to apply strategies. Metacognitive strategies refer to strategies learners use to become more aware of themselves as learners, and include the ability to monitor one's understanding through self-regulation; to plan, monitor success, and correct errors when appropriate; and the ability to assess one's readiness for high-level performance in the field one is studying (Bransford et al., 1999). Reflecting about one's own learning is a major component of metacognition, and does not occur naturally in the science classroom, possibly due to lack of opportunity, because instructors do not emphasize its importance, and because it develops slowly. It is common to hear during a one-on-one tutorial session with a student the comment, "I am stuck on this problem," but when asked for more specificity about this condition of "stuckness," students are at a loss to describe what it is about the problem that has them stuck, and often just repeat that they are just stuck and can't proceed. In this instance, the student has a metacognitive awareness of his or her level of understanding but is unable to bring conditional knowledge of learning strategies to bear on the task. Without knowledge of learning strategies and an ability to fit their use to the needs of the current situation, a student will be unable to make progress.

A second kind of metacognition is learning to reflect on the types of problem-solving strategies one has learned in the past. While the first type of meta-cognition discussed above is focused upon *oneself as a self-monitoring learner,* there is also a meta-level to understanding how to select problem-solving strategies. That is, thinking about strategies and how strategies are selected for problem solving relates to students' deeper understanding of the possibilities—it is thoughtful behavior geared toward selection and application. Kuhn (2000) has shown that understanding *why* a particular strategy is preferable over others plays a critical role in determining whether an available strategy will be used. Kuhn believes that such meta-level understanding plays a critical role in students sustaining their own learning management and problem solving once the teacher and other supports (peers in groups) are no longer present. *What makes learning last* is the ability to monitor one's thinking, including selecting from the knowledge base of strategies one has learned in the past. Failure to transfer is the major limitation of many educational approaches because they do not focus on deep understanding and applying strategies or on how to develop such knowledge in students (Kuhn, 2000).

Promoting the habit that students should reflect on their learning is also pivotal in science courses that deviate from the norms of pedagogical practice in these courses. Despite research evidence demonstrating that students learn best when actively engaged, the norm in most science courses, especially at the college level, is the lecture, in which most students are passively taking notes. Courses that attempt to get students to work collaboratively, or that try other techniques to engage them, are often viewed by students as being deviant from the norm, and they simply tolerate the course rather than becoming engaged in more active learning. It is also the case that students' epistemological beliefs about the nature of science knowledge and how it might be acquired may be in conflict with their experiences in more activity-based courses (Hofer, 2000). In cases such as these, instructors should communicate with students why the course is being taught the way it is, and explain how research on learning suggests that the approach being used is superior to the teach-by-telling approach. Only by getting students to reflect about their learning, and by accruing evidence that in fact the "active learning" approach is making them learn more than a lecture approach, will students begin to buy into the approach and become active participants rather than simply tolerant participants. One possible way to engage students in reflecting on knowledge construction is journal writing. However, journals are hard to grade. A way to simplify this task is to use structured journal writing, for example "Weekly Reports" (Etkina, 2000), described in detail in the Assessment section. Research shows that there is a correlation between students' ability to reflect on the construction of knowledge and the quality of their questions with their conceptual gains (Harper, Etkina & Lin, in press; May & Etkina, 2001).

Assessment in the Service of Learning

Numerous studies involving a variety of disciplines and age groups (Gunstone & Mitchell, 1998; Hake, 1998) demonstrate that for many students a constructivist approach to teaching works better than a traditional one. However, a truly constructivist teaching philosophy requires assessment that is consistent with constructivist epistemology. If we want our students to construct and reflect on their knowledge, ask questions, and plan their own learning, we need to devise assessment strategies that evaluate and encourage these aspects of learning (Black & Wiliam, 1998; Glaser & Baxter, 1994; Zoller, Tsaparlis, Fatsow, & Lubezky, 1997). Formative assessments (assessment to help guide instruction and learning) and summative assessments (assessment to evaluate student performance) that are used in science courses send messages to the students about what the instructor considers important, as well as provide feedback on whether their efforts are successful. By changing assessment tasks we can shift the attention of our students to what we consider important (Etkina, 2002a). Black and Wiliam (1998) showed that the learning gains from systematic attention to formative assessment followed by feedback are larger than most of those found for any other educational intervention (effect sizes of 0.4 to 0.7). Although such findings provide impressive evidence of classroom practices that improve learning, Black and Wiliam also report that such practices are currently underdeveloped in most classrooms.

What is needed is a way to help students understand the target concept or ability that they are expected to acquire and the criteria for good work relative to that concept or ability. Students also need to evaluate (metacognitively) their own efforts in light of the criteria. Finally, they need to share responsibility in taking action in light of the feedback. The quality of the feedback rather than its existence or absence is a central point. Research suggests that the feedback should be descriptive, criterion-based feedback as opposed to numerical scoring or letter grades without clear criteria (Black & Wiliam, 1998). A "scoring rubric" can also help students assess their own work. The rubric might contain criteria for the grading. A student or a group of students can use the rubric to assess her or their own work. An instructor can use the rubric to evaluate a student's response and to provide feedback to the students.

Formative assessment also provides feedback to the instructor. After receiving this feedback, the instructor can modify her or his instruction to better achieve the course goals. One formative assessment approach that has shown promise for providing useful feedback to instructors is journal writing. Writing a journal encourages students to reflect on their learning (a metacognitive strategy), to put their thoughts into coherent sentences, and to communicate them to a reader (Eisen, 1996; Liss & Hanson, 1993; Moore, 1997; Moscovici & Gilmer, 1996). Yet, journal writing is not used in high school science courses and especially in college courses (Lester, Lambdin, & Preston, 1997), which means that preservice teachers do not experience it in their own education and consequently do not use it when they teach themselves. Perhaps the major obstacle to the use of journal writing in science courses is that it is time-consuming for both students (to write) and instructors (to evaluate) (Moscovici & Gilmer, 1996).

One alternative to journal writing is the Weekly Report concept (Etkina, 2000; Etkina & Harper, 2002). A Weekly Report is a structured journal written by students each week, in which they answer three questions: (1) What did I learn this week and how did I learn it? (2) What remained unclear? (3) If I were the professor, what questions would I ask my students to find out if they understood the material? Weekly Reports provide benefits for both students and instructors. For the students, these benefits include developing metacognitive skills, such as reflecting upon their own learning process and upon their own particular knowledge state at a given moment, which tend to foster in a student a sense of responsibility for his or her own learning and move him or her away from being the traditional passive learner. The instructor gets the benefit of frequent feedback from students, finding out for all students (not just the vocal ones) which topics were learned thoroughly and which need some extra attention. Similar to portfolio assessment, Weekly Reports determine "what the student does know" as opposed to the tests that reveal "what the student does not know" (Slater, 1997, p. 315).

Other instructional strategies have emerged for incorporating formative assessment strategies in both large and small college lectures. For example, classroom communication systems now allow instructors to conduct large lecture classes in a "workshop" format where students are given questions to work on in groups, then students submit their answers electronically, and following a histogram display of the class' responses, a class-wide discussion ensues where students present and evaluate the reasoning that led to different answers (Dufrense, Gerace, Leonard, Mestre, & Wenk, 1996; Mazur, 1997; Mestre, Gerace, Dufresne, & Leonard, 1997). Other "workshop-style" approaches to instruction (Laws, 1991; McDermott, 1991), where students actively participate in their own learning with instructors continuously probing for understanding and serving as learning coaches, have also proven effective in college science teaching.

Summative assessment in the sciences at the post-secondary level have tended to focus on problem solving. Although problem solving is crucial for doing science, there are many other skills that scientists possess and value that are not evaluated in students. Recent research within the domain of physics education suggests that construction of summative assessment strategies to probe skills that we value in experts is possible. For example, research reviewed earlier (Chi et al., 1981) indicates that experts categorize problems on the basis of the underlying principle needed to solve the problem, whereas novices cue on problems' surface features. Assessment tasks have been devised that probe whether students are cueing on surface features or underlying principles when categorizing problems (Hardiman, Dufresne, & Mestre, 1989; Leonard, Dufresne, & Mestre, 1996; Royer, Carlo, Dufresne, & Mestre, 1996; Van Heuvelen, 1995). Furthermore, Chi and colleagues (1981) also demonstrated that when asked to state an approach they would use to solve specific problems, experts discuss the major principle they would apply, the justification for why the principle can be applied to the problem, and a procedure for applying the principle, whereas novices provide the equations they would manipulate. An assessment technique termed "strategy writing" (Leonard et al., 1996), where physics novices were asked to write strategies for solving problems that contained the three attributes contained in experts' discussions of solutions (principle, justification, procedure), has shown promise both as an instructional tool and as a summative assessment tool. Finally, it is clear that most of the research in the sciences entails the posing of interesting problems. Recently, "problem posing" has been used as a means of probing students' conceptual understanding and ability to link and apply conceptual knowledge flexibly across problem contexts (Mestre, 2002).

INSTRUCTIONAL PRINCIPLES IMPLIED BY THE SCIENCE OF LEARNING

In this section we discuss instructional principles that are suggested by the review of research on learning from the previous section.

Construction and Sense-Making of Science Knowledge Should be Encouraged

Although teachers can facilitate learning, research evidence indicates that students must construct the understanding themselves. Science ideas are not parcels that can be sent by a teacher to a student (Reddy, 1979) in ready-to-use form. They are complex constructions that the teachers (or book authors) code into words or symbols and that students then have to decode, process, and accommodate into useful understanding. The effort made by a student into decoding the information and making sense of it cannot be substituted by even the most excellent presentation given by a teacher. Furthermore, construction of knowledge is also a social enterprise. Just as scientists work collaboratively in research groups and discuss their findings at professional meetings and in refereed journal articles, students should have an opportunity to work collaboratively on science problems, to discuss the solutions with peers, to argue their point of view, and to ask questions and to challenge the views presented by others.

The process of construction of new knowledge by an individual student depends heavily on his or her prior experiences as well as knowledge derived from these experiences. A teacher should be aware of the knowledge state of her or his students (through questioning, pretests, and other activities) and design the instruction to incorporate students' ideas, keeping in mind that these ideas are context dependent and often hard to change. Authoritatively telling students the correct scientific view does not produce good results since students first have to engage in the difficult process of reconstructing knowledge, a process that is effortful and time-consuming. However, if incorporated into instruction through experimentations, discussions of everyday phenomena, or practical applications, students' ideas can serve as a productive foundation (Minstrel, 1999).

Hypothetico-Deductive Reasoning Should be Encouraged

How do scientists come to believe in different explanations of natural phenomena that students are asked to learn during their science classes? A scientific explanation serves not only to explain findings from experiments that scientists have already performed but also to predict new phenomena that scientists have not observed before. It is its predictive power that distinguishes a scientific from a nonscientific explanation, even if the prediction does not match the experimental outcome. To make a prediction, scientists follow hypothetico-deductive reasoning: If "such and such" is correct, and we do "this and that," the following will happen. Then they perform the experiment, compare the results with the prediction, and decide whether the explanation should be revised (Lawson, 2000).

Similar reasoning can be used by students to test their ideas. For example, many students believe that an object thrown upward continues to carry the force of the throw, a force that they believe diminishes as the object ascends until eventually dying out when the object reaches its apex, at which point gravity "takes over," forcing the object to fall back to Earth. A constructivist approach would not simply inform students that this is a misconception, but rather encourage them to use hypothetico-deductive reasoning to test their hypothesis. Students might come up with the following reasoning: If an object continues to move in the direction of the force of the throw until the force dies out, then if an object is thrown at an angle, it should move in a straight line in the direction of the throw, then stop when the force dies out, and then fall straight to the ground due to the pull of the Earth. Then they perform the experiment and observe that the object neither moves in a straight line if thrown at an angle, nor falls straight down after it reaches the highest point. This creates the need for a new idea to explain the actual observed motion of the object. Lawson and colleagues found that students who are able to perform hypothetico-deductive reasoning are better in acquiring new concepts (Lawson, Baker, DiDonato, Verdi, & Johnson, 1993).

Ample Opportunities Should be Available for Learning "the Processes of Doing Science"

Students should construct meaning of science concepts in ways that make sense to them. However, it is possible for them, under a teacher's guidance, to use processes similar to those used by scientists to construct knowledge (Etkina & Van Heuvelen, 2001; Etkina, Van Heuvelen, Brookes, & Mills, 2002). These processes include: observing natural phenomena or laboratory events; classifying, recording, identifying patterns; devising models to explain patterns; testing the explanations in new experiments; and applying them to design simple devices or solve problems. Experimental testing of the students' devised explanations means that students use the model that they constructed to explain the outcomes of one experiment to predict the outcomes of an undone experiment through hypothetico-deductive reasoning. An unsuccessful prediction means that the model should be revised. For example, students observe how a wet water spot dries out. Their explanation of the phenomenon is that the air "absorbs" water. It works well for many phenomena—blowing air makes hair or paint dry faster. How can the students find out whether this is a good explanation? Using hypothetico-deductive reasoning, they can reason that "if the air is responsible for drying then if we remove the air, then wet spots should not dry." They can then test this prediction using two equally wet pieces of paper that are placed inside and outside a vacuum jar. According to students' hypothesis, the paper inside a vacuum jar should dry slower. However, the experiment shows that it dries faster. Thus students conclude: "When we did the experiment, the paper in the absence of air dried faster than in the presence of air, therefore the air does not absorb water." Now the teacher can ask students whether they had similar experiences before. Many students will remember that in the mountains their lips become dry, supporting the testing experiment.

If similar processes are used throughout instruction, they will help students develop the explanations of phenomena consistent with scientists' current models for how the physical and biological worlds work. Classroom environments in which students are actively engaged and the instructor plays the role of learning coach (cooperative group learning, problem-based learning, Socratic dialogue, inquiry laboratory investigations instead of cookbook labs, minilabs) are helpful in achieving this goal.

Ample Opportunities Should be Provided for Students to Apply Their Knowledge Flexibly Across Multiple Contexts

In the physical sciences, it is usually the case that a handful of concepts can be applied to solve problems across a wide range of contexts. The transfer research literature suggests that when people acquire knowledge in one context they can seldom apply this knowledge to situations in related contexts that look superficially different from the original context, but which are related by the major idea that could be applied to solve or analyze them. The implication is that students should learn to apply major concepts in multiple contexts in order to make the knowledge "fluid." Other sciences that have larger sets of concepts also require practice for students to relate the concepts to new and varied situations. Providing practice exercises across a variety of contexts and situations is what makes learning last—it is the way to promote *transfer of learning*. An example from physics is using vectors to solve problems. Students learn vectors at the very beginning of the typical introductory physics course and apply them immediately in kinematics while dealing with two-dimensional motion. Later they can use them to draw force diagrams when solving problems in dynamics, or to determine complex electric and magnetic fields in electromagnetism, or to explain interference patterns in optics. Frequent opportunities to apply vectors in different contexts is a necessary condition for mastering vectors. If teachers emphasize vectors at the beginning of a course but later do not provide ample opportunities for using them for problem solving, students will not be able to transfer the ability to apply vectors to problems in new contexts such as electricity, magnetism, or optics.

Qualitative Reasoning Based on Concepts Should be Encouraged

Much of the knowledge that scientists possess is referred to as "tacit knowledge"; it is frequently used knowledge that is seldom made explicit or verbalized (e.g., when applying conservation of mechanical energy, one must make sure that there are no nonconservative forces doing work on the system). Working with principles tacitly is fine for experts, but tacit knowledge should be made explicit to novice learners so that they recognize it, learn it, and apply it. One way of making tacit knowledge explicit is by constructing qualitative arguments using the science that is being learned. By both constructing qualitative arguments and evaluating others' arguments, students can begin to appreciate the role of conceptual knowledge in "doing science."

Qualitative reasoning is strongly enhanced by using multiple representations of the same process. For example while solving a problem of a final speed of a skier going downhill, students can draw a picture, a motion diagram, a force diagram, and an energy bar chart (Van Heuvelen & Xou, 2001). All these representations should agree with each other. If the motion diagram shows that the skier is speeding up, the force diagram should have the resultant force pointed downhill and the energy bar chart should indicate that the kinetic energy of the skier at the bottom is higher than on top. If all representations do not agree with each other, then qualitative reasoning can be used to find the source of the disagreement. In chemistry, students can solve ideal gas problems by drawing a picture of the container with the gas in the initial and final states; or graphs of the processes plotted against coordinates P(ressure) and V(olume), V(olume) and

T(emperature), and P and T; or a picture of the molecules inside the container in different states accompanied by a verbal description of the microscopic explanation of the situation. These representations should also agree with each other. For example, suppose students are discussing what happens to the pressure of air in a syringe after its volume decreased by two-thirds by moving a plunger. They could start by drawing molecules inside the syringe before and after the plunger was moved and consider whether the molecules get closer together or farther apart as the volume allowed for them decreases, whether they push the walls of the syringe more or less often, and whether the overall effect is to increase or decrease the pressure. In short, we should keep in mind that mathematics is just one of the languages that science uses, and that other, more qualitative forms of reasoning from representations such as verbal, pictorial, and graphical, are equally useful.

Helping Students Organize Content Knowledge According to Some Hierarchy Should be a Priority

To learn lots of things about a topic, to recall that knowledge efficiently, and to apply it flexibly across different contexts requires a highly organized *mental* framework. A hierarchical organization, in which the major principles and concepts are near the top of the hierarchy, and ancillary ideas, facts, and formulas occupy the lower levels of the hierarchy but are linked to related knowledge within the hierarchy, is needed if a learner is to achieve a high level of proficiency in a field. There are several ways to incorporate this idea into instruction. One is to ask students to list major concepts that they need to solve a particular problem without actually solving the problem. Another one is to use hierarchical charts during instruction to show students the place of newly learned knowledge in the knowledge system. Later, the same chart can be used to decide what concept is needed for a particular problem (Van Heuvelen, 1995).

Metacognitive Strategies Should be Taught so that Students Learn How to Learn

Students should learn to predict not only their ability to perform tasks but also their current levels of mastery and understanding. Helping students to be self-reflective about their own learning will assist them in *learning how to learn* more efficiently. For example, when stuck trying to solve a problem, asking oneself questions such as "What am I missing or what do I need to know to make progress here?", "In what ways is this problem similar to others I've seen before?", "Am I stuck because of a lack of knowledge or because of an inability to identify or implement some procedure for applying a principle or concept?" are often helpful in deciding on a course of action. After solving a problem, reflecting on the solution by asking questions such as "What did I learn that was new by solving this problem?"; "What were the major ideas that were applied and what is their order of importance?"; "Why did the instructor give this particular problem to us?"; "Am I able to pose a problem in an entirely different context that can be solved with the same approach?" helps one monitor mastery and understanding of the topics being learned. Teachers should implement these self-reflective strategies in problem-solving exercises by having students engage in post-problem-solving summaries that address these kinds of questions. In this way, students' own learning progress becomes more evident to them.

Students should also reflect on the process of construction of knowledge. If they are accustomed to such questions as "What did you learn today?" "How did you learn it?" "Why do you believe in this particular idea?" "How is what we learned today connected to what we learned yesterday?" they can engage in independent learning and develop scientifically appropriate epistemologies.

Encouraging students' questions also promotes metacognition. For a scientist, asking a good question is often more important than getting an answer. In the practice of schooling, these values are reversed. Students seldom are rewarded for asking a good question, especially since most questions come from the teacher. A classroom atmosphere where profound questions are rewarded the same way as profound answers would resemble the practice of science more closely.

Formative Assessment Should be Used Frequently to Monitor Students' Understanding and to Help Tailor Instruction to Meet Students' Needs

Assessment provides feedback to both students and instructors, but the kind of assessment that would be useful in guiding teaching may be quite different from the kind of assessment used to determine students' competence at the end of instruction. Formative assessment helps students realize what they don't understand (an online monitoring of the learner's progress, so to speak), and formative assessment helps teachers craft tailored instructional strategies to help students achieve necessary and appropriate understanding in a particular learning exercise. It is important that students get assessed on a regular basis not only on mastering problem solving but also on an ability to collect and analyze data, devise hypotheses explaining patterns in the data, test hypotheses experimentally, reason qualitatively, and represent a concept in multiple ways. Other examples of formative assessment tasks can be problem design, evaluation of other students' work, and review of the section in their textbook. Self-assessment is the most productive variation of formative assessment (Black & Wiliam, 1998). To be successful at these tasks, students need to be provided with guidelines and examples of work at different levels of competence, or so-called "scoring rubrics," which greatly facilitate self-assessment for students.

Motivation is an Important Factor

Young children are curious about the world surrounding them. They constantly ask their parent "why" questions. Bugs, rainbows, and magnets fascinate them. What happens to all this interest when it is time for them to learn physics or chemistry? Many factors contribute to the decline in enthusiasm and motivation (e.g., social comparison, experiences of failure, etc.). One important contributing factor to low levels of motivation is the lack of connection between what they are learning in a classroom and their everyday experiences. For example, problems encountered in a physics class deal with frictionless situations, which are unrealizable in the "real" world; in chemistry class chemical reactions occur with materials they never see or use in day-to-day experience. Thus to students there often seems to be no practical benefit of learning "school science" except to get a good grade. The focus on performance and grades can be deleterious. Performance-oriented learners use strategies that are less productive than strategies of interest-oriented learners (Thorkildsen & Nicholls, 1998). The question is how to maintain students' curiosity and motivation. A study by Nolen (2003) showed that "students in science classrooms where teachers were perceived to endorse independent thinking and to desire deep understanding of science concepts had higher achievement and greater satisfaction with their science learning" (p. 363). Thus motivation can be increased by encouraging students' questions, focusing on the explanations of phenomena, promoting collaboration of students, and making students' grades independent of each other (no curving). A natural way to achieve these goals is to turn the students into young scientists at every step of their learning (see using the processes of science above) and help them develop the ownership of ideas and the application of the ideas to real-world problems. The experience of autonomy that is necessitated by the student as investigative scientist is strongly associated with motivational gains (Stipek, 2002). In addition, teachers can use homework assignments to encourage students to identify phenomena at their homes similar to those learned in class. For example, when students are learning about acids and bases they can use red cabbage juice as an indicator to find out which substances used in cooking at home are basic and which are acidic. Then they can test their saliva and decide if there is a relationship between the food taste and the properties of the saliva.

EXAMPLES OF SCIENCE INSTRUCTION BASED ON THE SCIENCE OF LEARNING

In this section we describe several approaches to teaching science that are based on the principles discussed in the previous section.

Science Learning Cycle

A science learning cycle approach has been developed for elementary and middle school science (Benbow & Mably, 2002; Martin, Sexton, Wagner, & Gerlovich, 1998). There are several forms of the cycle, but most of them include three phases: (1) exploration/term introduction/explanation, (2) application, and (3) expansion. During the exploratory phase students conduct experiments and collect data. During the second stage students explain their observations under the teacher's guidance. The teacher helps refine the scientific explanation of what students observed and the appropriate terminology. After this students use the new knowledge to explain findings from new experiments. Then they discuss and expand on the implications of the concepts learned.

A science learning cycle for heat transfer by radiation might look as follows. Students receive two identical containers filled with cold water. They cover the surface of one container with black paper and the other with aluminum foil, and place thermometers in the water for each. Then they position both containers below, and equidistant from, a high-wattage light bulb. They record the readings of the thermometers and observe that the temperature in the container covered with black paper rises faster. During the discussion the teacher introduces such terms as radiation, reflection, and absorption of radiation. Students use this knowledge to explain why we wear white clothes in the summer. A similar cycle can be used in college (Zollman, 1990) in large-enrollment classes. Students perform experiments before they come to the lecture where the professor helps students in the concept-building, term introduction, and application phases.

The investigative science learning environment (ISLE) is a more constructivist extension of the learning cycle. The instructor does not provide the explanations for the observed experiments; rather, students are encouraged to devise them and later test the explanations in new experiments (Etkina & Van Heuvelen, 2001). Students can devise multiple explanations for the same observations, and testing of the explanations with the help of hypothetico-deductive reasoning will rule some out. More specifically, the sequence proceeds as follows. Students first observe several experiments (called observational experiments). They describe their observations in words using familiar terms and try to identify patterns. Next they devise several plausible explanations for the patterns and use the explanations to predict the outcome of a new experiment (testing experiment). An ISLE cycle for projectile motion might look as follows. First, students observe a person walking or running who throws a basketball in the vertical direction. They observe that, independent of the speed of the person, the ball lands in her hands. This is a pattern. They explain this pattern by saying that the ball continues to move horizontally independently from its vertical motion. Then they use this explanation to predict the outcome of the following experiment. Two balls placed about 2 m from the floor are connected with a spring that pushes one of them horizontally while the other one just drops down. Which ball will land first? According to the model, the vertical motion of each ball is independent from the horizontal motion. Thus the vertical motions are identical and so the balls should reach the floor simultaneously. When students make this prediction following similar reasoning, they often say that the prediction does not make much sense since the ball shot sideways travels a longer path. This doubt makes the testing experiment even more exciting, especially since the balls do land at the same time.

This approach can be used in middle-school classes as well as in large-enrollment introductory science courses both in lectures and labs. For every concept that a teacher wishes students to construct, he or she needs to come up with at least two experiments—one to make students observe a phenomenon, the second one to make them test the explanation that they used to explain the findings from the first experiment to predict the outcome of the second (Etkina et al., 2002). After using the ISLE approach for a while, students move into designing their own testing experiments for invented concepts.

Modeling instruction (Hestenes, 1992, 1997; Wells, Hestenes, & Swackhamer, 1995) is another variation of the learning cycle approach. This method focuses on a limited number of mathematical models that students infer from the data and then apply to explain physical phenomena. In this method

> students are actively engaged in understanding the physical world by constructing and using scientific models to describe, explain, predict, and to control physical phenomena. Course content is organized around a small set

> of basic models. Instruction is organized into ways which move students systematically through all phases of model development, evaluation, and application in concrete situations—thus developing skill and insight in the procedural aspects of scientific knowledge. (Wells et al., 1995, p. 606)

Interactive lecture demonstrations is another type of learning cycle approach that also takes advantage of active engagement in both small and large classes (Sokoloff & Thornton, 1997). Before students are shown a demonstration, they are asked to predict the outcome and to provide an explanation for their prediction. They then view the demonstration. If the outcome does not match the prediction, they need to revise their explanations so that it can explain the observed outcome. The difference between this approach and ISLE is that students make their predictions based on intuition. This approach takes into consideration students' prior knowledge. For example, when studying projectile motion, students first are asked to predict the outcome of the experiment with the two-ball apparatus described above. As students can visualize the motion of the balls, they predict that the one that starts with a horizontal velocity will land last, as it travels a longer distance. However, when they observe the experiment, they see that the balls land simultaneously. This is an unexpected outcome. Students need to revise their explanation to account for why the balls land at the same time. Students now analyze the motion of the balls and see that the vertical component is the same, thereby coming to a conclusion that vertical and horizontal motions are independent of each other. Similar strategies can be used in K–12 classrooms (Chinn & Brewer, 1993; Vosniadau, 1994).

Promoting Conceptual Understanding in Introductory University Physics

Two studies in the domain of physics illustrate how research on learning can be used to promote a deeper conceptual understanding. In two studies (Dufresne, Gerace, Hardiman & Mestre, 1992; Mestre, Dufresne, Gerace, Hardiman, & Touger, 1993), high-performing students who had finished an introductory calculus-based physics course with a grade of B or better were asked to analyze problems using a computer-based tool called the hierarchical analysis tool (HAT). The HAT allowed a hierarchical analysis of the problem, starting with the major principle that could be applied to solve a problem, followed by specification of ancillary concepts and procedures needed for a solution. Once the analysis was completed, an equation would be provided that was specific to the analysis performed, and that (presumably) could be applied to solve the problem; if the conceptual analysis was carried out appropriately, then the equation would be tailor-made for solving the problem. On the other hand, if the analysis was erroneous, then the equation provided was consistent with the choices made in the conceptual analysis but was not useful for solving the problem. The design features of the HAT were motivated by research by Chi and colleagues (1981), who found that when asked to state an approach to solving a problem, experts discuss principles and procedures, whereas novices discuss equations and variables. Hence, the study's focus was to ascertain whether students using the HAT to analyze problems would exhibit traits of expertise following treatment.

After spending 10 hours solving problems over several weeks gaining proficiency with this approach, pre- and post-assessments were used to compare students undergoing the HAT regimen to two other control groups of similar high-performing novices who also spent 10 hours solving the same problems as the HAT group did but using more traditional approaches. Findings revealed that the HAT students outperformed the control groups. More specifically, HAT students performed significantly better than control students on pre–post problem-solving performance. Furthermore, the HAT students also significantly outperformed the control students in ability to categorize problems according to similarity of solution, with the HAT students better able to categorize according to principles, which has been shown to be a trait of expertise (Chi et al., 1981; Hardiman et al., 1989).

A second study explored the ability to scale up the HAT approach in a large lecture format (Leonard et al., 1996). In this study students enrolled in an introductory calculus-based physics course were required to write qualitative "strategies" for solving problems to accompany solutions. This was also based on the Chi and colleagues (1981) study, which found that when asked to provide an approach for solving a problem, experts discussed the principles needed, the justification for why the principles applied to the problem, and a procedure for applying it. Similarly, the strategies that students were taught to generate were qualitative descriptions for solving problems that contained those three components: the major physics principle(s), the justification, and a procedure for applying it (them). Students in the strategy-based course did significantly better than students in a traditionally taught course in both ability to categorize problems according to physics principles, and ability to recall the major concepts covered in the course months after finishing the course.

Overview of Case Study Physics (Van Heuvelen, 1991)

In this approach to learning physics, students first encounter a conceptual overview of a big chunk of the material (e.g., a chapter on kinematics or energy) in a qualitative form. They observe natural phenomena related to the concepts, learn to reason about them qualitatively without mathematics under the guidance of the teacher who focuses on multiple representations of the concepts (e.g., pictorial representations, motion diagrams, free body diagrams, energy bar charts, and ray diagrams). After students are facile with qualitative descriptions of phenomena, a quantitative description is introduced, and students revisit the same material but now the list of multiple representations is enriched with mathematics. The last step in mastering the new material is working on case studies—complicated multistep problems that require the application of material from several topics (e.g., kinematics, dynamics, energy conservation, etc.). Before exams students perform problem categorization exercises (such as those described above) in which students identify which major principles or concepts would be required in the solution of a problem; they do not actually solve the problems, just access their memory to find appropriate major ideas that could be applied to solve them. The instructor helps students to find cues in the problems that might steer them in the right direction, and to discard cues that might steer them to superficial views of the problems. This approach was proven to be very successful even with at-risk students (Gautreau & Novemsky, 1997).

Cheche Konnen: Inquiry Science with Minority Students

An important question in educational equity is how to teach science to younger children or to students who are considered to be educationally "at risk." One approach that has been especially useful in science teaching was developed with language-minority grade-school children. The approach, termed "Cheche Konnen," which in Haitian Creole means "search for knowledge" (Rosebery, Warren, & Conant, 1992), stresses how discourse is a primary means for the search for knowledge and scientific sense-making. The Cheche Konnen approach to teaching began by creating "communities of scientific practice" in language-minority classrooms in a few Boston and Cambridge, Massachusetts, public schools. "Curriculum" emerges in these classrooms from the students' questions and beliefs and is shaped in ongoing interactions that include both the teacher and students. Students explore their own questions; students design studies, collect information, analyze data, and construct evidence; and they then debate the conclusions that they derive from their evidence. In effect, the students build and argue about theories. Question posing, theorizing, and argumentation form the structure of the students' scientific activity. Within this structure, students explore the implications of the theories they hold, examine underlying assumptions, formulate and test hypotheses, develop evidence, negotiate disagreements in belief and evidence, argue alternative interpretations, and offer conclusions. The process as a whole provides a rich, scientifically grounded experience that is often missing in the conventional textbooks or laboratory experiences. The teacher guides and supports students as they explore problems and define questions that are of interest to them. In challenging one another's thoughts and beliefs, students must be explicit about their meanings; they must negotiate conflicts in belief or evidence; and they must share and synthesize their knowledge to achieve understanding (Brown & Palincsar, 1989; Inagaki & Hatano, 1987).

Cheche Konnen demonstrates the power of a sense-making approach to science learning that builds on the knowledge that students bring with them to school from their home cultures, including their familiar discourse practices. Students learned to think, talk, and act scientifically, and their first and second languages mediated their learning in powerful ways. They learn the language of the scientific discipline so that they learn the importance of communication and of a common set of terms for their discourse. Using Haitian Creole, they designed their studies, interpreted data, and argued theories; using English, they collected data from their mainstream peers, read standards to interpret their scientific test results, reported their findings, and consulted with experts. (For a fuller discussion, see Bransford et al., 1999, Chap. 7.)

CLASSROOM PRACTICES AND TEACHER PREPARATION: REVOLUTION OR SCATTERED SKIRMISHES?

In the previous sections we outlined several principles that are important for high-quality science teaching and showed how these principles can be implemented in science instruction. However, the key in the implementation of these principles is the teacher in the classroom. Many excellent curriculum innovations have not succeeded because their designers did not take into account the level of knowledge and skills of teachers. Teacher-proof curricula do not work. The goal of this section is to discuss what knowledge and skills are necessary for successful teaching, how to prepare prospective teachers, and how to arrange lifelong professional development. Simultaneously we will discuss the present situation in K–16 science classrooms in terms of whether the cognitive revolution has altered teaching and learning significantly.

Shulman (1987) identified two critical dimensions of the professional knowledge base needed for good teaching: content knowledge and pedagogical content knowledge (PCK). Content knowledge is composed of knowledge of *substantive* and *syntactic* structures of a discipline. Substantive knowledge includes knowledge of facts, concepts, and principles within the content area, and knowledge of the relationship among these elements. Syntactic knowledge refers to knowledge of the principles of inquiry and values inherent in a particular field, and of the methods with which new ideas are added and deficient ones replaced by those who produce knowledge in that field. When applied to the sciences, syntactic knowledge corresponds to the nature of science.

Pedagogical content knowledge (PCK) refers to a "blending of content and pedagogy into an understanding of how particular topics, problems, or issues are organized, represented, and adapted to the diverse abilities of the learners, and presented for instruction" (Shulman, 1987, p. 8). PCK includes the most commonly taught topics in one's subject area, and alternative ways of representing these topics to make them more understandable to students (Van Heuvelen, 1991). These alternative representations include analogies, illustrations, examples, explanations, and demonstrations. In addition, PCK includes student difficulties with the most commonly taught topics and the instructional methodologies needed to aid students in their accommodation of a new scientific conception. Shulman and Sykes (1986) list eight categories of PCK that form the knowledge base of science teaching:

1. General/liberal education including basic skills of reading, mathematics, writing, and reasoning.
2. Content knowledge in the domains in which teaching will occur.
3. Content-specific pedagogical knowledge.
4. General knowledge of pedagogical principles and practice.
5. Curriculum knowledge.
6. Understanding of student diversity and individual differences.
7. Performance skills.
8. Foundations of professional understanding.

As we see the knowledge of content that the teacher is supposed to teach is pivotal, so is content-specific teaching knowledge. Unfortunately, most preservice teachers do not learn the science content the way they are expected to teach it (McDermott, 1974). Let us now look at the state of science instruction in the past 20 years, after the beginning of cognitive revolution.

Levels of Instruction

1. Science Teaching at the Elementary Level. According to many studies (American Association for the Advancement of Science, 1989; Bethel, 1984; Daehler, & Shinohara, 2001; DeBoer, 1991; Heikkinen, 1988; Tamir, 1983), many elementary school teachers do not have even a rudimentary education in science. Because of this, they often neglect the teaching of science or place too much emphasis on textbook learning and seldom use meaningful inquiry (Beiswenger, Stephans, & McClurg, 1998). In contrast, current emphases in science education focus on introducing students to the process of scientific inquiry, and encourage teachers to provide students with opportunities to investigate questions of interest to them. However, few elementary school teachers have been taught science in this way (Tilgner, 1990). In addition, few elementary school teachers have majored, or minored, in a science area in college.

Although arguments for emphasizing subject-matter preparation as the curricular priority in preparing teachers have been made (Anderson, 1988), an emphasis on content alone is insufficient. Content knowledge alone may not be linked directly to an ability to communicate an understanding of this content (Yager, Hidayat, & Penick, 1988), or to building the pedagogical content knowledge needed for effective instruction. Research in teacher education has shown that pedagogical content knowledge has an influence on the content and process of instruction (Grossman, 1988). Only the combination of science content knowledge and pedagogical content knowledge produces successful instruction (Bellamy, Borko, & Lockard, 1992; Shulman, 1987). One of the possible solutions for the problem in elementary science instruction is to enroll preservice elementary teachers in content courses taught through inquiry. If they experience this method of teaching while simultaneously learning new content themselves there is an increased likelihood that they will build both content knowledge and pedagogical content knowledge, and that they will use student-centered instructional approaches when they eventually become teachers.

2. Science Teaching at the High School Level. Traditionally, science teachers have, and continue, to implement didactic approaches to teaching that are strongly lecture-centered and largely based on textbook readings (Stofflett & Stoddart, 1994). Many science teachers adhere to traditional, logical positivist views of their roles in the classroom. The work of Abd-El-Khalick and BouJaoude (1997) demonstrates that teachers "failed to see that most science concepts can be taught in one form or another at any grade level, . . . teachers were not knowledgeable about their students' alternative conceptions regarding the target topics and concepts" (p. 691).

For example, many science teachers do not challenge their students to explain phenomena, and they even demonstrate a lack of consistency in their use of the terms *describe* and *explain* (Horwood, 1988; Winograd & Hare, 1988). Students confuse observational data and models that explain patterns in the data (Grosslight, Unger, Jay, & Smith, 1991) and use authority as a warrant of their knowledge (Hofer, 2001; Ritchie, Tobin, & Hook, 1997). At the same time, there is a great deal of evidence that students are capable of not only developing explanations of the observed phenomena, but also of testing them by predicting the results of new experiments in traditional settings (Etkina, Matilsky, & Lawrence, in press; Zachos, Hick, Doane, & Sargent, 2000; Zollman, 1990). This would suggest that it would prove fruitful to combine content courses taught by inquiry with subject-specific education methods courses. For example, preservice biology teachers need to learn how to teach biology by acquiring both biology knowledge and PCK on the teaching of biology. The absence of content-specific methods courses in the schools of education makes these courses too general and of limited value for learning PCK. When young teachers start teaching their subjects they usually revert to the traditional methods of instruction that they experienced while being students themselves.

3. College Science Courses. College professors and teaching assistants in science courses usually do not have much pedagogical training in their background. Most of them transmit their own knowledge of the subject by mimicking their own instructors. Future teachers, in their college science courses, are exposed primarily to the traditional didactic pedagogy that is associated with university science courses. As a result, they leave these courses with a very limited knowledge of (1) the structure, function, and development of their scientific disciplines; (2) pedagogical content knowledge, and (3) the nature of science (Duschl, 1983; Gallagher, 1991; Gess-Newsome & Lederman, 1993). College professors' exposure to a primarily didactic pedagogy strongly influences their own pedagogical conception of science teaching. Although a great deal of effort has been invested in physics education research (McDermott, 1974, 1991; McDermott & Redish, 1999; Redish, 1994, 2000; Wilson, 1997), which in turn has led to many notable examples of successful changes in undergraduate instruction (Laws, 1991; Mazur, 1997; Sokoloff & Thorton, 1997; Tobias, 1992), in most universities, physics is still taught the way it has been taught for years.

Preparation of Science Teachers

Having just discussed both the knowledge base needed for high-quality science instruction and the situation in K–16 classrooms, we now focus on how teacher preparation programs could be structured so that teachers graduate ready to teach in ways consistent with research on teaching and learning. There are many ways to prepare teachers—through teacher colleges where students get an undergraduate teaching degree, through graduate schools of education where students receive a graduate degree in education, and through alternate routes where teachers are certified in the process of teaching. The recommendations below should prove beneficial to all three of these approaches to teacher education.

1. Science Content and Pedagogy Should be Integrated. When pedagogy and content are taught separately, they are seldom integrated. An ideal science course for prospective teachers integrates the subject content with effective ways of teaching that specific content, the goal being to develop pedagogical content knowledge (Shulman, 1987). Just as we have argued that what we know about students' learning comes from the science of learning, pedagogy needs to be grounded in scientific investigations as well, and subject content and pedagogy need to work together in science courses for future teachers.

Clearly, any science course for prospective teachers has to be based on specific science content, but at the same time it should not be so laden with content details that it becomes a race to cover as many topics as possible. The emphasis should be on in-depth understanding in a few major topics rather than the memorization of facts about many topics; the former has lasting value, the latter is quickly forgotten after the course is over. The former is more likely to transfer; the latter is rarely related to other learning contexts. Another important aspect is that the teaching methods that the college instructor uses to help future teachers master science content should be based on the instructional principles discussed in Section III. Even if few topics are covered in great detail through lecturing and cookbook labs, preservice teachers will not experience the learning process that they are supposed to build and refine for use with their future students. Programs that educate prospective science teachers should provide opportunities for their prospective teachers to experience: hypothetico-deductive reasoning when constructing scientific concepts; organizing knowledge using hierarchical charts or concept maps; making links between new science concepts learned and problem-solving contexts for applying them to facilitate transfer; reflecting on their own learning and the construction of knowledge to promote the development of metacognitive strategies; and formative and summative assessment strategies that are consistent with research on learning.

2. Real Science Should be Practiced. It is easier to advocate inquiry teaching that models authentic science practice if preservice teachers are familiar with it. To learn about authentic practices in doing science, preservice teachers can work as interns in scientific laboratories, attend science seminars, or take part in research summer programs for high school students

(Etkina, Lawrence, & Charney, 1999). The most important part of these experiences is preservice teachers' reflection on how scientists collect and analyze data, identify patterns, devise explanations, and test hypotheses. Working in a science lab helps dispel the notion that answers to scientific questions come from "authorities" who know answers before investigating the appropriate questions through experimentation.

With the help of modern technology, authentic science can be brought into a science classroom. A great deal of real-time data are available today on the Web. For example, X-ray astronomical data collected by X-ray satellites (the most famous being Chandra) can be downloaded from the NASA website and analyzed using the software that the site provides (Etkina et al., in press). Weather monitoring is another option. If preservice teachers learn one or more approaches to real-time data collection and analysis via the Internet, they can bring cutting-edge science into the classroom.

3. Clinical Practice Should Accompany Methods Courses. When asked to rate the relative usefulness of particular program elements, teachers cite student teaching (and other forms of clinical practice) as most useful (Kennedy, 1999). Integrating science learning with instructional practices in order to build pedagogical content knowledge is facilitated by extensive use of clinical practice, and not just at the end of the prospective teacher's training. Preservice teachers should visit local schools (urban and suburban), observe different types of science lessons, and possibly even teach mini-lessons. Preservice high school biology, chemistry, and physics teachers can also work as instructors in recitations or labs in interactive engagement undergraduate science courses. This practice is very successful at Rutgers University (Etkina, 2002b) where future physics and chemistry teachers have an opportunity to practice the methods that they are learning, to improve their content knowledge, and to see in action how students respond to an inquiry setting.

In most teacher preparation programs student teaching takes place during a single semester. If this "one-shot" approach remains the norm, it is very important that preservice teachers are placed with cooperating teachers who reflect sound pedagogical practices, such as using an inquiry approach to science teaching; this ensures that preservice teachers have ample opportunities to use what they learned in methods courses. Thus a close interaction between university faculty, cooperating teachers, and science supervisors is necessary for adequate progress. It also is very important to have a parallel seminar course for student teachers to share their classroom experiences, to plan lessons, and to design assessments.

4. Professional Development. Now, imagine a hypothetical teacher preparation program in which all of the principles above are implemented. Would this guarantee that the "ideal" teacher will emerge and she or he will be equipped for lifelong teaching? In a modern society there is no job that does not require constant retraining, and the job of a teacher requires considerable and continuous "upgrading." New technology, new students, new scientific discoveries, new curriculum materials, new learning research, and many other factors make lifelong professional development a part of any teacher's life. Billions of dollars have been invested in professional development workshops, summer programs, classes, and so on. Some of them were successful but most did not bring the desired change in teaching practice. What are the secrets of successful professional development practice? Professional development that was inquiry based was shown to be successful (Supovitz & Turner, 2000). In addition, when teachers chose to participate, long-term maintenance of new strategies was more likely to be successful (Bainer & Wright, 1998).

The literature on teacher professional development suggests that teachers will not change and new practices will not be sustainable if teachers participate in single workshops or training sessions for the implementation of curriculum materials designed by some research group or commercial company (Cohen & Ball, 1990; Hall & Hord, 1987; Joyce & Showers, 1988; Kaestle, 1993). These training sessions do not take into account the realities of a particular group of teachers nor their knowledge base, and do not provide in-class support to address problems that teachers may encounter in the classroom when implementing the purchased materials (Guskey, 1985; Richardson, 1990; Sarason, 1993).

Corcoran (1995) suggests that for professional development to be effective, it should:

- Stimulate and support site-based initiatives
- Support teacher initiatives
- Be grounded in knowledge about learning and teaching
- Model desirable teaching
- Offer intellectual, social, and emotional engagement with ideas, materials, and colleagues
- Demonstrate respect for teachers as professionals and adult learners
- Provide sufficient time and follow-up support for teachers to master new content and strategies and to integrate them into their practices

Research regarding effective staff development suggests that the most successful programs are those in which the participants become involved in literature, instruction, assessment, and discussions that are specifically designed to facilitate the implementation of a particular topic. These programs employ the model of presentation of information coupled with demonstrations, practice, feedback, and in-classroom coaching (Joyce & Showers, 1982). An excellent example of such a professional development program is Japanese Lesson Study (Lewis, 1995). In this approach a group of same-grade or same-subject teachers meets at the beginning of the school year and chooses two or three "research lessons" that they will work on. During several months the teachers plan the unit to which a lesson belongs, identifying those foundational concepts that are necessary for the student to understand the lesson as well as the ideas from the lesson that will be used in later lessons. Then they investigate the in-depth content of the lesson, decide on the materials and order of instruction, and assessment instruments. When the lesson is ready, one of the teachers in the group teaches it in her or his classroom, and the rest observe. Then the teachers meet and discuss the lesson, revise it, and another teacher teaches the revised version. After one cycle is complete, the group moves to the next lesson. These lessons later are published, and other teachers can use them.

In sum, professional development programs should be based on the same cognitive principles for teaching science discussed in Section III, and address Corcoran's recommendations listed above. Combined together, they lead to four major goals for professional development:

1. A focus on both the science content that teachers are/will be teaching and the pedagogical techniques for teaching that content (to develop pedagogical content knowledge).
2. A focus on making the methods used to teach pre- and in-service teachers and the methods used by them in their science classrooms seamless and reflective of cognitive principles.
3. A focus on having preservice teachers engaged in the clinical practice of teaching throughout their entire preservice training.
4. A focus on fostering metacognitive habits in teachers so that they reflect on their own teaching and learning experiences, and then collaboratively work with other teachers in refining instructional strategies (thereby forming communities of teaching and learning).

CONCLUSIONS

Ideas supportive of constructivist theory have been around a long time but it is only in recent years that there is a consensus on the view of learning described above. The principles of learning that are now clearly described by APA's *Learner Centered Principles* (CPSE, 2003) and the emerging science of learning (Bransford et al., 1999) portray the learner as active, engaged, motivated and self-regulated, deploying metacognitive skill in solving problems, and utilizing prior knowledge and experience to make sense of new experiences. This portrayal of the general learner is consistent with the vision of students learning science that has been advanced by the National Research Council (1996) and AAAS (1989). In this chapter, we have shown that examples of instruction in science that support such learning are available and successful. We have also shown that the professional preparation and development of teachers is crucial in making meaningful learning experiences in science learning available to a very broad range of students. Much remains to be done in developing strategies for how best to provide this preparation and continuous development.

NOTE

1. Work supported in part by NSF Grant No. REC-0106771 to Jose P. Mestre.

REFERENCES

Abd-El-Khalick, F., & BouJaoude, S. (1997). An exploratory study of the knowledge base for science teaching. *Journal of Research in Science Teaching, 34*(7), 673–699.

American Association for the Advancement of Science (1989). *Science for all Americans: A project 2061 report on literacy goals in science.* Washington, DC: Author.

Anderson, C. (1988). The role of education in academic disciplines in teacher education. In A.Woofolk (Ed.), *Research perspectives on the graduate preparation of teachers* (pp. 88–107). Englewood Cliffs, NJ: Prentice Hall.

Anderson, C. W. (1987). Strategic teaching in science. In B. F. Jones, A.S. Palincsar, D. S. Ogle & E. G. Carr (Eds.), *Strategic teaching and learning: cognitive instruction in the content areas* (pp. 73–91). Alexandria, VA: Association for Supervision and Curriculum Development.

Anderson, J. R. (1990). *Cognitive psychology and its implications.* San Francisco: Freeman.

Bainer, D. L., &. Wright, D. (1998). *Teacher choices about their own professional development in science teaching and learning.* (ERIC Document Reproduction Service No. ED420517)

Barnett, S. M. & Ceci, S. J. (2002). When and where do we apply what we learn?: A taxonomy for far transfer. *Psychological-Bulletin, 128*(4), 612–637.

Bassok, M. (1990). Transfer of domain-specific problem-solving procedures. *Journal of Experimental Psychology: Learning, Memory and Cognition, 16,* 522–533.

Bassok, M., & Holyoak, K. J. (1989). Interdomain transfer between isomorphic topics in algebra and physics. *Journal of Experimental Psychology: Learning, Memory and Cognition, 15,* 153–166.

Bellamy, M., Borko, H., & Lockhard, D. (1992, March). *The roles of three types of teacher knowledge—content knowledge, pedagogical knowledge, and pedagogical content knowledge—in the teaching of high school Mendelian genetics.* Paper presented at the annual meeting of the National Association for Research in Science Teaching, Boston.

Beiswenger, R. E., Stepans, J. I., & McClurg, P. A. (1998). Developing science courses for prospective elementary teachers. *Journal of College Science Teaching, 27*(4), 253–257.

Benbow, A., & Mably, C. (2002). *Science education for elementary teachers.* Wadsworth, Thomson Learning.

Bethel, L. (1984). Science teacher preparation and professional development. In D. Holdzkom & P. Lutz (Eds.), *Research within reach: Science education.* Washington, DC: National Science Teachers Association.

Bjork, R.A., & Richardson-Klavhen, A. (1989). On the puzzling relationship between environment context and human memory. In C. Izawa (Ed.), *Current issues in cognitive processes: the Tulane Flowerree symposium on cognition.* Hillsdale, NJ: Erlbaum.

Black, P., & Wiliam, D. (1998). Assessment and classroom learning. *Assessment in Education, 5* (1), 2–57.

Blanchette, I., & Dunbar, K. (2001). Analogy use in naturalistic settings: The influence of audience, emotions, and goals. *Memory and Cognition, 29,* 730–735.

Bransford, J. D., Brown, A. L., & Cocking, R. R. (1999). *How people earn: Brain, mind, experience, and school.* Washington, DC: National Academy Press.

Bransford, J. D., & Schwartz, D. (1999). Rethinking transfer: A simple proposal with multiple implications. In A. Iran-Nejad & P. D. Pearson (Eds.), *Review of research in education* (Vol. 24, pp. 61–100). Washington, DC: American Educational Research Association.

Bransford, J. D., Stein, B. S., Vye, N. J., Franks, J. J., Auble, P. M., Mezynski, J. J., & Perfetto, G. A. (1983). Differences in approaches to learning: An overview. *Journal of Experimental Psychology: General, 3,* 390–398.

Brown, A. L. (1975). The development of memory: Knowing, knowing about knowing, and knowing how to know. In H. W. Reese, (Ed.), *Advances in child development and behavior* (Vol. 10). New York: Academic Press.

Brown, A. L. (1980). Metacognitive development and reading. In R. J. Spiro, B. C. Bruce, & W. F. Brewer (Eds.), *Perspectives from cognitive psychology, linguistics, artificial intelligence, and education.* Hillsdale, NJ: Erlbaum.

Brown, A. L. (1990). Domain-specific principles affect learning and transfer in children. *Cognitive Science, 14,* 107–133.

Brown, A. L., Bransford, J. D., Ferrara, R. A., & Campione, J. S. (1983). Learning, remembering, and understanding. In J. H. Flavell & E. M. Markman (Eds.), *Handbook of child psychology: Vol 3. Cognitive development* (4th ed, pp. 78–166). New York: Wiley.

Brown, A. L., & Campione, J. (1998). Designing a community of young learners: Theoretical and practical lessons. In N. M. Lambert & B. L. McCombs (Eds.), *How students learn: Reforming schools through learner-centered education* (pp. 153–186). Washington, DC: American Psychological Association.

Brown, A. L., & Kane, M. J. (1988). Preschool children can learn to transfer: Learning to learn and learning from examples. *Cognitive Psychology, 20,* 493–523.

Brown, A. L., & Palinscar, A. S. (1989). Guided, cooperative learning, and individual knowledge acquisition. In L. B. Resnick (Ed.), *Knowing, learning, and instruction: Essays in honor of Robert Glaser* (pp. 393–451). Hillsdale, NJ: Erlbaum.

Bruning, R. H., Schraw, G., & Ronning, R. R. (1999). *Cognitive psychology and instruction* (3rd ed). Columbus, OH: Merrill.

Carey, S., & Smith, C. (1993). On understanding the nature of scientific knowledge. *Educational Psychologist, 28,* 235–251.

Carey, S., & Spelke, E. (1994). Domain-specific knowledge and conceptual change. In L. A. Hirshfeld & S. Gelman (Eds.), *Mapping the mind: Domain specificity in cognition and culture* (pp. 169–200). New York: Cambridge University Press.

Carraher, T. N. (1986). From drawings to buildings: Mathematical scales at work. *International Journal of Behavioral Development, 9,* 527–544.

Center for Psychology in the Schools and Education. (2003). *Learner-centered psychological principles: A framework for school reform and redesign.* Retrieved May 5, 2003, from http://apa.org/ed/lcpnewtext.html

Chi, M. T. H. (2000). Self-explaining: The dual process of generating inference and repairing mental models. In R. Glaser (Ed.), *Advances in instructional psychology: Educational design and cognitive science* (Vol. 5, pp. 161–238) Mahwah, NJ: Erlbaum.

Chi, M. T. H., Feltovich, P. J., & Glaser, R. (1981). Categorization and representation of physics problems by experts and novices. *Cognitive Science, 5,* 121–152.

Chi, M. T. H., & Glaser, R. (1981). The measurement of expertise: Analysis of the development of knowledge and skills as a basis for assessing achievement. In E. L. Baker & E. S. Quellmalz (Eds.), *Design, analysis and policy in testing* (pp. 37–47). Beverly Hills, CA: Sage.

Chinn, C. A., & Brewer, W. F. (1993). The role of anomalous data in knowledge acquisition: A theoretical framework and implications for science instruction, *Review of Educational Research, 63,* 1–49.

Cohen, D. K., & Ball, D. L. (Eds.). (1990). Policy and practice [Special issue]. *Educational Evaluation and Policy Analysis, 12,* 249–353.

Corcoran, T. B. (1995, June). *Helping teachers teach well: Transforming professional development* (CPRE Policy Briefs). Brunswick, NJ: Consortium for Policy Research in Education, Rutgers, the State University of New Jersey.

Daehler, K. R., & Shinohara, M.(2001). A complete circuit is a complete circle: exploring the potential of case materials and methods to develop teachers' content knowledge and pedagogical content knowledge of science. *Research in Science Education, 31,* 267–288.

DeBoer, G. (1991). *A history of ideas in science education.* New York: Teachers College Press.

De Lisi, R., & Golbeck, S. L. (1999). Implications of Piagetian theory for peer learning. In A. M. O'Donnell & A. King (Eds.), *Cognitive perspectives on peer learning* (pp. 3–37). Mahwah, NJ: Erlbaum.

Dufresne, R. J., Gerace, W. J., Leonard, W. J., Mestre, J. P., & Wenk, L. (1996). Classtalk: A classroom communication system for active learning. *Journal of Computing in Higher Education, 7,* 3–47.

Dufresne, R. J., Gerace, W. J., Hardiman, P. T., & Mestre, J. P. (1992). Constraining novices to perform expert-like problem analyses: Effects on schema acquisition. *Journal of the Learning Sciences, 2,* 307–331.

Duschl, R. (1983). The elementary level science methods course: Breeding ground of an apprehension towards science? A case study. *Journal of Research in Science Teaching, 20,* 745–754.

Eich, E. (1985). Context, memory, and integrated item/context imagery. *Journal of Experimental Psychology: Learning, Memory, and Cognition, 11,* 764–770.

Eisen, A. (1996). "Disease of the week" reports: Catalysts for writing and participation in large classes. *Journal of College Science Teaching, 25*(5), 331–333.

Ericsson, K. A., & Kintsch, W. (1994). Long-term working memory. *Psychological Review, 102,* 211–245.

Etkina, E. (2000). Weekly reports: a two-way feedback tool. *Science Education, 84,* 594–605.

Etkina, E. (2002a). Formative and summative assessment in a physics class: time to change. *Proceedings of the 2002 Physics Education Research Conference* (pp. 25–31). New York: PERC Publishing.

Etkina, E. (2002b). *How can we help physics teachers acquire pedagogical content knowledge?* Paper given the National Convention of the American Association of Physics Teachers, Boise, ID.

Etkina, E., & Harper, K. (2002). Weekly reports: student reflections on learning. *Journal of College Science Teaching, 31*(7), 476–480.

Etkina, E., Lawrence, M., & Charney, J. (1999). Introducing astrophysics research to high school students. *Physics Education, 34*(5), 300–305.

Etkina, E., & Van Heuvelen, A. (2001). Investigative science learning environment: Using the processes of science and cognitive strategies to learn physics. *Proceedings of the 2001 Physics Education Research Conference* (pp. 17–21). New York: PERC Publishing.

Etkina, E., Van Heuvelen, A., Brookes, D., & Mills, D. (2002). Role of experiments in physics instruction—A process approach. *The Physics Teacher, 40*(6), 351–355.

Etkina, E., Matilsky, T., & Lawrence, M. (in press). What can we learn from pushing to the edge? Evolution of student physics knowledge, understanding of the nature of science, and independent research ability in the context of the Rutgers Astrophysics Institute. *Journal of Research in Science Teaching.*

Flavell, J. H. (1973). Metacognitive aspects of problem solving. In L. B. Resnick (Ed.), *The nature of intelligence.* Hillsdale, NJ: Erlbaum.

Fullan, M. (2001). *Leading in a culture of change.* San Francisco: Jossey-Bass.

Gallagher, J. (1991) Prospective and practicing secondary school science teachers' knowledge and beliefs about the philosophy of science. *Science Education, 75,* 121–133.

Gautreau, R., & Novemsky, L. (1997). Concepts first—A small group approach to physics learning. *American Journal of Physics, 65,* 418–428.

Gelman, R., & Lucariello, J. (2002). Role of learning in cognitive development. In H. Pashler & R. Gallistel (Eds.), *Steven's handbook of experimental psychology: Vol. 3. Learning, motivation, and emotion* (3rd ed., pp. 395–443). New York: Wiley.

Gelman, R., & Williams, E. (1998). Enabling constraints for cognitive development and learning: Domain specificity and epigenesis. In D. Kuhn & R. Siegler (Eds.), *Handbook of child psychology: Vol. 2. Cognition, perception and language* (5th ed., pp. 575–630). New York: Wiley.

Gess-Newsome, J., & Lederman, N. G. (1993). Preservice biology teacher's knowledge structures as a function of professional teacher education: A year long assessment. *Science Education, 77,* 24–45.

Gick, M.L., & Holyoak, K.J. (1980). Analogical problem solving. *Cognitive Psychology, 12,* 306–355.

Glaser, R. (1992). Expert knowledge and processes of thinking. In D. Halpern (Ed.), *Enhancing thinking skills in the sciences and mathematics* (pp. 63–75). Hillsdale, NJ: Erlbaum.

Glaser, R., & Baxter, G. P. (1994). *A cognitive framework for evaluating innovative science assessments.* Paper presented at the annual meeting of the National Association of Research in Science Teaching, Anaheim, CA.

Grosslight, L., Unger, C., Jay, E., & Smith, C. (1991). Understanding models and their use in science: conception of middle and high school students and experts. *Journal of Research in Science Teaching, 28*(9), 799–823.

Grossman, P. L. (1988). *A study in contrast: Sources of pedagogical content knowledge for secondary English.* Unpublished doctoral dissertation, Stanford University.

Gunstone, R. F., & Mitchell, I. J. (1998). Metacognition and conceptual change. In J. J. Mintzes, J. H. Wandersee, & J. Novak (Eds.), *Teaching science for understanding: A human constructivist view* (pp.133–163). San Diego, CA: Academic Press.

Guskey, T. R. (1985). Staff development and teacher change. *Educational Leadership, 4*(7), 57–60.

Hake, R. R. (1998). Interactive engagement versus traditional methods: A six-thousand-student survey of mechanics test data for introductory physics courses. *American Journal of Physics, 66,* 64–74.

Hall, G. E., & Hord, S. M. (1987). *Change in schools: Facilitating the process.* Albany: State University of New York Press.

Hardiman, P.T., Dufresne, R. J. & Mestre, J. P. (1989). The relation between problem categorization and problem solving among experts and novices. *Memory and Cognition, 17,* 627–638.

Hartnett, P., & Gelman, R. (1998). Early understandings of numbers: Paths or barriers to the construction of new understandings? *Learning and Instruction, 8*(4), 341–374.

Harper, K., Etkina, E., & Lin, Y.-F. (in press). Encouraging and analyzing student questions in a large physics course: meaningful patterns for instructors. *Journal of Research in Science Teaching.*

Hayes, J. R., & Simon, H. A. (1977).† Psychological differences among problem isomorphs. In N. J. Castellan, Jr., D. B. Pisoni, & G. R. Potts (Eds.), *Cognitive theory* (Vol. 2, pp. 21–41). Hillsdale, NJ: Erlbaum.

Heikkinen, M. W. (1988). The academic preparation of Idaho science teachers. *Science Education, 72,* 63–71.

Henry, N. (Ed.). (1947). *The Forty-Sixth Yearbook of the National Society for the Study of Education, Part. 1: Science Education in the American Schools.* Chicago: University of Chicago Press.

Hestenes, D. (1992). Modeling games in the Newtonian world, *American Journal of Physics, 60,* 732–748.

Hestenes, D. (1997). Modeling methodology for physics teachers. In E. Redish & J. Rigden (Eds.), *The changing role of the physics department in modern universities* (pp. 935–957). Woodbury, NY: American Institute of Physics.

Hunter, G. W. (1934). *Science teaching at junior and senior high school levels.* New York: American Book Company.

Hofer, B. (2000). Dimensionality and disciplinary differences in personal epistemology. *Contemporary Educational Psychology, 25,* 378–405.

Hofer, B. (2001, April). *Personal epistemology research: Implications for learning and instruction.* Paper presented at the annual meeting of the American Educational Research Association, Seattle, WA.

Horwood, R. H. (1988). Explanation and description in science teaching. *Science Education, 72,* 41–49.

Huxley, T. H. (1898). *Science and education.* New York: D. Appleton.

Inagaki, K., & Hatano, G. (1987). Young children's spontaneous personification as analogy. *Child Development, 58,* 1013–1020.

Kaestle, C. R. (1993). The awful reputation of educational research. *Educational Researcher, 22*(1), 26–31.

Kennedy, M. M. (1999). The problem of evidence in teacher education. In R.A. Roth, (Ed.), *The role of the university in the preparation of teachers* (pp. 87–107). Philadelphia: Falmer Press.

Klahr, D., & Carver, S. (1988). Cognitive objects in logo debugging curriculum: Instruction, learning, and transfer. *Cognitive Psychology, 20,* 362–404.

Kuhn, D. (2000). Why development does (and doesn't) occur: Evidence from the domain of inductive reasoning. In R. Siegler & J. McClelland (Eds.), *Mechanisms of cognitive development: Neural and behavioral perspectives.* Mahwah, NJ: Erlbaum.

Joyce, B. R., & Showers, B. (1988). *Student achievement through staff development.* New York: Longman.

Larkin, J. H. (1979). Information processing models in science instruction. In J. Lochhead & J. Clement (Eds.), *Cognitive process instruction* (pp. 109–118). Hillsdale, NJ: Erlbaum.

Lave, J. (1988). *Cognition in practice: Mind, mathematics and culture in everyday life.* Cambridge, UK: Cambridge University Press.

Laws. P. (1991). Calculus-based physics without lectures. *Physics Today, 44*(12), 24–31.

Lawson, A. (2000). The generality of hypothetico-deductive reasoning: making scientific thinking explicit. *The American Biology Teacher, 62*(7), 482–495.

Lawson, A. E., Baker, W. P., DiDonato, L, Verdi, M., & Johnson, M. (1993). The role of hypothetico-deductive reasoning and physical analogies of molecular interaction in conceptual change. *Journal of Research in Science Teaching, 30*(9), 1073–1087.

Leonard, W. J., Dufresne, R. J., & Mestre, J. P. (1996). Using qualitative problem-solving strategies to highlight the role of conceptual knowledge in solving problems. *American Journal of Physics, 64,* 1495–1503.

Lester, F. K., Lambdin, D. V., & Preston, R. V. (1997). A new vision of the nature and purposes of assessment in the mathematics classroom. In G. D. Phye (Ed), *Handbook of classroom assessment: learning, adjustment, and achievement* (pp. 287–319). San Diego, CA: Academic Press.

Lewis, C. C. (1995). *Educating hearts and minds.* Cambridge, MA: Cambridge University Press.

Liss, J. M., & Hanson, S. D. (1993). Writing-to-learn in science. *Journal of College Science Teaching, 22*(4), 342–345.

Littlefield, J., Delclos, V. R., Lever, S., Clayton, K. N., Bransford, J. D., & Franks, J. J. (1988). Learning LOGO: Method of teaching, transfer of general skills, and attitudes towards school and computers. In R. E. Mayer (Ed.), *Teaching and learning computer programming: Multiple research perspectives* (pp. 111–135). Hillsdale, NJ: Erlbaum.

Mandler, J. M., & Orlich, F. (1993). Analogical transfer: The roles of schema abstraction and awareness. *Bulletin of the Psychonomic Society, 31,* 485–487.

Martin, R., Sexton, C., Wagner, K., & Gerlovich, J. (1998). *Science for all children*. Boston: Allyn & Bacon.

May, D., & Etkina, E. (2002). College physics students' epistemological self-reflection and its relationship to conceptual learning. *Physics Education Research: A Supplement to the American Journal of Physics, 70*(12), 1249–1258.

Mazur, E. (1997). *Peer instruction: A user's manual.* Upper Saddle River, NJ: Prentice Hall.

McDermott, L. C. (1974). Combined physics course for future elementary and secondary school teachers. *American Journal of Physics, 42,* 668–676.

McDermott, L. C. (1991). Millikan lecture 1990: What we teach and what is learned—Closing the gap. *American Journal of Physics. 59*(4), 311–315.

McDermott, L. C., & the Physics Education Group at the University of Washington. (1996). *Physics by Inquiry.* New York: Wiley.

McDermott, L. C., & Redish, E. F. (1999). Resource letter: PER1: Physics education research. *American Journal of Physics, 67,* 755–767.

Mestre, J. P. (1991). Learning and instruction in pre-college physical science. *Physics Today, 44*(9), 56–62.

Mestre, J. P. (1994, February). Cognitive aspects of learning and teaching science. In S. J. Fitzsimmons & L. C. Kerpelman (Eds.), *Teacher enhancement for elementary and secondary science and mathematics: status, issues and problems* (pp. 3.1–3.53) (NSF 94-80). Washington, DC: National Science Foundation.

Mestre, J. P. (2002). Probing adults' conceptual understanding and transfer of learning via problem posing. *Journal of Applied Developmental Psychology, 23,* 9–50.

Mestre, J. P., & Cocking, R. (Eds.). (2000). Special issue on the science of learning, *Journal of Applied Developmental Psychology, 21*(1).

Mestre, J. P., & Cocking, R. R. (2002). Applying the science of learning to the education of prospective science teachers. In R. W. Bybee (Ed.), *Learning science and the science of learning* (pp. 13–22). Arlington, VA: NSTA Press.

Mestre, J.P., Dufresne, R. J., Gerace, W. J., Hardiman, P. T., & Touger, J. S. (1993). Promoting skilled problem solving behavior among beginning physics students. *Journal of Research in Science Teaching, 30*, 303–317.

Mestre, J. P., Gerace, W. J., Dufresne, R. J., & Leonard, W. J. (1997). Promoting active learning in large classes using a classroom communication system. In E. F. Redish & J. S. Rigden (Eds.), *The changing role of physics departments in modern universities: Proceedings of the International Conference on Undergraduate Physics Education* (pp. 1019–1036). Woodbury, NY: American Institute of Physics.

Minstrel, J. (1999). Expertise in teaching. In R. J. Sternberg, & J. A. Horvath (Eds.), *Tacit knowledge in professional practice: researcher and practitioner perspectives.* Mahwah, NJ: Erlbaum.

Moore, R. (1994). Writing to learn biology. *Journal of College Science Teaching, 23*(5), 289–295.

Moscovici, H., & Gilmer, P. J. (1996). Testing alternative assessment strategies—the ups and the downs for science-teaching faculty. *Journal of College Science Teaching, 25*(5), 319–323.

Moshman, D. (1982). Exogenous, endogenous, and dialectical constructivism. *Developmental Review, 2,* 371–384.

National Research Council. (1996). *National Science Education Standards.* Washington, DC: National Academy Press.

Nolen, S. (2003). Learning environment, motivation, and achievement in high school science. *Journal of Research in Science Teaching, 40,* 347–369.

Phillips, D. (1995). The good, the bad and the ugly: The many faces of constructivism. *Educational Researcher, 24*(7), 5–12.

Reddy, M. J. (1979). The conduit metaphor: A case of frame conflict in our language about language. In A. Ortony (Ed.), *Metaphor and thought* (2nd ed.). Cambridge, UK: Cambridge University Press.

Reed, S. K., Dempster, A., & Ettinger, M. (1985). Usefulness of analogous solutions for solving algebra word problems. *Journal of Experimental Psychology: Learning, Memory, and Cognition, 11,* 106–125.

Reed, S. K., Ernst, G. W., & Banerji, R., (1974). The role of analogy in transfer between similar problem states. *Cognitive Psychology, 6,* 436–450.

Redish, E. F. (1994). Implications of cognitive studies for teaching physics. *American Journal of Physics, 62,* 796–803.

Redish, E. F. (2000). Discipline-based education and education research: The case of physics. *Journal of Applied Developmental Psychology, 21,* 85–96.

Resnick, L. B. (1983). Mathematics and science learning: A new conception. *Science, 220,* 477–478.

Resnick, L. B. (1987). *Education and learning to think.* Washington, DC: National Academy Press.

Resnick, L. B., & Resnick, D.P. (1992). Assessing the thinking curriculum: New tools for educational reform. In B. R. Gifford & M. C. O'Connor (Eds.), *Changing assessments: Alternative views of aptitude achievement and instruction*_(pp. 37–75). Boston: Kluwer.

Richardson, V. (1990). Significant and worthwhile change in teaching practice. *Educational Researcher, 19*(7), 10–18.

Ritchie, S. M., Tobin, K., & Hook, K. (1997). Teaching Referents and the warrants used to test the viability of students' mental models: Is there a link. *Journal of Research in Science Teaching, 34*(3), 223–239.

Rosebery, A., Warren, B., & Conant, F. R. (1992). Appropriating scientific discourse: Findings from language minority classes. *Journal of the Learning Sciences, 2,* 61–94.

Royer, J. M., Carlo, M. S., Dufresne, R., & Mestre, J. (1996). The assessment of levels of domain expertise while reading. *Cognition and Instruction, 14,* 373–408.

Sarason, S. B. (1993). *The case for change: Rethinking the preparation of educators.* San Francisco: Jossey-Bass.

Saxe, G. B. (1989). Transfer of learning across cultural practices. *Cognition and Instruction, 6*(4), 325–330.

Schauble, L. (1990). Belief revision in children: The role of prior knowledge and strategies for generating evidence. *Journal of Experimental Child Psychology, 49,* 31–57.

Shulman, L. S. (1987). Knowledge and teaching: Foundations of the new reform, *Harvard Educational Review, 57,* 1–22.

Shulman, L. S., & Sykes, G. (1986). *A national board for teaching? In search of a bold standard. A report for the task force on teaching as a profession.* New York: Carnegie Corporation.

Singley, K., & Anderson, J.R., (1989). *The transfer of cognitive skill.* Cambridge, MA: Harvard University Press.

Slater, T. F. (1997). The effectiveness of portfolio assessments in science. *Journal of College Science Teaching, 26*(5), 315–318.

Sokoloff, D., & Thornton, R. (1997). Using interactive lecture demonstrations to create an active learning environment, *The Physics Teacher, 35,* 340–346.

Stipek, D. (2002). *Motivation to learn: Integrating theory and practice.* Boston: Allyn & Bacon.

Supovitz, J. A., & Turner, H. M. (2000). The effects of professional development on science teaching practices and classroom culture. *Journal of Research in Science Teaching, 37,* 963–980.

Stofflett, R., & Stoddart, T. (1994). The ability to understand and use conceptual change pedagogy as a function of prior learning experience. *Journal of Research in Science Teaching, 31,* 31–51.

Tamir, P. (1983) Inquiry and the science teacher. *Science Education, 67,* 657–672.

Tilgner, P. (1990). Avoiding science in the elementary school. *Science Education, 74*(4), 421–431.

Tobias, S. (1992) *Revitalizing undergraduate science.* Tuscon, AZ: Research Corporation.

Thorkildsen, M., & Nicholls, J. G. (1998). Fifth graders' achievement orientations and beliefs: Individual and classroom differences. *Journal of Educational Psychology, 90,* 179–201.

Twiss, G. R. (1929). *A textbook in the principles of science teaching,* New York: MacMillan.

von Glasersfeld, E. (1989). Cognition, construction of knowledge, and teaching. *Synthese, 80,* 121–140.

von Glasersfeld, E. (1992, March). A constructivist's view of learning and teaching. In R. Duit, F. Goldberg, & H. Niedderer (Eds.), *The Proceedings of the International Workshop on Research in Physics Education: Theoretical issues and empirical studies.* Kiel, Germany: IPN.

von Glasersfeld, E. (1997). Amplification of a constructivist perspective. *Issues in Education: Contributions from Educational Psychology, 3,* 203–210.

Van Heuvelen, A. (1991). Overview, Case Study Physics. *American Journal of Physics, 59,* 898–907.

Van Heuvelen, A. (1995). *Active Learning Problem Sheets* (the ALPS Kits). Plymouth, MI: Hayden McNeil.

Van Heuvelen, A., & Xou, X. (2001). Multiple representations of work-energy processes, *American Journal of Physics, 69*(2), 184–194.

Vosniadou, S. (1994). Capturing and modeling the process of conceptual change. *Learning and Instruction, 4,* 45–69.

Vosniadou, S., & Brewer, W.F. (1992). Mental models of the earth: A study of conceptual change in childhood. *Cognitive Psychology, 24,* 535–585.

Wells, M., Hestenes, D., & Swackhamer, G. A. (1995). Modeling method for high school physics instruction. *American Journal of Physics, 63,* 606–619.

Wilson, J. M. (Ed.). (1997). *Conference on the introductory physics course.* New York: Wiley.

Winograd, P., & Hare, V.C. (1988). Direct instruction of reading comprehension strategies: the nature of teacher explanation. In C.E. Weinstein, E. T. Goetz, & P.A. Alexander (Eds.), *Learning and study strategies: issues in assessment, instruction, and evaluation* (pp. 121–139). San Diego, CA: Academic Press.

Woodhull, J. F. (1918). *The teaching of science.* New York: MacMillan.

Yager, R., Hidayat, E., & Penick, J. (1988). Features that separate least effective from most effective science teachers. *Journal of Research in Science Teaching, 25,* 165–177.

Zachos, P., Hick, L. T., Doane, W. E. J., & Sargent, S. (2000). Setting theoretical and empirical foundations for assessing scientific inquiry and discovery in educational programs. *Journal of Research in Science Teaching. 37*(9), 938–962.

Zoller, U., Tsaparlis, G., Fatsow, M., & Lubezky, A. (1997). Student self-assessment of higher-order cognitive skills in college science teaching. *Journal of College Science Teaching, 27*(2), 99–101.

Zollman, D. (1990). Learning cycles for a large enrollment class. *The Physics Teacher, 28*(1), 20–25.

CHAPTER 7

THE SELF AND ACADEMIC MOTIVATION

Theory and Research after the Cognitive Revolution

Frank Pajares
Emory University
Dale H. Schunk
University of North Carolina at Greensboro

ABSTRACT

Our aim in this chapter is to trace the development of research and theory on the self and motivation as a result of the cognitive revolution. Because other chapters in this volume focus with some specificity on the nature of the revolution itself, we describe how this important event in the history of psychology—and for our purposes the history of educational psychology—influenced the field of academic motivation. Subsequent to the cognitive revolution, motivation came to be viewed as a psychological process that cannot be divorced from processes related to the self. To be motivated is to invest one's self in a task or activity. Whatever forces impinge on an individual, the agent ultimately responsible for engaging that goal-directed activity

The Impact of the Cognitive Revolution on Educational Psychology, pages 165–198

is the individual's sense of self. Hence, motivation cannot be understood without understanding the self-processes from which it emanates. As we show, motivation theory, particularly after the cognitive revolution, has come to be defined nearly exclusively by self-constructs. For these reasons, both our brief history of academic motivation and the motivation constructs we describe focus on the self-beliefs that individuals hold about themselves and their potentialities. We close our chapter with some thoughts regarding the future of motivation research and theory, particularly in regard to continued attention to self-beliefs.

Simply casting a casual glance toward the area of achievement motivation these days one becomes immediately aware of the fact that the near totality of constructs and theories in this area share one focus in common: the discipline's taken-for-granted assumption that individuals' self-beliefs are a critical component of human striving. In their expansive overview of the current state of knowledge related to theories and principles of motivation for the 1996 *Handbook of Educational Psychology*, Graham and Weiner noted just how pervasive that focus has become, suggesting that current research "reflects what is probably the main new direction in the field of motivation—the study of the self." The almost exclusive focus on self-constructs as key indexes of achievement motivation caused Graham and Weiner to suggest that "it is evident that the self is on the verge of dominating the field of motivation" (p. 77). In the world of education, this assumption is grounded on the belief that the perceptions that students create, develop, and hold to be true about themselves and about their academic capabilities are vital forces in their success or failure in school.

So taken-for-granted is the assumption of today's theorists and researchers that people's self-beliefs are a critical component of their motivation that one might reasonably conclude it has always been so. In fact, one would think that educational psychology has always made research on the self a focus of educational research. But this has not been the case. For myriad reasons, from psychology's birth during the time of William James to the modern day, interest in self-beliefs as motivational agents has waxed and waned.

During much of the past century, attention to self-constructs in educational and psychological research, and in motivation theorizing in particular, was scrupulously avoided. There are a number of reasons for the haphazard attention self-constructs received throughout psychology's first century. For long periods, attention was scrupulously avoided because psychologists viewed all internal mental events with suspicion. At other times, rigor in empirical research left much to be desired, and researchers became bewildered by the inconsistent and sometimes contradictory findings that investigations produced. Still at other times, the application of research findings and scholarly thinking on self-beliefs produced mixed results in schools when insights regarding the correlates of self-beliefs made the transition from research to intervention practices. As a consequence, they fell prey to disinterest and ultimately ridicule (see Kohn, 1994). What, then, has been responsible for the resurgence of interest in self-constructs? And where is this research leading?

Although there are varied reasons for the shifting fortunes of self-constructs in psychology and education, there is a fairly clear reason for its resurgence—the renewed interest in mental events and human cognition that was shepherded into psychology by what has come to be called the cognitive revolution. Soon after cognition's victory in the fight for the heart of the discipline, motivation came to be viewed as a psychological process deeply entwined with processes related to the self. Our aim in this chapter is to trace the development of research and theory on the self and motivation as a result of this changing worldview, call it a paradigm shift if you like, in psychology. Because other chapters in this volume focus with some specificity on the nature of the revolution itself, we limit our task to describing how this important event in the history of psychology—and for our purposes the history of educational psychology—influenced the field of academic motivation.

No doubt, however, readers will be aware of the fact that numerous excellent texts and expansive chapters and articles have been devoted to fully describing the theories and constructs currently at play in the area of achievement motivation (e.g., Alderman, 1999; Eccles & Wigfield, 2002; Graham & Weiner, 1996; Pintrich, 2003; Pintrich & Schunk, 2002; Stipek, 2002). Naturally, it is not our desire to replicate or rehash the information provided by these excellent sources. How, then, should we structure our exploration?

Pintrich and Schunk (2002) define motivation as "the process whereby goal-directed activity is instigated and sustained" (p. 5). The dominance of self-beliefs in motivation research and theory suggests that modern theorists concur with the view that, whatever forces impinge on an individual, the agent ultimately responsible for engaging that goal-directed activity rests, as we noted earlier, within the beliefs that individuals come to hold about themselves and about their capabilities. As a consequence, current motivation scholars take it as self-evident that motivation cannot be understood without understanding the self-processes from which it emanates. For these reasons, both our brief history of academic motivation and the motivation constructs we later describe will focus on the self-beliefs that individuals hold about themselves and their potentialities. We close our chapter with some thoughts regarding the future of motivation research and theory, particularly in regard to continued attention to self-beliefs.

MOTIVATION BEFORE THE COGNITIVE REVOLUTION

As students of the history of psychology well know, from very nearly the start of the 20th century until the early 1970s, colleges of education throughout the United States were dominated by three divergent lines of thought, often called "forces." The first force had been ushered in by the work of Wilhelm Wundt and the Russian school of reflexology that flowered at the end of the 19th century. This movement, which had a profound influence on European elementist psychologists, was an antimentalist view of human functioning in which only observable experience was deemed worthy of scientific scrutiny. This radical positivist perspective traveled to the United States by way of structuralist Edward Titchener and others, so that by the early 1900s the intellectual precursors of John Watson's and B. F. Skinner's brand of radical behaviorism were well on their way to capturing the discipline. In this way of looking at human functioning and learning, self-perceptions and other internal mental states played no meaningful role in scientific reasoning and theorizing. As for motivation, radical behaviorists viewed motivated activity as governed nearly exclusively by a process of stimulus–response and subsequently understood by exploring an individual's reinforcement history (see Chance, 1988).

The second force in psychology, and the only real opposition to the behaviorist movement until the late 1940s, was initiated by the voluminous psychoanalytic writings of Sigmund Freud. Unlike behaviorists, psychoanalysts paid heed to internal mental states. Anna Freud's (1946) writings on ego development during adolescence provided the foundation for future explorations of self and identity by theorists such as Erik Erikson and James Marcia. Yet, as Gay (Freud, 1960) and Munroe (1955) have observed, Freudians, and even neo-Freudians, generally hesitated to make the self a primary psychological unit or to give it central prominence in their theoretical formulations. In part, this was due to the psychodynamic emphasis of the important role that biological processes play in human development and functioning, hence the oft-repeated Freudian maxim that "biology is destiny." As for motivation, psychodynamic thinkers wrote of it in terms of an intrapsychic struggle between id and superego forces moderated and resolved by an individual's ego.

The third force in psychology was ushered in by what has often been referred to as the "humanistic revolt" in psychology, which came about as a result of discontent with the intellectual currents of the day (Crain, 1992). Dissatisfied with the direction that psychodynamic theories were taking and apprehensive about what they considered the narrow and passive view of human functioning that behaviorism represented, during the 1940s and 1950s a group of psychologists called for renewed attention to inner experience, internal processes, and self-constructs. This group of scholars also consisted of followers of theorists critical of traditional psychoanalysis, including Alfred Adler, Carl Jung, Otto Rank, and other neo-Freudians (Maslow, 1968). In concert with Gestalt, existential, and phenomenological movements of the day, the writings of these new theorists (e.g., Gordon Allport, Kurt Goldstein) caught the attention of scholars and researchers, and the humanistic movement was born.

Perhaps the most powerful voice in this movement was that of Abraham Maslow, now generally recognized as the father of modern humanistic psychology. Maslow began his academic career as a staunch behaviorist. In 1943, however, with the publication of "A Dynamic Theory of Human Motivation," Maslow outlined a motivational process based on the human desire to fulfill certain needs and culminating in the need to become self-actualized, that is, to achieve one's potentialities, capacities, and talents. Maslow's (1954) theory held that individuals fulfill a series of needs hierarchically ordered and that the goal of each individual is to become self-actualized, which entails developing one's full potential as a human being and reaching self-fulfillment, inner peace, and contentment. The movement toward self-actualization, or "full-humanness," progresses through the fulfillment of psychological needs that include beliefs about one's own competence and sense of self-value.

Another eloquent and significant voice of this new humanistic movement was that of Carl Rogers. In an influential series of articles, books, and lectures, Rogers (e.g, 1951) presented a system of psychotherapy called "client-centered," which was built around the importance of the self in human adjustment and motivation. So powerful was the voice of this third force of psychology, Diggory (1966) observed, that "the fact that the new self psychologists were able to argue substantive matters of learning theory and motivation with the heirs of the behaviorists made the latter pay attention and finally to agree that there might be something to the idea of self after all" (p. 57).

This is not to say that these three forces were the only prominent psychological movements in education before the cognitive revolution. At the turn of the 20th century, when American psychology began to take its place among the other academic disciplines, there was a great deal of interest both in the self and in the role that academic motivation plays in human conduct. For example, when Williams James (1981) published the *Principles of Psychology* in 1890, his chapter on "The Consciousness of Self" was prominent in his volumes. James was one of the first writers to use the term "self-esteem," which he argued could be raised either by succeeding in our endeavors or, in the face of incessant disappointments, by lowering our sights and giving up certain pretensions. And, although he never wrote directly about motivation—the term is never used in *Talks to Teachers*—James (1958) provided a rather clear blueprint regarding how students should be motivated. He wrote with some conviction about the role of interest, attention, association, and habit formation in fostering academic achievement. Most notably, James argued that key self-regulatory practices are at the heart of academic progress.

Psychological theories have always had a strong influence on education. Through the years, teachers have followed the prescriptions of psychologists, from William James with his emphasis on habit, to Sigmund Freud with his focus on unconscious motivations, to J. B. Watson and B. F. Skinner with their stress on observable and measurable behavior. And, of course, those who attended colleges of education during the 1960s and 1970s well remember the powerful influence that humanistic psychology exercised during those decades, with its focus on the importance of students' self-esteem.

Prior to the arrival of the cognitive revolution, students in colleges of education engaging in the study of motivation could choose from five prominent lines of thought (see Graham & Weiner, 1996, for a detailed discussion of four of these lines). Three of those currents we have already identified—the associationist/behaviorist theories that at that time were spearheaded by the work of B. F. Skinner, the psychodynamic views of Freud and many of his followers, and the humanistic psychology of Maslow, Rogers, and others. A fourth movement had also drawn notice, and this was a movement that focused rather exclusively on motivation. Drawing from the early work of associationists such as E. L. Thorndike, a group of psychologists led by Clark Hull provided a view of motivation based on what came to be called *drive theory*.

Hull (1943, 1951) had first proposed a drive theory of motivation in the early 1940s. According to this view, individuals have physiological deficits, or needs, that seek to be fulfilled. Along lines similarly proposed by Jean Piaget and other cognitive psychologists, Hull argued that the biological concept of homeostasis—the metabolic equilibrium all organisms require in order to flourish—could be applied to psychological processes, such that human beings could be viewed as requiring not only physiological but cognitive equilibrium. Hull also borrowed from James the concept of *habit*—which he viewed as the behavioral patterns that humans developed as a result of conditioning—and he argued that the habits that human beings develop provide the direction that an individual's behavior would take. For Hull, however, habit had no agentic or self-propelling properties and hence could not generate behavior on its own. Both habit and drive were required to generate human behavior, hence Hull's famous formula that Behavior is the product of Drive times Habit. If either is lacking in an individual, no motivation is possible.

The fifth line of thought prominent in achievement motivation before the cognitive revolution was Kurt Lewin's *field theory*. Lewin (1935) viewed the self as a central and relatively permanent organization that gave consistency to the entire personality. According to his field theory, which was deeply influenced by Gestalt psychologists interested in human perception, human beings act within a "life space" that contains all of the psychological forces influencing an individual at any specific time. People organize their life space by separating it into regions, or fields, each of which has associated with it a "valence," which is the result of the positive versus negative perceptions that a particular feature of the life space has for the individual. Motivation results from the interplay among three factors: the magnitude of an individual's physiological and psychological needs, the goal desired by the individual, and the perceived psychological distance between the individual and the goal. For Lewin, unfulfilled needs or unattained goals create tension in an individual, such that the individual must take action to reduce that tension. Before making such actions, people formulate a subjective probability that a particular behavior will result in the attainment of a particular goal. This probability multiplied by the valence of a particular field is seen as the force that provides the motivation toward a particular action. This way of conceptualizing the likelihood of successfully or unsuccessfully navigating one's environment influenced early motivation scholars who put forth what came to be called expectancy-value theories.

As we have briefly sketched, the five lines of thought about motivation that were prevalent before the cognitive revolution focused on different behavioral mechanisms, but they also had a number of features in common. Note, for example, that a view of humans as having basic needs, both physiological and psychological, was prominent in four of them. The psychodynamic view of a tension between id and superego was mirrored by Lewin and Hull in their conceptualizations of drives and conflict, respectively. And, for both Lewin and Hull, as for the behaviorists, the direction of motivation was defined in terms of an association between a stimulus and a response. An emphasis on self-processes was central to humanistic psychology and to Lewin's field theory, albeit in a markedly different manner.

Due to the efforts of theorists such as Lewin, Maslow, and Hull, by the mid-20th century it was clear that interest in the self in the study of human functioning was on the rise. Enough so that, in a presidential address before the American Psychological Association, Ernest Hilgard (1949) defended the thesis that the self could be a unifying concept in problems of motivation. This renewed attention to the self would coincide handsomely with the onset of the cognitive revolution.

MOTIVATION AFTER THE COGNITIVE REVOLUTION

As Mayer (1992) elegantly traces it, the cognitive revolution in psychology was born in the mid-1950s, when George Miller (1956) published a seminal paper on the limits of the information-processing abilities of humans. It was organized in the 1960s, particularly as a result of Neisser's (1967) and Atkinson and Shiffrin's (1968) descriptions of the human information-processing system. By the 1970s it had matured, as exemplified by Newell and Simon's (1972) computer simulation of human problem solving. And during the 1980s and 1990s it evolved toward a constructivist view of cognition in an effort to explain learning in ecologically valid, real educational settings (see Joplin & Neisser, 1997; Neisser, 1988, 1991, 1993).

At its inception and for two decades afterward, the cognitive revolution proffered a view of learning that emphasized the acquisition of knowledge (see Mayer, 1992, 1996). Beginning in the late 1970s, and in keeping with emerging constructivist notions of the time, the revolution gradually abandoned the focus on knowledge acquisition of learners for one in which learners were perceived as constructing knowledge. Motivation theorists followed suit, abandoning mechanistic and formulaic understandings of academic motivation and pursuing metacognitive views of motivational processes. Because psychologists began to explore how students' selection and interpretation of information influenced their learning, motivation theorists began to focus on the nature and meaning of those interpretations. Out of these efforts to better understand the interpretations that students make as they endeavor to make sense of their learning, the new theories of motivation emerged.

Several prominent lines of inquiry regarding academic motivation rose to prominence following the cognitive revolution (see Graham & Weiner, 1996). One focused on constructs concerned with students' perceptions of their own abilities, or competence beliefs, such as self-efficacy, self-concept of ability, self-worth, and helplessness beliefs. Another line explored the role played by achievement goals in human decision making and task selection. These included achievement goal orientations, intrinsic motivation, and cooperative versus competitive goal beliefs and structures. Some lines of inquiry, such as expectancy-value theory and attribution theory, predated the revolution but retained their relevance to the present day.

Examination of recent academic texts on motivation in education (e.g., Alderman, 1999; Pintrich & Schunk, 2002; Stipek, 2002) suggests that these are indeed the theoretical lines that generate the greatest interest in researchers, not to mention the ones to which most teachers in training and students of academic motivation are currently exposed. Each of the texts devotes chapters or large sections to expectancy-value theory; social cognitive theory, focusing on self-efficacy and self-regulation; attribution theory; achievement goal orientations; intrinsic motivation and rewards, including self-determination theory; and self-concept or self-worth beliefs. Some also include interest and affect (Pintrich & Schunk, 2002). In some fashion all include discussions of teacher and classroom influences, school culture, and sociocultural influences on academic motivation. Stipek (2002) includes key ideas from reinforcement theory as well.

Alternatively, in their overview of where the field currently stands, Eccles and Wigfield (2002) identify four major movements at play in the area of academic motivation. The first has focused on competence beliefs, most prominently self-efficacy, self-concept, and theories of perceived control. The second movement emphasizes the value of a task or activity, and these include intrinsic motivation, self-determination, flow, interest, and achievement goal orientations. A third movement integrates competence perceptions and value, and these include attributions of success and failure and theories focusing on expectancy value and self-worth. Finally, they identify theories that integrate motivation and cognition, such as social cognitive theories of self-regulation and motivation and of motivation and volition.

As is evident from these overviews, a key motivation construct prominent both before and after the cognitive revolution refers to the beliefs that individuals hold about their own self. These and conceptually similar self-beliefs, such as beliefs about one's own competence, have become central to most modern theories and approaches in the study of motivation. We begin our exploration of these constructs with self-concept beliefs, for they are, after all, the self-constructs that have proved most hardy in the study of achievement motivation and human functioning.

Self-Concept

Self-concept is a psychological construct with a long ancestry. In 1890, William James (1981) became one of the first writers to use the term "self-esteem," which he described as a self-feeling that "in this world depends entirely on what we back ourselves to be and do" (p. 310). He even provided a formula for "self-esteem" showing that how individuals feel about themselves depends on the success with which they accomplish those things they wish to accomplish. Self-esteem may be raised, James argued, either by succeeding in endeavors or, in the face of failure, by lowering one's sights and surrendering certain aims.

Although the construct of self-concept predates the cognitive revolution, this movement had a pronounced influence on how this self-belief has come to be conceptualized and investigated. Prior to the revolution, self-concept beliefs were primarily viewed in affective terms, in keeping with the emphasis on self-esteem that had been chartered by Carl Rogers and other humanistic psychologists. It is for this reason that students and researchers continue to have difficulty navigating the conceptual difference between the constructs of self-esteem and self-concept. Before the cognitive revolution, researchers tended to view self-concept as having two characteristics, one simply descriptive and the other affective. To distinguish between these characteristics, theorists began to draw a distinction between self-concept, the totality of self-knowledge that one possesses about oneself, and self-esteem, which is considered the evaluative component of the self-concept. For example, Combs (1962) argued that an individual's self-concept is, quite simply, "what an individual *believes* he is" (p. 52). Coopersmith (1967) defined self-esteem as involving an attitude of approval or disapproval and "indicates the extent to which the individual believes himself to be capable, significant, successful, and worthy. In short, self-esteem is a personal judgment of worthiness that is expressed in the attitudes the individual holds toward himself" (pp. 4–5).

After the emphasis on metacognitive processes wrought by cognitive psychology, researchers began to define the construct primarily in cognitive terms. For example, Hattie (1992) defined self-concept beliefs as "cognitive appraisals expressed in terms of expectations, descriptions, and prescriptions integrated across various dimensions" (p. 37) that individuals attribute to their own selves. After careful investigation, modern researchers concluded that descriptive and evaluative perceptions of self-concept have not been empirically separated in research studies and may not be empirically separable (Shavelson & Bolus, 1982). For this reason, these days the terms are used interchangeably, although most prefer the term "self-concept."

Cooley (1902) used the metaphor of the self as mirror, or looking-glass self, to illustrate the idea that individuals' sense of self is primarily formed as a result of their perceptions of how others perceive them. That is, the appraisals of others act as mirror reflections that provide the information that individuals use to define their own sense of self. This conception of self brought to the forefront of psychological thought an emphasis on the importance of social comparisons in the development of self-beliefs, an emphasis that continues to be made. As Coopersmith (1967) wrote, "each person's self-concept, to a considerable extent, is a mirror reflection of how he has been (and is) seen by others who are important to him" (p. 201).

Today, self-concept researchers posit that self-concept judgments are based on social- and self-comparisons, which they described as "frame of reference effects." Individuals use external and internal comparisons to determine their self-worth. By comparing one's own performance with those of others ("I am a better reader than most of my friends") and also one's own performance in related areas ("I am better at reading than at mathematics"), an individual develops a self-concept (Marsh, Walker, & Debus, 1991).

Researchers have identified seven features critical to a definition of self-concept: organized, multifaceted, hierarchical, stable, developmental, evaluative, and differentiable (Byrne, 1984, 1986; Shavelson, Hubner, & Stanton, 1976; Shavelson & Marsh, 1986). The hierarchical feature has received the most attention. Marsh and Shavelson (1985) differentiated between the self-perceptions one has about oneself as an individual that involve the totality of one's self-knowledge and the self-perceptions that one has in regards to specific areas or domains in one's life. General self-perceptions comprise the global self-concept, whereas the more bounded self-perception can comprise self-concepts about academic, social, emotional, or physical facets of the self. The hierarchy progressively narrows into more discrete types of self-concepts. Academic self-concepts can be subject-specific, such as language arts, history, mathematics, science, art, or music self-concepts; social self-concepts can include self-perceptions regarding family, peers, or significant others. People become increasingly aware of their differing domain-specific self-concepts as they grow older.

The hierarchical nature of self-concept is rooted in the observation that, when an individual makes a self-appraisal, this appraisal is contextually bounded. Whereas theorists once defined and operationalized self-concept in broad terms as global perceptions of self-worth, modern authors contend that how individuals perceive themselves in one area of their life may be unrelated to how they perceive themselves in another. For example, how a young woman perceives herself as a student may differ markedly from how she perceives herself as a daughter, sister, or girlfriend. Even as a student, she may perceive herself quite differently in differing academic areas. She may view herself positively in mathematics but negatively in reading. This is not to argue that self-concept beliefs do not generalize through an individual's self-system or that they do not relate to each other in important ways, nor does it mean that one does not possess a general view of oneself. Rather, it means that self-conceptions can differ across differing domains of functioning, and it is the self-views in discrete and specific areas of one's life that are most likely to guide and inform behavior in those areas. Numerous studies have provided support for this hierarchical model (see Bong & Clark, 1999; Marsh, 1993).

An analysis of 128 studies conducted up to the late 1970s revealed that researchers had reported relationships between self-concept and academic achievement that ran the gamut from a strong negative correlation to nearly perfect positive correspondence (Hansford & Hattie, 1982; see also Byrne, 1984). Over 90% of the studies reported moderate to weak correlations. In most studies during those years, however, researchers compared general, or global, self-concept with academic achievement. In studies in which academic self-concept was measured, correlations were moderately positive, a finding that has been supported by self-concept researchers during the last 20 years (see Bong & Clark, 1999). Assessing global self-concept and comparing it to academic achievement in early self-concept research had the effect of lowering the statistical relationship between the two constructs.

As we have pointed out, Shavelson and colleagues (1976) introduced a hierarchical model that differentiated between general, academic, social, emotional, and physical self-concepts. Academic self-concepts were further differentiated as English, history, science, and mathematics self-concepts. This conceptualization represented an important step in the study of self-concept. The hierarchical nature is now widely accepted, and researchers warn that using global indexes of self-concept can provide limited value (Bong & Skaalvik, 2003; Byrne, 1984, 1986; Marsh, 1990b, 1993; Marsh & Shavelson, 1985; Shavelson & Bolus, 1982; Shavelson & Marsh, 1986).

According to the hierarchical model, subject-specific self-concepts are distinguishable from each other and from academic and global self-concepts. Relations among self-concept dimensions are themselves hierarchically structured. The relationship between subject-specific self-concept (e.g., mathematics self-concept) and related performance (mathematics performance) is stronger than that between academic self-concept and academic achievement, which, in turn, is stronger than that between global self-concept and achievement (Marsh 1990a; Marsh, Barnes, Cairns, & Tidman, 1984; Marsh, Byrne, & Shavelson, 1988). For example, mathematics self-concept has been found to be related to mathematics achievement, but the strength of the relationship decreased as mathematics achievement is compared with academic self-concept, and it decreased even further when compared with verbal self-concept (Marsh & O'Neill, 1984). It is clear that self-concept becomes more empirically sensitive to, and more predictive of, achievement outcomes the more specifically that it is conceived and assessed.

When domain-specific self-concept is compared with achievement in the same domain (e.g., mathematics self-concept with mathematics achievement), the relationship is positive and strong (Marsh, 1993). Marsh (1990b) reported on a number of studies in which correlations between mathematics self-concept and mathematics achievement indexes ranged from .17 to .66 with a median of .33. Other studies report higher correlations, generally ranging from .40 to .70 (e.g., Marsh, 1992; Skaalvik & Rankin, 1990; see also Bong & Clark, 1999; Bong & Skaalvik, 2003; Skaalvik, 1997, for an overview of findings regarding self-concept and achievement outcomes).

Self-Efficacy Beliefs

In 1941, Miller and Dollard proposed a theory of social learning and imitation that rejected behaviorist notions of associationism in favor of drive reduction principles. It was a theory of learning, however, that failed to take into account the creation of novel responses or the processes of delayed and nonreinforced imitations. In 1963, Bandura and Walters wrote *Social Learning and Personality Development*, broadening the frontiers of social learning theory with the now-familiar principles of observational learning and vicarious reinforcement. By the 1970s, however, and with the cognitive revolution exercising a strong influence on psychology, Bandura was becoming aware that a key element was missing not only from the prevalent learning theories of the day but from his own social learning theory. In 1977, with the publication of "Self-Efficacy: Toward a Unifying Theory of Behavioral Change," he identified the important piece of that missing element: self-beliefs.

With the publication of *Social Foundations of Thought and Action: A Social Cognitive Theory*, Bandura (1986) further advanced a view of human functioning that accords a central role to cognitive, vicarious, self-regulatory, and self-reflective processes in human adaptation and change. People are viewed as self-organizing, proactive, self-reflecting, and self-regulating rather than as reactive organisms shaped and shepherded by environmental forces or driven by concealed inner impulses. From this theoretical perspective, human functioning is viewed as the product of a dynamic interplay of personal, behavioral, and environmental influences. How people interpret the results of their own behavior informs and alters their environments and the personal factors they possess, which, in turn, inform and alter subsequent behavior. This interplay forms the foundation of Bandura's conception of reciprocal determinism, the view that (a) personal factors in the form of cognition, affect, and biological events, (b) behavior, and (c) environmental influences create interactions that result in a triadic reciprocality.

It is a testament to the influence of the cognitive revolution in Bandura's thinking that he altered the label of his theory from social learning to social "cognitive" both to distance it from prevalent social learning theories of the day and to emphasize the powerful role played by cognition in people's capability to construct reality, self-regulate, encode information, and perform behaviors. Because human lives are not lived in isolation, Bandura expanded the conception of human agency to include collective agency. People work together on shared beliefs about their capabilities and common aspirations to better their lives. This conceptual extension makes the theory applicable to human adaptation and change in collectivistically-oriented societies as well as individualistically-oriented ones.

Bandura's (1986) social cognitive theory stands in clear contrast to theories of human functioning that had overemphasized the role that environmental factors play in the development of human behavior and learning. Behaviorist theories, for example, show scant interest in self-processes because theorists assume that human functioning is caused by external stimuli. Because inner processes are viewed as transmitting rather than causing behavior, they are dismissed as a redundant factor in the cause-and-effect process of behavior and unworthy of psychological inquiry. Bandura argued that a psychology without introspection cannot aspire to explain the complexities of human functioning. It is by looking into their own conscious mind that people make sense of their own psychological processes. To predict how human behavior is influenced by environmental outcomes, it is critical to understand how the individual cognitively processes and interprets those outcomes. More than a century ago, William James (1890/1981) had argued that "introspective observation is what we have to rely on first and foremost and always" (p. 185). For Bandura (1986), "a theory that denies that thoughts can regulate actions does not lend itself readily to the explanation of complex human behavior" (p. 15).

In this sociocognitive perspective, individuals are viewed as proactive and self-regulating rather than as reactive and controlled by biological or environmental forces. Also in this view, people are understood to possess self-beliefs that enable them to exercise a measure of control over their thoughts, feelings, and actions. Of all the self-beliefs that affect human functioning, and standing at the very core of social cognitive theory, are self-efficacy beliefs, which Bandura (1986) defined as "people's judgments of their capabilities to organize and execute courses of action required to attain designated types of performances" (p. 391). According to Bandura's social cognitive theory, self-efficacy beliefs provide the foundation for human motivation, well-being, and personal accomplishment. This is because unless people believe that their actions can produce the outcomes they desire, they have little incentive to act or to persevere in the face of difficulties. Much empirical evidence now supports Bandura's (1997) contention that self-efficacy beliefs touch virtually every aspect of people's lives—whether they think productively, self-debilitatingly, pessimistically, or optimistically; how well they motivate themselves and persevere in the face of adversities; their vulnerability to stress and depression; and the life choices they make. Self-efficacy is also a critical determinant of self-regulation. In fact, according to Bandura (1986, 1997), how people behave can often be better predicted by the beliefs they hold about their capabilities than by what they are actually capable of accomplishing, for these self-perceptions help determine what individuals do with the knowledge and skills they have.

Self-efficacy beliefs are posited to influence the choices people make and the courses of action they pursue. Individuals tend to engage in tasks about which they feel competent and confident and avoid those in which they do not. Efficacy beliefs also help determine how much effort people will expend on an activity, how long they will persevere when confronting obstacles, and how resilient they will be in the face of adverse situations (see Bandura, 1997; Pajares, 1996, 1997; Schunk, 1991, 1995). The higher the sense of efficacy, the greater the effort, persistence, and resilience. Efficacy beliefs also influence the amount of stress and anxiety individuals experience as they engage in an activity (Pajares & Miller, 1994). As a consequence, self-efficacy beliefs exercise a powerful influence on the level of accomplishment that individuals ultimately realize.

Beliefs of personal competence also help determine the outcomes one expects. Individuals who are confident anticipate successful outcomes. Students confident in their writing capabilities anticipate high marks on writing assignments and expect the quality of their work to reap academic benefits. Conversely, students who doubt their writing ability envision low marks before they even begin to write. The expected results of these imagined performances will be differently envisioned: academic success and greater options for the former, academic failure and curtailed possibilities for the latter.

A strong sense of efficacy enhances human accomplishment and well-being in countless ways. Confident individuals approach difficult tasks as challenges to be mastered rather than as threats to be avoided. They have greater interest and deep engrossment in activities, set themselves challenging goals and maintain strong commitment to them, and heighten and sustain their efforts in the face of failure. They more quickly recover their confidence after failures or setbacks, and attribute failure to insufficient effort or deficient knowledge and skills that are acquirable. High self-efficacy helps create feelings of serenity in approaching difficult tasks and activities. Conversely, people who doubt their capabilities may believe that things are tougher than they really are, a belief that fosters stress, depression, and a narrow vision of how best to solve a problem. Not surprisingly, confidence in one's academic capability is a critical component of school success.

There is ample empirical evidence showing that self-efficacy beliefs are related to and influence numerous academic outcomes. Researchers have also been successful in demonstrating that these self-beliefs mediate the effect of skills, previous experience, mental ability, or other motivation constructs on subsequent achievement, which is to say that they act as a filter between prior determinants and academic indexes. Bandura (1997) provides extensive evidence that self-efficacy beliefs are powerful determinants of achievement outcomes in varied fields of human endeavor. In a meta-analysis, Stajkovic and Luthans (1998) found that the average weighted correlation between self-efficacy and work-related performance was (G)r = .38, which transforms to an impressive 28% gain in task performance. In education, a meta-analysis of studies published between 1977 and 1988 revealed that efficacy beliefs were positively related to academic achievement (Multon, Brown, & Lent, 1991). Self-efficacy beliefs were related to academic outcomes ($r_u = .38$) and accounted for approximately 14% of the variance. Effects were stronger for high school and college students than for elementary students. Stronger effects were obtained by researchers who compared specific efficacy judgments with cognitive skills measures of performance or classroom-based indexes such as grades than with global, standardized achievement tests. Effect sizes were also stronger in studies in which researchers developed highly concordant self-efficacy/performance indexes and administered them at the same time. Self-efficacy beliefs have also been found highly predictive of college students' selection of majors and career choices (Hackett, 1995; Lent & Hackett, 1987).

Correlations between self-efficacy and academic performances in investigations in which self-efficacy is analyzed at the item- or task-specific level and corresponds to the criterial task have ranged from .49 to .70; direct effects in path-analytic studies have ranged from β = .349 to .545 (see Pajares, 1996, 1997). Results tend to be higher in studies of mathematics than of other academic areas such as language arts, but even in these areas relationships are considerably higher than previously obtained if the criteria by which students rate their self-efficacy judgments is used as the criteria for scoring essays or assessing reading comprehension (see Pajares, 2003). Schunk (1991) provided an overview of research on the effect of self-efficacy beliefs on academic motivation, specifically the role played by variables such as perceived control, outcome expectations, value, attributions, and self-concept. He concluded that all are a "type of cue" (p. 211) used by individuals to assess their efficacy beliefs (see Schunk, 1991).

Because of conceptual similarities, researchers have been eager to tease out the differences between self-concept and self-efficacy. Bong and Skaalvik (2003) thoroughly examined the empirical properties of the two constructs and concluded that they differ in important ways. Self-efficacy beliefs are cognitive, goal-referenced, relatively context-specific, and future-oriented judgments of competence that are relatively malleable due to their task dependence. Self-concept beliefs, on the other hand, are primarily affective, heavily normative, typically aggregated, hierarchically structured, and past-oriented self-perceptions that are relatively stable due to their sense of generality. According to Bong and Skaalvik, self-efficacy acts as an active precursor of self-concept development. This is consistent with Covington's (1984, 1992) self-worth theory, which puts forth a view of self-concept in which competence beliefs are central to the self-system, such that individuals are motivated by a need to perceive themselves as competent. Because modern society places a powerful premium on achievement, individuals want strongly to be perceived as capable, and they define self-worth in terms of such capability. According to Covington (1984), the need to safeguard this mental perception of competence often gives rise to external attributional beliefs and self-handicapping strategies that serve to protect the individual from potential feelings of incompetence.

Self-efficacy and self-concept theorists have each emphasized the need to keep the contextual nature of these self-beliefs in mind when conducting investigations. Bandura (1997) argued that, to predict academic outcomes from students' efficacy beliefs, "self-efficacy beliefs should be measured in terms of particularized judgments of capability that may vary across realms of activity, different levels of task demands within a given activity domain, and under different situational circumstances" (p. 6). In a similar vein, Marsh (1993) cautioned that "research clearly demonstrates that self-concept and its relation to other variables cannot be adequately understood if its multidimensional, domain-specific nature is ignored" (p. 92). And both have cautioned that the self-beliefs assessed should always correspond with the achievement index with which they are compared.

It bears noting that Graham and Weiner (1996) concluded that, after 25 years of research it cannot easily be disputed that self-efficacy beliefs have been "a much more consistent predictor of behavior and behavior changes than have any of the other closely related expectancy variables" (p. 75). Currently, self-efficacy is generating research in areas as diverse as psychology, medicine, sports, politics, and business, and it has become a staple of educational research (see Bandura, 1997).

Expectancy Value

At the very cusp of the cognitive revolution, John Atkinson (1957, 1964) proposed an expectancy-value theory in which motivation is viewed primarily as a result of individuals' beliefs about the likely outcomes of their actions (expectancy) and of the incentive value they place on those outcomes (see also Vroom, 1964). Atkinson's (1957) early views regarding motivation were originally focused on the effect of motivation on risk-taking behavior. In essence, Atkinson argued that individuals would risk a particular behavior depending on the product of the perceived probability of success (or of failure) times the incentive value attached to the outcome expected. Performance, persistence, and choice of behavior are linked directly to the beliefs that individuals hold about their expectancy and the value of the task at hand. Individuals will be motivated to engage in tasks when they value the outcome expected; they will be less predisposed to perform tasks whose outcomes they do not value.

Modern expectancy-value theories differ from those earlier conceptions (Eccles & Wigfield, 2002). Expectancy and value beliefs are more specifically defined and are linked to myriad psychological and sociocultural factors. Atkinson (1964) had posited that expectancy and value beliefs can interact in such a way that they can be inversely related, in the sense that expectations to succeed at difficult tasks is valued, whereas expectations to succeed at easy tasks is not. Today, theorists contend that expectancy and value are positively related. They also define expectancies for success as "individuals' beliefs about how well they will do on upcoming tasks, either in the immediate or longer-term future" (Eccles & Wigfield, 2002, p. 119), and they assess them in a manner similar to that used by self-efficacy researchers.

Eccles and her colleagues have provided an expectancy-value model in which human behavior is viewed as influenced both by the positive and negative features of a particular task or activity and in which the choices that people make have costs associated with them because one choice can eliminate others (Eccles et al., 1983). In this model, the relative value and probability of success of various options are key determinants of choice, and individuals' expectancies for success are influenced by self-beliefs such as self-efficacy. The expectancies themselves directly influence performance, persistence, and task choice. Competence beliefs are construed as domain-specific judgments of competence, in contrast to expectancies, which are operationalized as relatively specific expectations to succeed on a specific upcoming task. Expectancy-value theorists contend that, even though the two constructs are conceptually distinct, they are not empirically separable, and they report that children and adolescents do not easily distinguish between domain- and task-specific competence beliefs (Eccles & Wigfield, 2002).

Task value has four components—attainment value, intrinsic value, utility value, and cost—and these components are typically included when assessing value (see Eccles et al., 1983; Eccles & Wigfield, 2002). Attainment value is defined as an individual's perceived importance of succeeding on a task or activity. Intrinsic value is the enjoyment that people experience as they engage in the task or activity, as well as the interest they experience. This type of value is similar to the idea of intrinsic motivation as elaborated by Deci and his colleagues (e.g., Deci, 1975; Deci & Ryan, 1985). Utility value refers to the relationship between a task or activity and the individual's current and future goals, which can include social, academic, or career goals. Tasks have positive utility value to the degree that they promote desired goals, even if the task or activity has no inherent intrinsic value for the person. People often engage in actions they do not find intrinsically interesting or enjoyable, but which they view as necessary to meeting desired goals such as pleasing others or being well-perceived. Cost is conceptualized in terms of the negative feelings or perceptions that can be involved in engaging in a particular task or activity. For example, cost can be viewed in terms of anxiety and apprehension, fear of failure, the effort it will take to succeed, or the opportunities that are lost when one choice is made over another.

Expectancy-value researchers report that, even after controlling for previous performance, competence beliefs and expectancies each predict academic performance in various academic areas, whereas task values predict course plans and enrollment decisions as well as involvement in sport activities (Eccles, 1987; Eccles, Adler, & Meece, 1984; Eccles et al., 1983; Meece, Wigfield, & Eccles, 1990). Expectancies and values each predict career choices (Wigfield & Eccles, 1992).

Possible Selves

Another important self-construct powerfully influenced by the emphasis on cognition that the cognitive revolution brought to psychology was provided by Hazel Markus and her associates (Markus, 1977; Markus & Nurius, 1986; Markus & Wurf, 1987). This was a perspective on the self that focuses on individuals' self-schemas and possible selves. Finding the traditional view of self-concept too "monolithic," Markus (1977) argued that an individual's self-system is composed of self-schemas, which are defined as "cognitive generalizations about the self, derived from past experience, that organize and guide the processing of self-related information contained in the individual's social experiences" (p. 64). Markus agreed that one's self-concept comprises the totality of information available to an individual as a result of past experience, but contended that an individual does not make use of this total structure when undergoing day-to-day activities. Instead, people selectively extract from their vast storehouse of self-knowledge the information required to deal with specific events or experiences. The information selected results in the construction of self-schemas that help a person determine the specific stimuli selected for attention, the information to be remembered and incorporated into the self-system, and the inferences and decisions to be made in a particular context. Seen this way, the concept of self-schemas is similar to Freud's description of the executive function of the ego.

Self-schemas have four dimensions. First, they have an affective dimension, in that some self-perceptions are positive, whereas others are negative. The efficacy dimension is composed of beliefs about one's capability to become a desired self or avoid an undesired self. They also have a value dimension in that individuals are aware of the importance and centrality of the beliefs that make up the schema. Finally, self-schemas possess a temporal dimension—experience results in past, present, and future conceptions of self. Of these dimensions, the reader will find the first three remarkably in concert with the model of belief systems we sketched out in the previous chapter. It is the temporal dimension that has sparked research into self-schemas, for this dimension speaks to the importance of an individual's possible selves as a motivational construct in influencing one's life course.

Possible selves are cognitions of the type of person one can become and the things one can accomplish. These cognitions also include the type of person one fears becoming. In a similar vein, Erikson (1959/1980) had earlier observed that one's self "demands to be synthesized with abandoned and anticipated selves" (p. 160). Because individuals' conceptions of their possible selves aid them in selecting information from the self-system and constructing the self-schemas necessary to function adaptively, Markus (1977) argued that they provide the essential link between the self-concept and motivation.

Self-schemas, then, are the pieces of the self-concept activated and brought to consciousness by an individual to deal with specific experiences. Because the self-concept includes all the self-knowledge available, an individual must carefully and creatively construct the self-schemas most relevant and appropriate to the experience at hand. In a sense, this is a working self-concept. Markus (1977) suggested that self-schemas are in part constructed from information and strategies that have proven effective in past similar situations and reflect significant and enduring personal concerns. They are also constructed to reflect the individual's enduring goals, hopes and aspirations, motives, fears, and threats. In constructing self-schemas, individuals are guided by visions of themselves in the future, visions of their possible selves.

Achievement Goal Orientations

Achievement goals are the reasons that individuals have for engaging in a particular task or activity. Typically, researchers have described these goals in terms of either task or performance orientations. Task goals (sometimes called learning or mastery goals) represent an individual's concern with mastering material and concepts, challenge-seeking, and learning as an end in itself. Performance goals (also called ego goals) represent a person's concern with social comparisons, doing better than others, appearing smart, and avoiding appearing unable. Results have typically demonstrated that having a task goal orientation has motivational benefits, whereas having a performance goal orientation can be detrimental and maladaptive (see Ames, 1992; Ames & Ames, 1989; Elliot & Dweck, 1988; Nicholls, 1984; Urdan, 1997; Urdan & Maehr, 1995).

In general, educational researchers have reported that task goals are positively related to motivation indexes such as perceived ability, use of deep processing strategies, task engagement, attributions of success to effort, and persistence in the face of difficulty (Anderman & Young, 1994; Dweck, 1986; Dweck & Leggett, 1988; Meece, Blumenfeld, & Hoyle, 1988; Midgley & Urdan, 1995; Schunk, 1996; see also Urdan, 1997, for a review). Task goals are also negatively associated with use of self-handicapping strategies and helplessness (Midgley & Urdan, 2001; Sideridis, 2003). Performance goals have been shown related to maladaptive behaviors such as lack of persistence, use of shallow cognitive strategies, avoiding help seeking, negative affectivity, and attributions of failure to lack of ability (Anderman & Young, 1994; Meece et al., 1988; Ryan & Pintrich, 1997).

Students have task goals to develop their ability and performance goals to demonstrate that ability. But theorists have argued that motivation has both an approach and an avoidance tendency (Elliot & Harackiewitcz, 1996). That is, individuals can strive to attain success or to avoid failure. In school, students may hold performance goals related to an approach tendency, such as wanting to do well so as to please teachers or parents, impress their classmates, or feel important. They may also hold performance goals related to an avoidance tendency, such as wanting to do well so as to avoid showing a lack of ability, being embarrassed, or looking stupid. Until recently, researchers tended to confound the approach and avoid tendency of goal orientations (Urdan, 1997).

Some researchers have reported that task goals and performance-approach goals are unrelated (Middleton & Midgley, 1997; Midgley & Urdan, 2001). Others have argued that task goals and performance-approach goals are each grounded in self-regulatory practices that lead to positive outcomes such as the attainment of competence or task mastery and that, as such, they represent an approach orientation and should be related (Elliot & Harackiewicz, 1996; Pajares, Britner, & Valiante, 2000; Pajares & Cheong, 2003). Some researchers have suggested that a developmental component may be at work in determining whether these goals serve a facilitative function in fostering motivation and that performance-approach goals may relate differently to motivation indexes at different ages (Pajares & Cheong, 2003). It may be that, as Middleton and Midgley (1997) suggest, performance-approach goals do not serve a facilitative function at early ages but that, as Elliot and Harackiewicz (1996) suggest, at later ages these goals may foster motivation. Investigations focusing on the development of achievement goal orientations from lower to higher academic levels would serve to bring greater clarity to this issue.

After a thorough review of findings related to performance-approach goals, Midgley, Kapland, and Middleton (2002) concluded that researchers should not reconceptualize goal theory to suggest that performance-approach goals are adaptive and should be encouraged. Other researchers have found that a strong performance-approach goal orientation is adaptive only when paired with a strong task goal orientation (Pintrich, 2000).

A number of theorists have rightly argued that individuals can have myriad goals as they contemplate their actions. For example, some have called for greater attention to performance goals that have social aims (Juvonen & Wentzel, 1996; Skaalvik, 1996; Urdan, 1997; Wentzel, 1994). Others have distinguished between within-person goals and person-environment goals (Ford, 1992; Ford & Nichols, 1987). Within-person goals are the desired consequences of our personal actions. These can include affective goals such as happiness and contentment, cognitive goals such as creative expression and discovery, and organization goals such as unity and harmony. Person-environment goals represent an individual's desired relationship with the world, and they include self-assertive goals such as maintaining a sense of autonomy and individuality, integrative social relationship goals such as belonging to a group or collective and exercising social responsibility, and task-related goals such as expressing task mastery and reaping the fruits of one's efforts.

Other Constructs Prominent in the Study of Motivation

One of the first casualties of the cognitive revolution was the idea that external inducements are the key to motivation. As soon as researchers began to investigate the mental events that take place as individuals select tasks and persevere at them, they found that a critical component of motivation was intrinsic rather than extrinsic. For example, White (1959) put forth the idea that individuals are motivated by an *effectance motive* that satisfies their inherent need to be competent and master their environment.

A number of contemporary theorists have used this idea of effectance motivation as a framework for motivation research. These have included Harter's (1978) extension of White's theory, Rotter's (1966) concept of locus of control, de Charm's (1968) description of pawns and origins, and Deci and his colleagues' (Deci, 1980; Deci & Ryan, 1985) self-determination theory. Each of these perspectives has in common the idea that intrinsic forces within the individual are typically more effective in fostering motivated behavior than are extrinsic forces. Intrinsic motivation is typically understood as having root causes in the individual, these causes often viewed in terms of needs. As did White (1959), for example, self-determination theorists posit that all individuals have a need for autonomy, as well as a need for competence and for relatedness. The human needs for self-determination and competence are viewed as motivating an individual to engage in activities that may be challenging but that can be mastered. Extrinsic motivation includes forces that the environment exercises on individuals. In school, this can take the form of coercion, punishment, or intimidation.

Effectance motivation is defined as the motivation to interact effectively with one's environment and control critical aspects (White, 1959). In young children it is diffuse and affects many interactions. As individuals grow, their effectance motivation becomes more specialized and manifests itself in achievement behaviors in various academic areas. As is the case with self-efficacy, effectance motivation includes perceived capabilities for influencing important aspects of one's life.

The notion of perceived control also continues to receive attention by motivation researchers. According to theories of locus of control (Rotter, 1966), people expect success to the degree that they feel in control of their behavior, often referred to as internal locus of control. Research evidence supports this contention (see Findley & Cooper 1983). Connell and Wellborn (1991) proposed that internal locus of control is related to competence beliefs. People who believe they can control what they learn and perform are more apt to initiate and sustain behaviors directed toward those ends than are individuals who hold a low sense of control over their capabilities (Schunk, 2000).

Bernard Weiner's (1972) theory of causal attributions, which can be defined as the reasons that people give themselves for their success and failure in achievement situations, has also been a prominent staple of motivation texts since he originally formulated it (Weiner, 1969, 1972). Attribution theorists make two key assumptions. The first is that people are motivated by a goal of understanding and mastering the environment and themselves. The second is that individuals try to understand the causal determinants of their own behavior as well as the behavior of others, that is, the reasons why things happen to themselves and to others. Attribution theory posits that people interpret the outcomes of their actions and that it is these interpretations, rather than the outcomes themselves, that determine their motivation and subsequent actions. The explanations that people give themselves for their successes and failures are in themselves motivation self-beliefs, and they have three motivational dimensions. First, causes of outcomes are perceived as stable (fixed across situations and over time) or unstable (variable across situations and over time). Causes of outcomes are also interpreted as internal (due to the individual) or external (due to forces outside the individual). Finally, causes of outcomes can be perceived as controllable (within the individual's control) or uncontrollable (beyond the individual's control).

Of all the attributions that individuals make, those related to ability, effort, task difficulty, and luck are made most frequently (Weiner, 1985, 1992). Different attributions produce different motivational results. For example, when students attribute their failure to an external cause such as lack of ability, which they may view as a fixed entity (see Dweck & Leggett, 1988), their motivation to expend greater effort is likely to be diminished by the belief that nothing they can do will help ensure success because they simply lack the innate ability to produce the outcome desired. The same would be true of attributing an outcome to luck, a magical entity over which mere mortals are thought to have no control. Alternatively, attributing failure to an internal cause such as having put forth insufficient effort may result in reengaging the task and putting forth greater effort the next time around. Each of the dimensions has affective consequences, with the locus dimension being particularly linked to emotional reactions. Attributing success to an internal cause such as effort increases pride and a sense of accomplishment, whereas attributing failure to lack of effort may inspire embarrassment or shame. Attributing failure to an external cause such as ability or luck is likely to lead to frustration and anger.

Before we close our brief description of motivation constructs at play after the cognitive revolution, one more deserves mentioning, and it is one to which William James (1958) dedicated a full chapter in his *Talks to Teachers*. In recent years, researchers have increased their attention to "interest," particularly in academic contexts. As Hidi and Ainley (2002) correctly noted, emotional processes and physiological factors that contribute to an individual's psychological state and that serve to energize behavior need to be included in the theories of motivation. Interest is precisely such a construct, incorporating affective and cognitive factors to explain how and why individuals choose to engage in certain activities (Hidi & Berndorff, 1998; Schiefele, 2001). Interest may also have strong and yet not fully explained associations with the selection and pursuit of goals (Hidi & Harackiewicz, 2000).

Researchers who have investigated the construct of interest differentiate between situational interest and individual interest. Situational interest refers to the psychological state of interest generated by particular conditions and/or objects in one's environment, which is to say the specific features of a particular task or activity. Situational interest varies as the task or activity changes. On the other hand, people develop a rather stable individual interest to a particular activity when they repeat that activity with positive results and experience affective feelings, when they notice that their knowledge is growing as a result of engaging in that activity, and when they come to value the activity. Research findings attest to the relationship between both situational and individual interest on the one hand and a host of adaptive motivational and achievement outcomes on the other (see Hidi & Ainley, 2002, for a review).

THE FUTURE OF ACADEMIC MOTIVATION

We know, then, what the cognitive revolution has wrought in the area of academic motivation, and we know why. The title of Bernard Weiner's (1972) book, *Theories of Motivation: From Mechanism to Cognition*, handsomely encapsulates precisely the influence that the cognitive revolution had on thinking about human motivation. The focus on cognitive processes moved psychologists away from positivist notions and invited them to revisit self-constructs and investigate them from a cognitive perspective. The revolution also was influential in moving research studies from the narrow confines of the laboratory to the more ecologically sound classrooms and schools (Mayer, 1992).

As we have traced, during the last century the field of psychology has wafted from philosophical psychology to William James's brand of introspective psychology, to the associationist movement and the accompanying onslought of behaviorism, to humanistic psychology with its emphasis on self-processes, to cognitive psychology and its focus on cognitive processes, memory, and problem solving, and now to an almost eclectic and divided psychology in which no multiple movements have taken a foothold. In educational psychology, self-processes and self-beliefs have had a renaissance of sorts, and nowhere is this more evident than in motivation research.

Andy Elliot and Carol Dweck are in the process of editing an ambitious new volume entitled *The Handbook of Competence and Motivation.* In that volume, they identify the currently prominent motivation constructs as effectance and intrinsic motivation, motives, achievement goals, attributions, competence perceptions, values, self-beliefs, and evaluation anxiety. In addition, they emphasize the importance of self-regulatory processes such as flow, creativity, defensive strategies, self-regulated learning, coping and disengagement, and social comparison. This seems to us to reflect a reasonable view of where the field is now and where it will remain for the near future.

But what can we glean from present-day trends that may help us conjecture as to where motivation theory and research are now heading? Clearly, these sorts of divination enterprises are always fraught with risk. In the 1960s, Decca Records and Columbia turned down a recording contract with the Beatles with the unprophetic evaluation, "We don't like their sound. Groups of guitars are on their way out." Beethoven's teacher once proclaimed with haughty certainty that his student was "hopeless as a composer."

Naturally, we cannot presume to know the direction that educational psychology or motivation research will take in the future. It would be short-sighted, however, not to admit the likelihood that during the 21st century many, if not most, of the powerful insights about psychological processes likely will not come from within mainstream psychology. Sigmund Freud (1961, 1991) conjectured that the future of psychology would ultimately lie with biology, that all of psychology's provisional ideas will at some time be likely based on an organic substructure and all psychological terms replaced with physiological or chemical ones. Steven Pinker (2002) envisions a similar future in his book, *The Blank Slate*. If this should prove to be the case, what will this mean for educational psychology in general and motivation theory in particular? for theories of learning? for development? for educational policy and school practices? for educational research? It seems almost unavoidable that the architecture of the mind and of the nature of learning is more likely to be crafted in the future by cognitive science, neuroscience, behavioral genetics, and evolutionary psychology than by either psychology or educational psychology. Nonetheless, educational psychologists, and motivation theorists in particular, can surely maintain their relevance by continuing to interpret biological processes and brain activity in circumstances in which individuals make judgments, learn, and engage in behaviors. As Pinker also points out, it would be a serious error to distance philosophical thinking from scientific research and theory, for it will be precisely such thinking that will guide investigators toward the metaphysical, epistemological, and ethical questions and methods that will be at the heart of scientific advancement.

Having said that, it bears noting that modern social cognitive theories such as Bandura's (1986) differ from theories of human functioning that overemphasize the influence of biological factors in human development and adaptation. Although these theories acknowledge the influence of evolutionary factors in human adaptation and change, they reject the type of evolutionism that views social behavior as the product of evolved biology but fails to account for the influence that social and technological innovations that create new environmental selection pressures for adaptiveness have on biological evolution (Bussey & Bandura, 1999). Instead, they posit a bidirectional influence in which evolutionary pressures alter human development such that individuals are able to create increasingly complex environmental innovations that, "in turn, create new selection pressures for the evolution of specialized biological systems for functional consciousness, thought, language, and symbolic communication" (p. 683). This bidirectional influence results in the remarkable intercultural and intracultural diversity evident on our planet.

As regards future efforts related to shedding additional light on the academic motivation of young people in our schools, our best sense is that we are nearly past the point of showing that differing motivation constructs are adaptive or maladaptive and related to and predictive of or unrelated to and unpredictive of academic attainment. Studies that continue to make these points in varied academic domains are becoming redundant. Rather, we are in need of putting into practice the policies, interventions, and schooling strategies that emanate from insights obtained from research findings. We would like to see motivation research dedicate itself to this path with greater vigor. All of which is to say that we already know a great deal about academic motivation. What is lacking is putting that knowledge into greater use by working to alter school and classroom structures, the content of teacher education programs, and educational policy. To accomplish these things, of course, we will need to overcome powerful political, cultural, and academic obstacles that do not easily budge.

In addition, we hope that researchers will continue to place a greater emphasis on ecologically grounded investigations and less on survey-type studies, and that they will give greater value to sound qualitative efforts and less to decontextualized quantitative efforts that serve little function other than to confuse lay readers and practitioners and to provide fodder for psychometricians in search of a problem. Also, and as Eccles and Wigfield (2002) point out, nearly all current motivation theories can be criticized for emphasizing rational cognitive processes leading to motivation and behavior, although such processes are often not used because people's "folk psychology" results in more fallible and optimistic decisions. Moreover, students' academic achievement is too complex a matter to permit the contention that such achievement is caused by any particular motivation construct (Schunk, 1984). These constructs are neither the *prima causa* of achievement in any particular case nor a magic elixir that can make all kids work to their full potential. Students perform differently in school for many reasons, including differences in aptitude, general mental ability, interest, perceived value, effort, perseverance, use of self-regulatory strategies, teaching and instruction, and availability of materials (see Gustaffson & Undheim, 1996; Keogh & MacMillan, 1996; Snow, Corno, & Jackson, 1996). Social and familial variables such as peer influence, family income, and parental expectations also play a hand in the academic outcomes that students will attain (see Steinberg, Brown, & Dornbusch, 1996). And it bears emphasizing that no amount of motivation can produce success when requisite skills and knowledge are absent (Schunk, 1984, 1991).

For these reasons, it goes without saying that we will have to give greater attention to the social (including political, home, and school) structures and practices that foster or impede the academic motivation of youngsters. As we have observed, it seems clear that we have quite a good handle on how motivation works, what its defining features are, and what things can be done to better motivate students. Looking at this from a "positive psychology" lens, what is required is greater attention to and deeper understandings of how to foster the psychological, sociocultural, and political processes and practices that work to maximize such motivation.

Finally, we hope that we will move away from efforts to provide universal prescriptions for educational problems and issues. The critical questions in education involve matters that cannot be settled by universal prescriptions. They demand attention to the cultural forces that shape our lives. Action, cognition, and motivation are always situated in a network of cultural particulars. However elegant the insight or telling the behavior, neither can be understood fully beyond the natural boundaries provided by the local, situated, cultural conditions in which they are embedded.

As the world shrinks and cultural diversity grows, attempting to understand how cultural variations influence academic motivation seems more critical than ever. Culturally attentive views of motivational processes can help us clarify how motivational and self-regulatory strategies are created and develop as a result of differing cultural practices or differing group membership, as well as how these practices influence our students' school success and enrich their lives.

REFERENCES

Alderman, M. K. (1999). *Motivation for achievement: Possibilities for teaching and learning.* Mahwah, NJ: Erlbaum.

Ames, C. (1992). Classrooms: goals, structures, and student motivation. *Journal of Educational Psychology, 84,* 261–271.

Ames, R., & Ames, C. (Eds.). (1989). *Research on motivation in education* (Vol. 3). New York: Academic Press.

Anderman, E. M., & Young, A. J. (1994). Motivation and strategy use in science: Individual differences and classroom effects. *Journal of Research in Science Teaching, 31,* 811–831.

Atkinson, J. W. (1957). Motivational determinants of risk-taking behavior. *Psychological Review, 64,* 359–372.

Atkinson, J. W. (1964). *An introduction to motivation.* Princeton, NJ: Van Nostrand.

Atkinson, R. C., & Shiffrin, R. M. (1968). Human memory: A proposed system and its control processes. In K. W. Spence & J. T. Spence (Eds.), *Advances in the psychology of learning and motivation research and theory* (Vol. 2, pp. 89–195). New York: Academic Press.

Bandura. A. (1977). Self-efficacy: Toward a unifying theory of behavioral change. *Psychological Review, 84,* 191–215.

Bandura, A. (1986). *Social foundations of thought and action: A social cognitive theory.* Englewood Cliffs, NJ: Prentice Hall.

Bandura, A. (1997). *Self-efficacy: The exercise of control.* New York: Freeman.

Bandura, A., & Walters, R. H. (1963). *Social learning and personality development.* New York: Rinehart & Winston.

Bong, M., & Clark, R. (1999). Comparisons between self-concept and self-efficacy in academic motivation research. *Educational Psychologist, 34,* 139–154.

Bong, M., & Skaalvik, E. M. (2003). Academic self-concept and self-efficacy: How different are they really? *Educational Psychology Review, 15,* 1–40.

Bussey, K., & Bandura, A. (1999). Social cognitive theory of gender development and differentiation. *Psychology Review, 106,* 676–713.

Byrne, B. M. (1984). The general/academic self-concept nomological network: A review of construct validation research. *Review of Educational Research, 54,* 427–456.

Byrne, B. M. (1986). Self-concept/academic achievement relations: An investigation of dimensionality, stability, and causality. *Canadian Journal of Behavioural Science, 18,* 173–186.

Chance, P. (1988). *Learning and behavior* (2nd ed.). Belmont, CA: Wadsworth.

Combs, A. W. (1962). *Perceiving, behaving, becoming* (Yearbook of the Association for Supervision and Curriculum Development). Washington, DC: Education Association.

Connell, J. P., & Wellborn, J. G. (1991). Competence, autonomy, and relatedness: A motivational analysis of self-system processes. In M. Gunnar & L. A. Sroufe (Eds.), *Minnesota symposia on child psychology* (Vol. 23, pp. 43–77). Hillsdale, NJ: Erlbaum.

Cooley, C. H. (1902). *Human nature and the social order.* New York: Scribner.

Coopersmith, S. (1967). *The antecedents of self-esteem.* San Francisco: W. H. Freeman.

Covington, M. V. (1984). The motive for self-worth. In R. Ames & C. Ames (Eds.), *Research on motivation in education* (Vol. 1, pp. 77–113). New York: Academic Press

Covington, M. V. (1992). *Making the grade: A self-worth perspective on motivation and school reform.* New York: Cambridge University Press

deCharms, R. (1968). *Personal causation: The internal affective determinants of behavior.* New York: Academic Press.

Deci, E. L. (1975). *Intrinsic motivation.* New York: Plenum Press.

Deci, E. L. (1980). *The psychology of self-determination.* Lexington, MA: Lexington Books.

Deci, E. L., & Ryan, R. M. (1985). *Intrinsic motivation and self-determination in human behavior.* New York: Plenum Press.

Diggory, J. C. (1966). *Self-evaluation: Concepts and studies.* New York: Wiley.

Dweck, C. S. (1986). Motivational processes affecting learning. *American Psychologist, 41,* 1040–1048.

Dweck, C. S., & Leggett, E. L. (1988). A social-cognitive approach to motivation and personality. *Psychological Review, 95,* 256–273.

Eccles, J. S. (1987). Gender roles and women's achievement-related decisions. *Psychology of Women Quarterly, 11,* 135–172.

Eccles (Parsons), J., Adler, T. F., Futterman, R., Goff, S. B., Kaczala, C. M., Meece, J. L., & Midgley, C. (1983). Expectancies, values, and academic behaviors. In J. E. Spence (Ed.), *Achievement and achievement motivation* (pp. 75–146). San Francisco: Freeman.

Eccles, J. S., Adler, T., & Meece, J. L. (1984). Sex differences in achievement: A test of alternate theories. *Journal of Personality and Social Psychology, 46,* 26–43.

Eccles, J. S., & Wigfield, A. (2002). Motivational beliefs, values, and goals. *Annual Review of Psychology, 53,* 109–132.

Elliott, E. S., & Dweck, C. S. (1988) Goals: An approach to motivation and achievement. *Journal of Personality and Social Psychology, 54,* 5–12.

Elliot, A. J., & Harackiewicz, J. M. (1996). Approach and avoidance goals and intrinsic motivation: A mediational analysis. *Journal of Personality and Social Psychology, 70,* 461–475.

Erikson, E. (1980). *Identity and the life cycle.* New York: Norton. (Original work published 1959).

Findley, M. J., & Cooper, H. M. (1983). Locus of control and academic achievement: A literature review. *Journal of Personality and Social Psychology, 44,* 419–427.

Ford, M. E. (1992). *Human motivation: Goals, emotions, and personal agency beliefs.* Newbury Park, CA: Sage.

Freud, A. (1946). *The ego and the mechanisms of defense.* New York: International Universities Press.

Freud, S. (1960). *The ego and the id.* New York: Norton.

Freud, S. (1961). *Beyond the pleasure principle.* New York: Norton.

Freud, S. (1991). *On narcissism: An introduction.* New Haven, CT: Yale University Press.

Graham, S., & Weiner, B. (1996). Theories and principles of motivation. In D. C. Berliner & R. C. Calfee (Eds.), *Handbook of educational psychology* (pp. 63–84). New York: Simon & Schuster Macmillan.

Gustafson, J., & Undheim, J. O. (1996). Individual differences in cognitive functions. In R. C. Calfee & D. C. Berliner (Eds.), *Handbook of educational psychology* (pp. 186–242). New York: Simon & Schuster Macmillan.

Hackett, G. (1995). Self-efficacy in career choice and development. In A. Bandura (Ed.), *Self-efficacy in changing societies* (pp. 232–258). New York: Cambridge University Press.

Hansford, B. C., & Hattie, J. A. (1982). The relationship between self and achievement/performance measures. *Review of Educational Research, 52,* 123–142.

Harter, S. (1978). Effectance motivation reconsidered: Toward a developmental model. *Human Development, 21,* 34–64.

Hattie, J. (1992). *Self-concept.* Hillsdale, NJ: Erlbaum.

Hidi, S., & Ainley, M. (2002). Interest and adolescence. In F. Pajares & T. Urdan (Eds.), *Adolescence and education: Academic motivation of adolescents* (pp. 247–276). Greenwich, CT: Information Age.

Hidi, S., & Berndorff, D. (1998). Situational interest and learning. In L. Hoffman, A. Krapp, K. Renninger, & J. Baumert (Eds.), *Interest and learning: Proceedings of the Seeon Conference on interest and gender* (pp. 74–90). Kiel, Germany: IPN.

Hidi, S., & Harackiewicz, J. M. (2000). Motivating the academically unmotivated: A critical issue for the 21st century. *Review of Educational Research, 70,* 151–179.

Hilgard, E. R. (1949). Human motives and the concept of the self. *American Psychologist, 4,* 135–142.

Hull, C. (1943). *Principles of behavior.* New York: Appleton-Century Crofts.

Hull, C. (1974). *Essentials of behavior.* New Haven, CT: Yale University Press.

James, W. (1958). *Talks to teachers: On psychology; and to students on some of life's ideals.* New York: Norton.

James, W. (1981). *The principles of psychology* (Vol. 1). Cambridge, MA: Harvard University Press.

Joplin, D., & Neisser, U. (Eds.). (1997). *Conceptual self in context: Culture, experience, self understanding.* Cambridge, MA: Harvard University Press.

Juvonen, J., & Wentzel, K. R. (1996). Social motivation: *Understanding children's school adjustment.* Cambridge, UK: Cambridge University Press.

Keogh, B. K., & MacMillan, D. J. (1996). Exceptionality. In R. C. Calfee & D. C. Berliner (Eds.), *Handbook of educational psychology* (pp. 311–330). New York: Simon & Schuster Macmillan.

Kohn, A. (1994). The truth about self-esteem. *Phi Delta Kappan, 76,* 272–283.

Lent, R. W., & Hackett, G. (1987). Career self-efficacy: Empirical status and future directions. *Journal of Vocational Behavior, 30,* 347–382.

Lewin, K. (1935). *A dynamic theory of personality.* New York: McGraw-Hill.

Markus, H. (1977). Self-schemata and processing information about the self. *Journal of Personality and Social Psychology, 35,* 63–78.

Markus, H., & Nurius, P. (1986). Possible selves. *American Psychologist, 41,* 954–969.

Markus, H., & Wurf, E. (1987). The dynamic self-concept: A social-psychological perspective. *Annual Review of Psychology, 38,* 299–337.

Marsh, H. W. (1990a). Influences of internal and external frames of reference on the formation of math and English self-concepts. *Journal of Educational Psychology, 82,* 107–116.

Marsh, H. W. (1990b). A multidimensional, hierarchical self-concept: Theoretical and empirical justification. *Educational Psychology Review, 2,* 77–172.

Marsh, H. W. (1992). Content specificity of relations between academic achievement and academic self-concept. *Journal of Educational Psychology, 84,* 35–42.

Marsh, H. W. (1993). Academic self-concept: Theory, measurement, and research. In J. Suls (Ed.), *Psychological perspectives on the self* (Vol. 4, pp. 59–98). Hillsdale, NJ: Erlbaum.

Marsh, H. W., Barnes, J., Cairns, L., & Tidman, M. (1984). Self-description Questionnaire: Age and sex effects in the structure and level of self-concept for preadolescent children. *Journal of Educational Psychology, 76,* 940–956.

Marsh, H. W., Byrne, B., & Shavelson, R. J. (1988). A multifaceted academic self-concept: Its hierarchical structure and its relation to academic achievement. *Journal of Educational Psychology, 80,* 366–380.

Marsh, H. W., & O'Neill, R. (1984). Self Description Questionnaire III: The construct validity of multidimensional self-concept ratings by late adolescents. *Journal of Educational Measurement, 21,* 153–174.

Marsh, H. W., & Shavelson, R. J. (1985). Self-concept: Its multifaceted, hierarchical structure. *Educational Psychologist, 20,* 107–125.

Marsh, H. W., Walker, R., & Debus, R. (1991). Subject-specific components of academic self-concept and self-efficacy. *Contemporary Educational Psychology, 16,* 331–345.

Maslow, A. H. (1943). A dynamic theory of human motivation. *Psychological Review, 50,* 370–396.

Maslow, A. H. (1954). *Motivation and personality.* New York: Harper & Row.

Maslow, A. H. (1968). *Toward a psychology of being* (2nd. ed.). New York: Van Nostrand Reinhold.

Mayer, R. E. (1992). Cognition and instruction: Their historic meeting within educational psychology. *Journal of Educational Psychology, 84,* 405–412.

Mayer, R. E. (1996). Learners as information processors: Legacies and limitations of educational psychology's second metaphor. *Educational Psychologist, 31,* 151–161.

Meece, J. L., Blumenfeld, P. C., & Hoyle, R. H. (1988). Students' goal orientations and cognitive engagement in classroom activities. *Journal of Educational Psychology, 80,* 514–523.

Meece, J. L., Wigfield, A., & Eccles, J. S. (1990). Predictors of math anxiety and its consequences for young adolescents' course enrollment intentions and performances in mathematics. *Journal of Educational Psychology, 82,* 60–70.

Middleton, M. J., & Midgley, C. (1997). Avoiding the demonstration of lack of ability: An underexplored aspect of goal theory. *Journal of Educational Psychology, 89,* 710–718.

Midgley, C., Kaplan, A., & Middleton, M. J. (2001). Performance-approach goals: Good for what, for whom, under what circumstances, and at what cost? *Journal of Educational Psychology, 93,* 77–86.

Midgley, C., & Urdan, T. (1995). Predictors of middle school students' use of self-handicapping strategies. *Journal of Early Adolescence, 15,* 389–411.

Midgley, C., & Urdan T. (2001). Academic self-handicapping and achievement goals: A further examination. *Contemporary Educational Psychology, 26,* 61–75.

Miller, G. A. (1956). The magic number seven plus or minus two: Some limits on our capacity for processing information. *Psychological Review, 63,* 81–97.

Miller, N. E., & Dollard, J. (1941). *Social learning and imitation.* New Haven, CT: Yale University Press.

Multon, K. D., Brown, S. D., & Lent, R. W. (1991). Relation of self-efficacy beliefs to academic outcomes: A meta-analytic investigation. *Journal of Counseling Psychology, 38,* 30–38.

Munroe, R. (1955). *Schools of psychoanalytic thought.* New York: Holt, Rinehart, & Winston.

Neisser, U. (1967). *Cognitive psychology.* New York: Appleton-Century-Crofts.

Neisser, U. (1988). Five kinds of self-knowledge. *Philosophical Psychology, 1,* 35–59.

Neisser, U. (1991). Two perceptually given aspects of the self and their development. *Developmental Review, 11,* 197–209.

Neisser, U. (Ed.). 1993. *The perceived self: Ecological and interpersonal sources of self-knowledge.* Cambridge, MA: Cambridge University Press.

Newell, A., & Simon, H. A. (1972). *Human problem solving.* Englewood Cliffs, NJ: Prentice Hall.

Nicholls, J. (1984). Achievement motivation: Conceptions of ability, subjective experience, task choice, and performance. *Psychological Review, 91,* 328–346.

Pajares, F. (1996). Self-efficacy beliefs in achievement settings. *Review of Educational Research, 66,* 543–578.

Pajares, F. (1997). Current directions in self-efficacy research. In M. Maehr & P. R. Pintrich (Eds.). *Advances in motivation and achievement.* (Vol. 10, pp. 1–49). Greenwich, CT: JAI Press.

Pajares, F. (2003). Self-efficacy beliefs, motivation, and achievement in writing: A review of the literature. *Reading and Writing Quarterly, 19,* 139–158.

Pajares, F., Britner, S., & Valiante, G. (2000). Writing and science achievement goals of middle school students. *Contemporary Educational Psychology, 25,* 406–422.

Pajares, F., & Cheong, Y. F. (2003). Achievement goal orientations in writing: A developmental perspective. *International Journal of Educational Research, 39,* 437–445.

Pajares, F., & Miller, M. D. (1994). The role of self-efficacy and self-concept beliefs in mathematical problem-solving: A path analysis. *Journal of Educational Psychology, 86,* 193–203.

Pinker, S. (2002). *The blank slate: The modern denial of human nature.* New York: Viking.

Pintrich, P. R. (2000). Multiple goals, multiple pathways: The role of goal orientation in learning and achievement. *Journal of Educational Psychology, 92,* 544–555.

Pintrich, P. R. (2003). Motivation and classroom learning. In W. M. Reynolds & G. E. Miller (Eds.), *Handbook of psychology: Vol. 4. Educational psychology* (pp. 103–124). New York: Jossey-Bass.

Pintrich, P. R., & Schunk, D. H. (2002). *Motivation in education: Theory, research, and applications* (2nd ed.). Englewood Cliffs, NJ: Prentice Hall.

Rogers, C. R. (1951). *Client-centered therapy.* Boston: Houghton Mifflin.

Rotter, J. B. (1966). Generalized expectancies for internal versus external control of reinforcement. *Psychological Monographs, 80,* 1–28.

Ryan, A. M., & Pintrich, P. R. (1997). "Should I ask for help?": Adolescent perceptions of costs and benefits of help-seeking in math class. *Journal of Educational Psychology, 89,* 329–341.

Schiefele, U. (2001). The role of interest in motivation and learning. In J. M. Collis & S. Messick (Eds.), *Intelligence and personality: Bridging the gap in theory and measurement* (pp. 163–194). Mahwah, NJ: Lawrence Erlbaum.

Schunk, D. H. (1984). Self-efficacy perspective on achievement behavior. *Educational Psychologist, 19,* 48–58.

Schunk, D. H. (1991). Self-efficacy and academic motivation. *Educational Psychologist, 26,* 207–231.

Schunk, D. H. (1995). Self-efficacy and education and instruction. In J. E. Maddux (Ed.), *Self-efficacy, adaptation, and adjustment: Theory, research, and application* (pp. 281–303). New York: Plenum Press.

Schunk, D. H. (1996). Goal and self-evaluative influences during children's cognitive skill learning. *American Educational Research Journal, 33,* 359–382.

Shavelson, R. J., & Bolus, R. (1982). Self-concept: The interplay of theory and models. *Journal of Educational Psychology, 74,* 3–17.

Shavelson, R. J., Hubner, J. J., & Stanton, G. C. (1976). Self-concept: Validation of construct interpretations. *Review of Educational Research, 46,* 407–441.

Shavelson, R. J., & Marsh, H. W. (1986). On the structure of self-concept. In R. Schwarzer (Ed.), *Self-related cognition in anxiety and motivation* (pp. 79–95). Hillsdale, NJ: Erlbaum.

Sideridis, G. D. (2003). On the origins of helpless behavior of students with learning disabilities: Avoidance motivation? *International Journal of Educational Research,* 39, 497–517.

Skaalvik, E. M. (1996). Dimensions of ego orientation (or performance goals). *International Journal of Psychology, 31,* 4744–4744.

Skaalvik, E. (1997). Issues in research on self-concept. In M. Maehr & P. R. Pintrich (Eds.), *Advances in motivation and achievement* (Vol. 10, pp. 51–97). Greenwich, CT: JAI Press.

Skaalvik, E. M., & Rankin, R. J. (1990). Math, verbal, and general academic self-concept: The internal/external frame of reference model and gender differences in self-concept structures. *Journal of Educational Psychology, 82,* 546–554.

Snow, R. E., Corno, L., & Jackson III, D. (1996). Individual differences in affective and conative functions. In R. C. Calfee & D. C. Berliner (Eds.), *Handbook of educational psychology* (pp. 243–310). New York: Simon & Schuster Macmillan.

Stajkovic, A. D., & Luthans, F. (1998). Self-efficacy and work-related performance: A meta-analysis. *Psychological Bulletin, 124,* 240–261.

Steinberg, L., Brown, B. B., & Dornbusch, S. M. (1996). *Beyond the classroom: Why school reform has failed and what parents need to do.* New York: Simon & Schuster.

Stipek, D. (2002). *Motivation to learn: Integrating theory and practice* (4th ed.). Boston: Allyn & Bacon.

Urdan, T. C. (1997). Achievement goal theory: Past results, future directions. In M. Maehr & P. R. Pintrich (Eds.), *Advances in motivation and achievement* (Vol. 10, pp. 99–142). Greenwich, CT: JAI Press.

Urdan, T. C., & Maehr, M. (1995). Beyond a two goal theory of motivation: A case for social goals. *Review of Educational Research, 65,* 213–244.

Vroom, V. H. (1964). *Work and motivation.* New York: Wiley

Weiner, B. (1969). Motivation. In R. Ebel (Ed.), *Encyclopedia of educational research* (4th ed., pp. 877–887). New York: Macmillan.

Weiner, B. (1972). *Theories of motivation: From mechanisms to cognition.* Chicago: Rand McNally.

Weiner, B. (1985). An attributional theory of achievement motivation and emotion. *Psychological Review, 92,* 548–573.

Weiner, B. (1992). *Human motivation: Metaphors, theories, and research.* Newbury Park, CA: Sage.

Wentzel, K. R. (1994). Relations of social goal pursuit to social acceptance, and perceived social support. *Journal of Educational Psychology, 86,* 173–182.

White, R. H. (1959). Motivation reconsidered: The concept of competence. *Psychological Review, 6,* 297–333.

Wigfield, A., & Eccles, J. S. 1992. The development of achievement task values: a theoretical analysis. *Developmental Review, 12,* 265–310.

CHAPTER 8

THE COGNITIVE REVOLUTION AND INSTRUCTIONAL DESIGN

Marcy P. Driscoll and Kerry J. Burner
Florida State University

ABSTRACT

In this chapter, we examine the impact of the cognitive revolution on instructional design. We begin by tracing the evolution in the thinking of Robert M. Gagné, who has arguably had the greatest impact on the instructional design field of any single individual to date. Then, we turn our attention to the Dick and Carey model of instructional design and consider how it has changed in over 40 years of research on human learning. The Dick and Carey model is often referred to as the field standard to which other models and approaches to instructional design are compared. Finally, we present some new directions in instructional design theory that appear to be growing out of constructivist and situated cognition views about learning.

FIRST AUTHOR'S FOREWORD

The assignment for this chapter was to examine the impact of the cognitive revolution on instructional design. It is, of course, impossible to do that without a careful inspection of Robert M. Gagné's work. He has arguably

The Impact of the Cognitive Revolution on Educational Psychology, pages 199–229

had the greatest impact on the instructional design field of any single individual to date. The late Gagné was both a colleague and a friend, someone with whom I had taught and published. Yet when I started this work, I was not sure I had anything new to say, having recently completed a chapter about him (Ertmer, Driscoll, & Wager, 2003) for *Educational Psychology: A Century of Contributions* (Zimmerman & Schunk, 2003). Furthermore, an entire book has been published on the subject of Gagné's legacy, wherein the contributing authors discuss the impact of his work on instructional theory, instructional design practice, military training, and new technologies for instruction (Richey, 2000b).

Scrutinizing Gagné's research from the perspective of how it may have been affected by the cognitive revolution, however, affords a fresh look at his legacy. Trained as an experimental psychologist, Gagné relied on psychological theory to inform his investigations. So it is natural to see how his own thinking shifted as beliefs about human learning evolved in educational psychology.

It is also a new angle to trace Gagné's theory development in relation to other prominent instructional design models, as those, too, have responded to the shifts in psychological learning theory over the past 40 years. It is neither an accident nor a surprise that I have chosen to focus on a particular instructional design (ID) model—Dick and Carey—for this analysis. Walter Dick was a colleague of mine and Gagné's for many years, one of those who helped to attract Gagné to Florida State University. As Gagné wrote, "When I came to Florida State University in 1969, I found a scholarly environment favorable to the establishment of an academic program that incorporated the idea of systematic application of the findings of learning research in the technology of instructional design" (Gagné, 1989). Beyond the personal connections between Gagné and Dick, however, the Dick and Carey model merits attention in its own right. "Without a doubt, the most widely cited ID model is the one originally published by Walter Dick and Lou Carey" (Gustafson & Branch, 2002).

Finally, to keep me honest in my biases and to add a contemporary perspective, I invited one of my students to work with me on this project. Together we have explored some of instructional design's history for the mind shifts in theoretical perspective that have occurred since its early days through to the present. Our analysis is by no means comprehensive, but we believe it is representative and presents a fair picture of how the cognitive revolution affected the field's evolution.

INTRODUCTION

Like most authors who write about instructional design, we feel compelled to begin by staking out the territory. Gustafson and Branch (2002) noted that despite efforts by many to set standard meanings for the terminology of the field, terms continue to be used inconsistently. Therefore, it is important for us to establish what we mean by "instructional design" and how this chapter differs from Wiley and colleagues' (Chapter 9, this volume) discussion of "instructional technology." In its broadest sense, instructional design concerns how "knowledge about human learning can be put to use in the school and in the workplace" (Gagné, 1989, Preface). Thus, it involves the systematic application of research findings to practical instructional problems. But instructional design is also the systematic process of, or set of procedures for, analyzing the problem and designing, developing, implementing, and evaluating a solution (Reiser, 2001). While this often involves the use of technology for creating and, sometimes, delivering instruction, the technology itself is not our focus. For purposes of this chapter, our focus is on the application of learning principles in the systematic process for designing instruction.

By all accounts, the field of instructional design emerged sometime during the 1960s, and formal programs of study in the field became established shortly thereafter. As one of the first, Florida State University formalized its instructional systems program in 1969 and graduated its first doctoral students in the early 1970s. Most faculty members at the time held degrees in experimental psychology with the strongly behavioral training of that era. Like Gagné, however, these were individuals who sought to solve practical problems in education and training, rather than focus on basic theoretical learning concepts. As such, they drew inspiration and theoretical foundations from not only psychology, but also from general systems theory, curriculum theory, communication, and personnel development. Many of the ideas from these areas remain deeply embedded in instructional design, even as evolving psychological theories have wrought changes.

We begin our analysis by examining the work of Robert M. Gagné. To review the entire corpus of his work, which includes more than 150 publications spanning 50 years, would go well beyond the scope of this chapter. Instead, we focused on the four editions of *The Conditions of Learning* (1965, 1970, 1977, and 1985) and Gagné's self-selected publications in his edited volume, *Studies of Learning: 50 Years of Research* (1989). To a lesser extent, we consulted the four editions of *Principles of Instructional Design* (with Briggs, 1974, 1979; with Briggs & Wager, 1988, 1992) and the two works recently published about Gagné (Ertmer et al., 2003; Richey, 2000b). We looked for evidence of how and when concepts of the cognitive revolution appeared to play a role in his thinking.

Then we turn to the Dick and Carey model as it is presented in *The Systematic Design of Instruction,* published first in 1978 and now in its fifth edition. Walter Dick acknowledges the impact of Gagné's work on his own, but it is interesting to note that others view Dick's text as explicating and applying Gagné's theories for practitioners. For example, Fields characterized the Dick and Carey editions as "'vintage systems theory' with the strong influence of Gagné in their application of instructional design theory to practice" (Fields, 2000). Gustafson and Branch (2002) referred to the Dick and Carey model as "the standard to which all other ID models (and alternative approaches to design and development of instruction) are compared" (p. 59). They, too, noted the connection to Gagné's ideas, specifically in the step of the model that calls for conducting an instructional analysis, which is "vintage hierarchical analysis as conceived by Gagné" (p. 61). As with Gagné's work, we looked to see how cognitive concepts became incorporated in the Dick and Carey model.

Finally, we present some new directions in instructional design theory and research that are prominent in the field today. In the past 10 years or thereabouts, so-called traditional views of instructional design as represented by the likes of Gagné and Dick have come under increasing criticism. Some have even made claims that classic instructional design is dead (Gordon & Zemke, 2000). Most agree, however, that instructional design and the practice of it are shifting now with the growing popularity of constructivism and situated learning.

THE INSTRUCTIONAL THEORIES OF ROBERT M. GAGNÉ

Gagné said that the focus of his interest in psychology throughout his long career was "the nature of learning as a phenomenon, and the conditions that bring it about" (Gagné, 1989). At first, this interest was driven by a desire to develop a mathematical description of simple learning events, and he published five papers in 1940 and 1941 on the external inhibition and disinhibition of conditioned operant responses in laboratory rats, a phenomenon observed by researchers such as Pavlov and Skinner. In these studies, rats were exposed to extraneous stimuli (such as a buzzer or the experimenter scratching lightly on the starting box of a runway maze) during acquisition and extinction trials of traversing the runway. Results indicated that performance was subject to both inhibitory and disinhibitory effects and supported the notion that any theory of conditioning must include the concept of inhibition (Gagné, 1941).

Gagné called the aim of this research "incompletely realized," most likely because World War II intervened and he enlisted. During the war, Gagné became involved in military training as part of the Aviation Psychology Program and he conducted research on conditions affecting transfer of training. After the war, he went to work for the Air Force Laboratories and began to apply that knowledge to the design of training devices and simulators. This research on Air Force training in the 1950s and 1960s reveals the first appearance of cognitive concepts in Gagné's work. In fact, it is probably fair to say that the major components of Gagné's instructional theory and his views on instructional design all took root during this period. He suggested that the outcomes of learning be described in categories, and he proposed a view of human functioning that is acknowledged as an early information-processing conception (Asher, 2003).

Varieties of Learning

In a review article on training devices and simulators, Gagné (Gagné, 1954) argued that perhaps the most important problem facing researchers was how to measure complex human performances. Decisions on how to design or use training devices such as flight simulators depend on what performance is expected after training. Once criterion performance could be sufficiently described, then it would be possible to determine how man[1] and machine interact in machine-based training devices and how such devices could be designed to better promote transfer. Gagné proposed in a subsequent address to the APA Division of Military Psychology that, to achieve reliability of measurement, learning performance should be described operationally at a behavioral level (Gagné, 1955, p. 101). He suggested five categories:

1. *Discrimination.* Examples of this are reading dials, aligning dials and instruments, distinguishing wave forms.
2. *Recall.* Examples of this category are the identification of components and parts, the remembering of procedures.
3. *The Use of Symbols.* Examples are reading or interpreting charts, diagrams, and tables and performing numerical operations.
4. *Decision making.* This means making a complex judgment in which several possibilities must be successfully or simultaneously weighed. It is a rather common type of behavior. In the maintenance situation, it is exemplified by systematic troubleshooting.
5. *Motor Skill.* This category consists of various kinds of tool-using, making fine adjustments of controls, tracking.

Gagné wrote that these are behavioral terms, and he maintained throughout his career the importance of specifying learning outcomes in behavioral terms. But it is also quite apparent that the categories he proposed embody cognitive ideas. Remembering, use of symbols, and complex judgment are cognitive terms, consistent with modern formulations of cognitive theory. The extent to which Gagné might have been influenced at this point by anyone writing about cognition is unclear. There is no evidence that he was.

In a report to the Air Force in 1957, Gagné elaborated his "principles of human functioning" to include "detection," "perception," "retention," and the notion of attitudes, which he distinguished from basic human capacities (reprinted in Gagné, 1989). By the mid-1960s and early 1970s, the learning taxonomy and conditions of learning for which Gagné is most known were truly taking shape. Gagné said that he originally described the domains of learning outcomes in a presidential address to the American Educational Research Association, which was later published in 1972. Yet he had already published the first two editions of *The Conditions of Learning* by then and both described different varieties of learning. However, the AERA address (Gagné, 1972) included the same five domains that appeared in the later editions of his text, so it is likely that Gagné was referring to this fact. Table 8.1 displays the varieties of learning that Gagné proposed as they evolved in his writing from the 1960s to the 1980s.

Interestingly, Gagné noted in the second edition of the *Conditions* book that "to distinguish a number of varieties of learning is not, of course, a completely novel idea. Learning types have been distinguished by many other writers" (Gagné, 1970, ,p. 64). He went on to cite Thorndike, Skinner, Hull, Tolman, Woodworth, Mowrer, and Harlow and concluded that "a detailed analysis would reveal many ideas within the present volume that are derived from these writers" (p. 65). By the third edition, Gagné (1977) was clearly familiar with cognitive research, because he incorporated the information-processing model as a foundation for his description of learning and the events of learning. However, there is no reference to this literature in his presentation of the five varieties of learning.

The connection of Gagné's taxonomy of learning outcomes to cognitive psychology became explicit in the 1980s. Shortly after the first author arrived at Florida State University in 1980, she and Bob Gagné were assigned to teach sections of the same course, which he had developed. The subject was learning theories, and assigned readings included articles by Skinner, Bruner, Ausubel, and Gagné himself. Driscoll cannot remember whose suggestion it was, but they soon adopted Roberta Klatzky's 1980 text on cognitive psychology to introduce students to research in that area. Informal discussions around the office also focused on what was to be learned from cognitive psychology research that would have practical implications for instructional design.

Table 8.1. Evolution of Gagné's Taxonomy of Learning Outcomes

Gagne (1965a)	*Gagne (1970)*	*Gagne (1972)*	*Gagne (1977)*	*Gagne (1985)*
Response differentiation	Type 1: Signal learning	1. Motor skills	1. Intellectual skills	1. Verbal information
Association	Type 2: Stimulus–response learning	2. Verbal information (also called "knowledge")	2. Discriminations	2. Intellectual skills
Multiple discrimination (identification)	Type 3: Chaining	3. Intellectual skills	3. Concepts[a]	3. Discriminations
Behavior chains (sequences)	Type 4: Verbal associations	4. Cognitive strategies	4. Rules	4. Concrete concepts
Class concepts	Type 5: Discrimination learning	5. Attitudes	5. Higher-order rules	5. Defined concepts
Principles	Type 6: Concept learning		6. Cognitive strategies	6. Rules
Strategies	Type 7: Rule learning		7. Verbal information	7. Higher-order rules
	Type 8: Problem solving		8. Attitudes	8. Cognitive strategies
			9. Motor skills	9. Attitudes
				10. Motor skills

[a]These are elaborated on later in the text as concrete and defined concepts, but when introduced, "concepts" encompass both types and simple rules as well.

When Gagné wrote again about categories of learning outcomes (Gagné, 1984), he began the article with a review of learning concepts stemming back to associationist philosophers and ending with the advent of cognitive psychology. Furthermore, he couched his presentation of the categories in cognitive terms. For instance, "as a category of learning outcome, intellectual skills have in recent times appeared to find their proper place in the scheme of things.... Perhaps the best-known synonym is *procedural knowledge*" (1984, reprinted in Gagné, 1989, p. 514, original emphasis). He also called verbal information declarative knowledge and discussed the idea of such knowledge being organized in networks of propositions, citing John Anderson's (1976, 1980) work extensively.

As Woolfolk (1995) defined conditional knowledge—knowing when and why to use one's declarative and procedural knowledge (Woolfolk, 1995)—she said that Gagné called it cognitive strategies. Whether Gagné agreed with this characterization is unknown; he seems to have preferred the use of strategic knowledge, as we have found no reference to conditional knowledge in any of his work. The meaning is clearly the same, though, as he discussed cognitive strategies as enabling learners to control when and how they will bring to bear what they know to tasks such as remembering or solving problems.

Conditions of Learning

Gagné was often asked, why five kinds of learning outcomes? As Driscoll has heard the question posed, the emphasis was always on the number, why *five* (or eight, if one was asking about an earlier version of Gagné's taxonomy). Gagné's answer, however, seemed more often than not to focus on the why: *why* is it important to differentiate domains at all? Put simply, he believed that different instructional treatments were needed for different learning outcomes. The reason to search for domains was to find out how widely generalizable conditions were that facilitated learning. In other words, "if the researcher has obtained a result that shows certain conditions to be facilitative of learning, he needs to know how widely this result can be generalized. Does it apply across subject-matter, across age levels, across classrooms?" (Gagné, 1972, reprinted in Gagné, 1989, p. 487).

In the 1984 paper on learning outcomes that is mentioned above, Gagné discussed the tension between trying to discover general principles of learning and distinguishing domains that seem to differ in the information-processing demands on learning. On one hand, he seems to foreshadow developments in situated learning when he called for an intensive look at what people learn:

> By this I mean, we need to gain an idea of what all kinds of people learn—not only school children or laboratory subjects but masons, carpenters, astronauts, politicians, housewives, and word-processing operators. Most of the overt behavior people engage in during each day, of course, is what they learned to do (1984, reprinted in Gagné, 1989, p. 513).

Yet examining what people learn in everyday life did not mean that learning was necessarily bound to the situation or context, as was later proposed by theorists in the situative perspective (Brown, Collins, & Duguid, 1989). Rather, Gagné asked, "How can we find principles of learning that can be generalized and not tied to specific subject matter?... I trust [learning psychologists] will continue to keep the appropriate goal in mind and will not be too seriously distracted by trendy issues suggested by neighboring disciplines" (p. 513). Gagné was well known in the office for his views on teacher education; he believed that "the view that we must discover and formulate principles of mathematics learning, science learning, automobile repair learning, and so on [is unacceptable]" (p. 523).

Gagné's conditions of learning, then, depended not on the subject matter or the situation in which a human capability may be acquired, but on the capability itself. And he looked for empirical evidence of what those conditions could be. As early as his research on Air Force training, there is evidence that Gagné believed in the need to consider internal as well as external conditions of learning. For example, in describing how men function within a machine environment, Gagné referred to "internal functional units" in man that may be called upon as the individual is interacting with the machine system, and he identified memory as one example. Discussing a specific example of machine operation, Gagné wrote:

> He [the machine operator] must have an *internal* means of matching the external display to one of the five classes in order that he can make five different responses, as required in our example. This means he must have previously acquired the "representative shapes" in his memory, by means of learning preceding the occasion when he tackles the job of operating the equipment. And this provides the basic reason for *training*.... (Gagné, 1962b, reprinted in 1989, pp. 108–109)

At this time, Gagné appears to have been influenced by psychological theories of perception as well as the ideas of Jerome Bruner, whose 1956 book, *A Study of Thinking*, is cited in Gagné's research and in *The Conditions of Learning*.

The research that Gagné conducted that led to his formulation of the learning hierarchy also demonstrates the importance he attributed to internal events in learning. Gagné was interested in what he called productive learning, or the ability of learners to respond appropriately to classes of tasks. Learning concepts and rules clearly fell into this arena, what he would later elaborate as intellectual skills. He argued that complex tasks such as solving linear equations or adding rational numbers could be analyzed into "a hierarchy of subordinate learning sets, which mediate positive transfer of learning in a unidirectional fashion from one to another, and ultimately to the final performance" (Gagné, Mayor, Garstens, & Paradise, 1962). Learning such tasks depended on the recallability of the relevant subordinate learning sets that comprised the complex final task.

At least initially, Gagné appeared to conceive of learning hierarchies as learning sets in much the same way as Harlow, to whose work with monkeys he referred (Gagné, 1962a). But in *The Conditions of Learning*, when Gagné distinguished between external and internal conditions, he said simply that "learning hierarchies pertain only to the *internal conditions*, in other words, to the capabilities that are to be learned and to the capabilities that are prerequisite to these" (Gagné, 1970, p. 241). He also emphasized the fact that learning hierarchies pertain to skill acquisition, not the acquisition of verbalizable knowledge (Gagné, 1968b). To this day, the procedures Gagné proposed for analyzing complex intellectual skills into learning hierarchies composed of their component skills are standard practice in instructional design.

While Gagné's own research clearly influenced his explication of internal learning conditions in the intellectual skill domain, he incorporated the work of many others across a wide spectrum of beliefs into learning conditions in the other domains. For basic forms of learning (signal learning, S–R learning, chaining, and verbal association), Gagné cited Pavlov and Skinner as well as verbal learning theorists such as Cofer and Underwood. For concept and rule learning, Gagné cited Ausubel and Bruner. To some of us who watched Gagné grow red in the face at the suggestion that Piaget's theory might be studied in a course on learning, it may seem surprising that he also cited Piaget in discussing the concept learning of young children. Yet he did so primarily to offer an alternative explanation to children's performance on conservation tasks over time. Gagné (1968a) argued that cumulative learning based on hierarchies of capabilities would prove a better model of human development than theories such as Piaget's, which are based on maturational readiness.

Finally, as mentioned previously, information-processing (IP) theory made its way into Gagné's third edition of the *Conditions* book (1977) in his discussion of learning conditions. Internal conditions for learning verbal information included a preexisting set of organized knowledge—a cognitive structure, as Ausubel called it, or a network of associations or propositions in the IP tradition—and encoding strategies. Internal conditions necessary for learning in the other domains amounted mostly to recall—of part skills (motor skills), concepts and information (attitudes), prerequisite skills (intellectual skills), and cognitive strategies (problem solving).

Information Processing and the Events of Instruction

In the *50 Years of Research* book, Gagné commented that a chapter he published in 1962 "may be viewed as incorporating an early information-processing conception of human performance" (Gagné, 1989). We have mentioned the work already, in that it seems to have been Gagné's first reference to internal events. However, he went on to describe processes of human functioning including long-term and short-term memory and mechanisms such as sensing, identifying, and interpreting. In this chapter, Gagné made explicit comparisons between human functioning and machine functioning, both as complex systems, in much the same way that information-processing theorists used the analogy of computer processing to describe mental functioning. The chapter clearly predates Atkinson & Shiffrin (1968), which Gagné cited as the basis for the general learning events he described in his later work. Moreover, from the references he cited in the chapter, Gagné appears to have had little knowledge of the research in cognitive psychology at that time.

From this chapter, Gagné's view of the learning process evolved to a sequence of four phases in the early editions of the *Conditions* book (Gagné, 1965, 1970):

1. *Apprehending phase*: this consisted of attending to a stimulus, perceiving of it, and coding it.
2. *Acquisition phase*: conceptually, this referred to the fact that a learner can execute a performance he or she could not do before learning.
3. *Storage phase*: this referred to the retention of learning. Gagné distinguished between short-term and long-term memory and suggested that long term could mean permanent in the sense of memory traces in the brain. Interestingly, he cited Penfield's and Lashley's work as evidence of long-term storage.
4. *Retrieval phase*: this was the performance of learned capabilities, involving, for example, the recall of information or the reinstatement of intellectual skills.

These phases were not, however, integrated into the rest of the book, despite Gagné's claim that they, along with the specific conditions for learning, enabled derivation of the nine events of instruction.

The events of instruction provided a means of control over external events of learning so as to ensure the proper timing and sequencing of internal events necessary for learning to occur. Driscoll remembers Gagné describing the nine events of instruction as "something good teachers do to make a difference in learning" and that he claimed to have derived them initially from observations of teachers. Indeed, Gagné spent considerable time observing a number of classes in all grades, the results of which he reported at a symposium at the American Psychological Association annual meeting (Gagné, 1973). There he described events that fell into four broad categories of introducing the lesson, guiding the initial learning process, making learning memorable and generalizable, and effecting reinforcement (p. 113).

Shortly after this, Gagné adopted the information processing model of Atkinson and Shiffrin (1968) as a theoretical foundation for describing learning processes, and he related those four categories (which he again called phases) to the learning processes they were presumed to affect (Gagné, 1974). From then on, Gagné's events of instruction were linked to learning processes as posited by the IP model, although the notion of instruction occurring in phases gradually disappeared (see Table 8.2). Once asked about the connection between the nine events and the IP model, which appears first in the third edition of the *Conditions* book, Gagné chuckled impishly and replied something to the effect that, "Yes, isn't it nice that the cognitive psychologists have found a way to explain why the nine events of instruction work!"

The Legacy Years

Gagné retired from Florida State University in 1986. He continued to publish, although mostly revisions of earlier work and reflections on the field. This included the 4th edition of the *Conditions* book (1985), the *50 Years* book (1989), the 4th edition of the *Principles of Instructional Design* book (1992), *The Conditions of Learning: Training Applications* (Gagné & Medsker, 1996), and conversations with David Merrill (Gagné & Merrill, 1990–1991).

In 1982, an interview with Gagné was published in *Educational Technology*, in which he discussed learning psychology as he perceived its influence on instructional design. He indicated that a major change had taken place in the shift from behaviorist, S–R psychology to cognitive learning psychology, specifically information processing and the notion that what gets stored in long-term memory is propositions. This transformation changed the question for instructional designers from how to present a stimulus to how to organize instructional material so as to support internal learning processes. When asked whether a cognitive approach would supplant behavioral theory in instructional design, Gagné said yes. "I do believe that the

Table 8.2. Evolution of Gagné's Events of Instruction and Their Relation to Learning

1965, 1970 Conditions		*1974—Essentials* *1977—Conditions*		*1985—Conditions* *1988—Essentials*	
Phases of Learning	*Events of Instruction*	*Phases of Learning*	*Events of Instruction*	*Learning Processes*	*Events of Instruction*
Apprehending Acquisition Storage Retrieval	1. Gaining and controlling attention 2. Informing the learner of expected outcomes 3. Stimulating recall of prerequisites 4. Presenting the stimulus 5. Offering guidance for learning 6. Providing feedback 7. Appraising performance 8. Making provisions for transferability 9. Insuring retention	Motivation Apprehending Acquisition Retention Recall Generalization Performance Feedback	1. Activating motivation 2. Informing learner of objective 3. Directing attention 4. Stimulating recall 5. Guiding learning 6. Enhancing retention 7. Promoting transfer 8. Eliciting performance 9. Providing feedback	Attention Expectancy Retrieval to working memory Selective perception Encoding: entry to LTM storage Responding Reinforcement Cueing, retrieval	1. Gaining attention 2. Informing the learner of the objective 3. Stimulating recall of prior learning 4. Presenting the stimulus 5. Providing learning guidance 6. Eliciting performance 7. Providing feedback 8. Assessing performance 9. Enhancing retention and transfer

cognitive approach will come to dominate, if it hasn't already" (Gagné, 1982, reprinted in 1989, p. 580). He went on to qualify his answer, though, by saying that "the ideas of Skinnerian behaviorism are very strong. They are very *sturdy*—and they *work*" (p. 580; original emphasis). Thus, Gagné did not believe that behaviorist principles would fall out of use altogether. Rather, instructional designers would continue to incorporate the "important parts" of behaviorism in their work.

In the interview, the conversation turned to the implications of computers, and Gagné lamented the commercial influence, with people turning out instructional programs for the computer that had no theoretical basis and thus little likelihood of effectiveness. He saw, however, the potential of computers to play a significant role in learning effectiveness and learning research:

> I think there is a convergence of developments here—a convergence in learning theory, research on learning, in computer science itself, and in the advent of the new microcomputer, which is just beginning to take over. There are many ways in which these things come together and give us new tools, new prospects for learning, and new ways of conducting instruction. (Gagné, 1982, reprinted in 1989, p. 585)

This comment is especially interesting for a couple of reasons. Once again, Gagné seems to foreshadow developments in learning theory. Nearly 20 years after Gagné made this comment, David Jonassen and Susan Land (2000) say much the same thing in the Preface to their edited volume on open-ended learning environments. "We have witnessed a convergence of learning theories never before encountered." Moreover, contributing authors, writing from the perspectives of constructivism and situated learning, refer in their chapters to technologies as tools to think and learn with.

Yet Gagné never subscribed to constructivism and, in fact, adamantly opposed it. Asking him what he thought about it would trigger a tirade. "Constructivism is ——!" (expletive deleted), he would bellow. Driscoll believes that the reason for Gagné's ire lay in the influence constructivism seemed to exert in science and math teacher education during the early 1990s. Poor student performance on nationwide assessment tests was attributed to a failure among schools to teach critical thinking. "Nonsense," Gagné would say. "You're talking about a lack of basic skills." He believed firmly that basic skills in reading, arithmetic, and other fields had to be mastered first before strategic thinking could be approached. In the 1982 interview, he said,

> No one says these [i.e., cognitive strategies] are not important, but the question is can you *start* from there and also *stop* there? I'm of the other school, and always have been. I say you have to start from the bottom up. Starting

> from the bottom up means you have to start with *skills.* I don't care whether you call them memorized or not. I think they have to be *firmly established* before one can free the human mind for the kind of thinking that has to be done, even in reading a page of text. (Gagné, 1982, reprinted in 1989, p. 583, original emphasis)

Given enough time and empirical evidence, Gagné might have found something of value in constructivism, as he apparently did in schema theory and some aspects of situated learning. Again, Driscoll recalls conversations around the office where Gagné expressed skepticism about the merits and implications of schema theory. Yet he incorporated the concept of schema when he and David Merrill wrote about how instructional designers could plan instruction that integrated multiple goals. They proposed the notion of an enterprise, which provides an overarching context and reason for learning particular knowledge and skills (Gagné & Merrill, 1990). "Each kind of enterprise is represented in memory by a schema that reflects the purpose or goal of the enterprise category, the various knowledges and skills required to engage in the enterprise, and a scenario which indicates when and how each piece of knowledge or skill is required by the enterprise" (Gagné & Merrill, 1990, p. 25). With a nod to situated learning, Gagné and Merrill suggested that real-world problems and situations could serve as useful enterprises in instruction.

Gagné's Impact versus Impact on Gagné

According to Pressley and Roehrig (2003, p. 356), Robert Gagné's writings "translated information processing theory into the basic research problems that were of greatest interest to educational psychologists of the [modern era]." Similarly, an information-processing view of learning has dominated instructional design practice for at least the past 20 years, largely due to the influence of Gagné's work (Richey, 2000a). Thus, the impact of Gagné's thinking on the field seems clear. As he incorporated cognitive concepts into the framework of his instructional theory and into his writings, these ideas led to instructional design conventions and procedures that were adopted widely.

The impact of the cognitive revolution on Gagné, however, is not as obvious in one respect. Although he came to make direct links between his theories and information-processing theory, for instance, Gagné appeared to think in cognitive terms even before he had such a theory to rely on. That he recognized this is apparent in the section introductions that he wrote for the *50 Years* book. There, he also provided a reasonable explanation for why a lag may have occurred between his thinking as it was evolving and the eventual utility that he found in cognitive psychology, especially information-processing theory. In describing a study on problem solving that he conducted in which subjects were asked to think aloud as they solved the target problem, Gagné said, "My behavioral training led me to neglect the idea of analyzing these verbal reports, however, as others would later do in 'protocol analysis'" (1989, p. 171).

The influence of his behavioral training notwithstanding, Gagné was persuaded by empirical evidence. It seemed as though he maintained a certain skepticism toward new ideas, kept developing his own ideas, and then sought reconciliation when those other ideas had the weight of evidence behind them. Most of the research he conducted himself was experimental, although we have already mentioned the observational study that Gagné conducted of teachers as he attempted to understand what they do that facilitates learning (Gagné, 1973). However, he was open to any research idea that would illuminate a question of interest.

Driscoll remembers that, during her job interview at Florida State, Gagné asked her a question about what people learn from television documentaries. It turns out that he was directing a doctoral study in which adults were shown an hour-long documentary and then asked to recall it. Their verbal protocols were compared with the documentary's narrative in terms of propositions, or idea units, with the result that approximately 50% of the narrative was recalled. Gagné appeared delighted with this finding, because he expected much less to be remembered. The result also led to other empirical questions, such as how the redundancy between the visual and verbal narratives and how the semantic structure of the verbal narrative (á la Kintsch) related to learning. We can only speculate as to whether Gagné would have embraced qualitative methodologies and analysis techniques that have gained in prominence during the last decade. We think so, provided he accepted the scientific rigor of such approaches and recognized their usefulness in "seeing a novel idea come to fruition in a network of empirical validations" (Gagné, 1989, Preface).

THE DICK AND CAREY INSTRUCTIONAL DESIGN MODEL

The Dick and Carey model as presented in the authors' text *The Systematic Design of Instruction* (Dick & Carey, 1978, 1985, 1990, 1996; Dick, Carey, & Carey, 2001) is the academic standard for students in the field of instructional systems design (Gustafson & Branch, 2002). The ADDIE model is commonly accepted as the foundational model for the field, and it calls for instructional designers to analyze, design, develop, implement, and evaluate instruction. The Dick and Carey model refines that process, clearly distinguishing 10 recursive steps that are meant to guide instructional designers through the process of developing efficient and effective instruction. These steps in the most current version are:

1. Assess needs and identify goals
2. Conduct instructional analysis
3. Analyze learners and contexts
4. Write performance objectives
5. Develop assessment instruments
6. Develop instructional strategy
7. Develop and select instructional material
8. Design and conduct formative evaluation of instruction
9. Revise instruction
10. Design and conduct summative evaluation.

The steps are not meant to be hard and fast directives nor should designers feel obliged to move through them in a lock-step fashion. Rather, the authors intended the model as a teaching tool to help students "understand the major concepts in the field of education," such as performance objectives and formative evaluation (Dick, Carey, & Carey, 2001, p. xxi). Now in its fifth edition, the text has evolved over the 25 years since its first publication, reflecting the movements in and influence on instructional design. The cognitive revolution appears to have had a twofold impact: (1) on the content of the model itself, and (2) on the communication of the model as instructional material in courses dedicated to teaching the instructional design process.

Changes to the Communication of the Model

A quick look at the table of contents across the five editions clearly demonstrates two of the many key concepts undergirding a cognitive approach to learning and instruction, schema theory, and advance organizers. In the 1978 and 1985 editions, each chapter contained the same headings with no subheadings. This very stylized approach changed in the 1990 edition to a format more consistent with implications of research on comprehension of text material. From that edition forward, the authors instead used descriptive headings and detailed subheadings, which clearly signaled the text organization (Armbruster, 1986). This articulation of the chapters' content functions along with the index to inform novice instructional designers of the locations of specific information for reference. More importantly, this approach offered learners an advance organizer and a schema for the new information with which they were engaging.

Graphic presentations of the Dick and Carey model also changed through the years, signaling what step of the model is the topic of each chapter, and including more comparative and procedural organizers to represent aspects of the model. Most chapters now provide tables and matrices that guide students through each step of the model, first showing an illustration of the step with a concrete example and then providing an empty shell for students to use in completing the step with their own projects.

Changes to the Content of the Model

These subtle changes in the presentation of the model belie the significant changes in the content of it. While the model, in essence, has not changed significantly, some of the model's steps have been renamed and modifications have been made in how some steps are conducted. These changes indicate both continued development of the model itself and a deeper understanding of the most effective way to communicate the model. Table 8.3 lists the 10 steps by year; the changes are in boldface. The most significant changes occur between the third edition (1990) and fourth edition (1996), with the changes to the first, third, and fifth steps. These steps are called—in the fifth edition—Assess Needs to Identify Goals, Analyze Learners and Contexts, Develop Assessment Instruments. In addition, the sixth step, Develop Instructional Strategy, appears to have undergone a change in content but not in name.

The first step was changed from *Identify Instructional Goals* to *Assess Needs to Identify Goals*. This change acknowledges that instructional goals are directly related to the needs assessment process, a process that identifies what knowledge, skills, and attitudes learners will need in order to perform the instructional goal. The third step is now known as *Analyze Learners and Contexts*, but it began as *Identifying Entry Behaviors, Characteristics*. The change reflects the influence of a cognitive psychological stance insofar as the authors are now looking beyond the learning goal as the central focus of the step. Previously the main focus of the chapter, the discussion of the target population and entry behaviors has been subsumed into a discussion of learner analysis. This subsection now includes discussions of learner motivation, attitude, and learning preferences, as well as the characteristics of the group of learners for which instruction will be designed. Additionally, the authors expanded the chapter to include an analysis of the performance and learning environments. They go further to discuss the compatibility of the instructional site with the instructional materials to be used.

Table 8.3. Changes to the Wording of the Steps in the Dick and Carey Model, 1978–2001

1978	Identify Instructional Goal	Conduct Instructional Analysis	Identify Entry Behaviors, Characteris-tics	Write Performance Objectives	Develop Criterion-Referenced Tests	Develop Instructional Strategy	Develop and Select Instruction	Design and Conducting Formative Evaluation	Revise Instruction	Design and Conduct Summative Evaluation
1985	**Identify Instructional Goal(s)**	**Conduct Instructional Analysis**	**Identify Entry Behaviors, Characteris-tics**	**Write Performance Objectives**	**Develop Criterion-Referenced Tests**	**Develop Instructional Strategy**	**Develop and Select Instructional Materials**	**Design and Conduct Formative Evaluation**	**Revise Instruction**	**Design and Conduct Summative Evaluation**
1990	Identify Instructional Goal(s)	Conduct Instructional Analysis	Identify Entry Behaviors, Characteris-tics	Write Performance Objectives	Develop Criterion-Referenced Tests	Develop Instructional Strategy	Develop and Select Instructional Materials	Design and Conduct Formative Evaluation	Revise Instruction	Design and Conduct Summative Evaluation
1996	**Assess Needs To Identify Goals**	Conduct Instructional Analysis	**Analyze Learners and Contexts**	Write Performance Objectives	**Develop Assessment Instruments**	Develop Instructional Strategy	Develop and Select Instructional Materials	**Design and Conduct Formative Evaluation of Instruction**	Revise Instruction	Design and Conduct Summative Evaluation
2001	Assess Needs To Identify Goal(s)	Conduct Instructional Analysis	Analyze Learners and Contexts	Write Performance Objectives	Develop Assessment Instruments	Develop Instructional Strategy	Develop and Select Instructional Materials	Design and Conduct Formative Evaluation of Instruction	Revise Instruction	Design and Conduct Summative Evaluation

The changes in these two steps reflect the impact of the cognitive revolution in that the steps are concerned with the acquisition of information by learners. Needs are assessed and learners are analyzed with a specific performance context in mind. Applications of cognitive psychology in instruction concern the individual learner's uniquely structured way of connecting new information to extant information. This is not to say that every learner's cognitive structure is thought to be different; rather, cognitive psychologists theorize that learners use common processes—like schema theory or cognitive information processing.

Factors influencing these processes go beyond the individual, however. Environment, especially the appropriateness of the environment to the subject matter, plays an important role in effective learning. Instructional goals must take into consideration who the learners are, what they will be able to do, where they'll be able to do it, and what they'll be using to perform the task (Dick & Carey, 1996). Taking this holistic approach to learning is evidence of the cognitive revolution. That is, instructional designers must bear in mind that learning that is not retained or transferred is not useful. Thus, instructional designers must consider more than the substeps of a task or behavior when creating instruction geared toward a specific instructional goal if they desire learners to make the instruction permanent.

Starting in the second edition of *The Systematic Design of Instruction,* the authors included Gagné's taxonomy of learning domains, using four of his five categories to help organize their discussions of instructional design—intellectual skill, verbal information, motor skill, and attitude. Subsequently, the step *Develop Assessment Instruments* was expanded to include a discussion about criterion-referenced test items relative to Gagné's domains of learning. This is significant in light of the cognitive revolution as it demonstrates the authors' deeper understanding of the differences between the thought processes behind the demonstration of learning. General statements about matching objectives to assessment do not help designers conceive of appropriate measurements for such vastly different skills as motor—riding a bike—and intellectual—writing a story. While this reliance on Gagné's taxonomy may not initially appear to be significant to the cognitive revolution, the fact that the authors now recognize differences in both learning and demonstrating that learning (behavior) signifies the influence of a cognitive approach. Different types of learning require different types of assessment.

This statement holds true for instructional strategies as well. Different types of learning require different types of instructional strategies. Although the authors did not rename the step *Developing Instructional Strategies,* by the fourth edition they did revise the chapter to include instructional strategies appropriate for the four categories of Gagné's taxonomy that they used to organize their overall discussion of instructional design, namely intellectual skill, verbal information, motor skill, and attitude. In doing this, they exhibit the impact of different theories, like cognitive psychology, on the development of their model.

For example, in their discussion (fifth edition) of instructional strategies for intellectual skills, they tell the reader: "The designer should be aware of both the way in which learners may have organized their entry knowledge in memory and the limits of their ability to remember new content. The strategy should provide ways in which the learner can link new content to existing prerequisite knowledge in memory" (Dick, Carey, & Carey, 2001, p. 198). The authors are discussing key concepts in cognitive psychology, namely schema theory, information processing, organization effects, and meaningful learning. The links are not made explicit, but this implicit inclusion of a cognitive approach is a louder statement of the role of cognitive psychology than a direct association made by the authors.

This tracing of cognitive psychology in the revisions to *The Systematic Design of Instruction* (Dick & Carey, 1978, 1985, 1990, 1996; Dick, Carey, & Carey, 2001) is by no means exhaustive; rather, the intent is to be illustrative of the kinds of subtle and important changes cognitive psychology has had on one of the most prevalent of all instructional design models.

RECENT DIRECTIONS IN INSTRUCTIONAL DESIGN

If Gagné's instructional theory and Dick and Carey's instructional design model represent traditional views of the field and illustrate the impact of the cognitive revolution as it has happened in the past, then what about the present and future? Van Merrienboer and Kirschner (2001) suggest that "different authors are performing their research and development work in different Worlds of Instructional Design" (p. 429). The distinctions they make among three different worlds provide a useful framework for our discussion.

First, the World of Knowledge is the world of learning outcomes, task and content analysis, and the selection of appropriate learning conditions based on the learning goals. It is the "traditional field of ID, where foundations were laid by Gagné" (van Merrienboer & Kirschner, 2001). According to the authors, this world is particularly useful for "medium-sized, integrated sets of knowledge components" (p. 432) that are typically associated with "traditional lesson design" (p. 433). The analytic approach characterizing the World of Knowledge can be seen in Merrill's component display theory (Merrill, 1983, 1994) and instructional transaction theory (Merrill, 1999) as well as Scandura's (2001) structural learning theory. Merrill's recent work has been an attempt to automate instructional design procedures so as to make it easier and more efficient for designers to build instruction based on sound principles of instructional design. Scandura has focused on building rule-based representations of complex behaviors that can serve as the basis for intelligent tutoring systems. Both approaches depend heavily on cognitive task analysis and diagnosis of learner prior knowledge.

The World of Learning, on the other hand, emphasizes analysis of learning processes rather than learning outcomes, and it is "rooted in educational and cognitive psychology" (van Merrienboer & Kirschner, 2001). Instructional design is focused on developing tools and systems to support "a particular, often predefined kind of learning process" (p. 431) rather than on developing lessons in a traditional sense. It is interesting that van Merrienboer and Kirschner do not mention Gagné's work in relation to the World of Learning, even though Gagné's events of instruction are presumed to support certain learning processes. The reason can perhaps be seen in the body of work that van Merrienboer and his colleagues have contributed on cognitive load theory (CLT), to which we turn our attention shortly. CLT focuses on working memory primarily, whereas Gagné's events related to all phases of information processing.

Finally, the World of Work is focused on professional task performance in the real world and, as such, involves authentic learning and the notion of apprenticeship. It is "best associated with constructivist views on learning" (van Merrienboer & Kirschner, 2001) and may involve complex problem solving in ill-structured domains.

The World of Learning and the World of Work clearly represent new directions for instructional design, and developments in each are worth further examination. While space does not permit an exhaustive review, we believe that research on CLT and mental models, as seen in the work of van Merrienboer and Jonassen, respectively, offers a glimpse of how the cognitive revolution is affecting instructional design today.

Cognitive Load Theory

CLT exemplifies van Merrienboer and Kirschner's (2001) World of Learning in that it "emphasizes working memory constraints as determinants of instructional design effectiveness" (Sweller, van Merrienboer, & Paas, 1998). It adopts the view that working memory involves a number of partially independent processors that are associated with different sensory modes. One processor may deal with visually based information, for instance, whereas another processor may deal with auditory information. According to CLT, the capacity of working memory, as well as the nature of material to be learned, can restrict the ease with which information is processed. This is what defines cognitive load.

CLT is clearly related to information-processing theory in general, but it also relies on schema theory and research conducted on memory differences between novices and experts (Chase & Simon, 1973). Possessing a large number of relevant and automated schemas reduces the load on working memory and allows for precious resources to be more efficiently allocated during problem solving. CLT researchers have systematically investigated variables that affect cognitive load and therefore have important implications for the design of instruction.

Sweller, van Merrienboer, and Paas (1998), for example, summarized research on CLT that has been conducted for the past 15 years in several different laboratories and described seven effects: goal-free, worked examples, completion problem, split-attention, dual modality, redundancy, and variability, with the last three existing under conditions of split-attention. In the goal-free effect, cognitive load is reduced when learners pursue a goal-free strategy as they solve problems. That is, they do not have to keep the goal in working memory nor the strategy by which they are trying to solve the problem. All their attentional resources can be directed at solving for values of variables that will lead ultimately to an overall solution to the problem. Problem-solving performance is enhanced as a result.

Providing worked examples and partially completed problems also appear to be effective instructional strategies for reducing cognitive load and improving problem-solving performance (Sweller et al., 1998). The split-attention effect, on the other hand, refers to the adverse impact on performance of instructional conditions wherein learners have to pay attention to information presented separately that must be integrated to solve a problem. For instance, a diagram could provide some of the information necessary to solve a problem, while a verbal description in the text provides the rest. This effect can be reduced by integrating the two sources of information for the learner, as in providing the problem description as part of the diagram.

Split-attention effect must be eliminated if at all possible, but when learners are learning under conditions of split-attention, then information (1) if ancillary, should be in a different mode from the primary presentation (the dual modality effect), (2) if redundant, should be easily pushed aside by learners (the redundancy effect), and (3) if germane, should be increased to maximize the positive learning effects of a high cognitive load (the variability effect).

Kalyuga, Chandler, and Sweller (2000) extended previous research into dual modality and laid the groundwork for future research forays into a new effect, the expert reversal effect. In two experiments, experienced learners in a domain performed better under opposite conditions than less experienced learners, hence the notion of "expertise reversal." That is, for less experienced learners, instructional guidance and redundancy reduced cognitive load because learners did not have to resort to general problem-solving strategies that tax working memory. By contrast, such guidance and redundancy impaired the performance of experienced learners. These strategies imposed cognitive load because the information was difficult to ignore even though the experienced learners were capable of solving the problem without it. The authors concluded that "differing levels of learner experience should be taken into account when selecting proper user-adapted instructional design" (Kalyuga et al., 2000).

The importance of considering the expertise of learners in the design of instruction can hardly be overrated. In their review of CLT literature, Kalyuga, Ayres, Chandler, and Sweller (2003) found that the expertise reversal effect has been replicated extensively across a range of instructional materials and learners. They argued that instructional designs, if not tailored to the expertise of learners, could be ineffective at best and potentially even counterproductive.

Van Merrienboer and his colleagues have now developed a four-component instructional design model for complex learning (4C/ID) that is rooted in CLT. The 4C/ID model "presupposes that well-designed learning environments for complex learning always consist of four components: (a) learning tasks, (b) supportive information, (c) procedural information, and (d) part-task practice" (van Merrienboer, Kirshener, & Kester, 2003). To promote positive cognitive load conditions, learning tasks should be sequenced in task classes that represent simple to complex versions of the whole task. A high level of embedded support should be provided at the beginning of instruction, with lower levels of support occurring as learners continue. Furthermore, supportive information should provide just-in-time help on nonrecurrent aspects of the task, while procedural information provides just-in-time help on consistent, recurrent aspects (van Merrienboer et al., 2003, p. 11). Finally, part-task practice may be provided when a high degree of automaticity on the learning task is desired.

The proponents of CLT clearly conceive of it as connected to both cognitive psychology and instructional design. Van Merrienboer and his colleagues (2003) working on the 4C/ID model indicated that cognitive load theory alone may not be sufficient to develop a useful instructional design model. They cited the reliance of 4C/ID as well on John Anderson's adaptive control theory (Anderson, 1996) to prescribe instructional methods within each of the four components. Paas, Renkl, and Sweller (2003) summed it up nicely when they concluded that "the existence of the applications generated by CLT validates not only CLT but also many of the constructs of cognitive psychology, such as schema construction and the distinction between working and long-term memory. ... CLT is continuing its role of using cognitive psychology principles to generate novel instructional design procedures."

Mental Models and Instructional Design for Ill-Structured Problems

While CLT certainly diverges from Gagné's theory in its focus, it rests on one of the same foundational assumptions, namely the information-processing view of cognition. However, the contrasting view of constructivism and situated learning in the cognitive revolution is also having an impact on instructional design. Van Merrienboer and Kirschner (2001) position these efforts in the World of Work because of researchers' belief that "instruction can and should be embedded in the real world" (p. 435) where people work together to solve complex problems and develop culturally situated knowledge.

Theories about mental models overlap the World of Learning and the World of Work in that they explore learning processes but they do so by looking at how these models function in complex, situated, social learning environments. Jonassen has written two articles four years apart exploring mental models and troubleshooting. In the first, Jonassen and Henning (1999) contrasted mental models as "knowledge in the head" (p. 37) and "knowledge in the world" (p. 40), and they identified six different kinds of internal knowledge or mental models and four kinds of external or socially based knowledge:

internal	**external**
structural knowledge	activity-based knowledge,
performance/procedural knowledge	social/relational knowledge
reflective/procedural knowledge	conversational/discursive knowledge
image of the system	artifactual knowledge
metaphors	
executive knowledge	

The results of their experiments and ethnographic analysis led them to conclude that (1) people create different mental models and those models have a direct effect on their ability to problem solve and (2) communal mental models exist and are socially negotiated (Jonassen & Henning, 1999).

Four years later, Jonassen (2003) applied what Jonassen and Henning learned about mental models from their 1999 studies to problem solving in the professional world by looking at troubleshooting and novice workers. "The troubleshooter's mental model consists of conceptual, functional, and declarative knowledge" (Jonassen, 2003). Jonassen presented seven intellectual activities that troubleshooting problem-solvers engage in and used those activities to guide his development of an "architecture for designing learning environments that engage novices in troubleshooting while supporting their construction of knowledge" (Jonassen, 2003, p. 38).

Jonassen further contributed to the research in the World of Work—the world of real-life tasks—with his regular investigation of problem solving and constructivist learning environments. In 1997, Jonassen distinguished well- and ill-structured problems, articulated the ways learners solve these problems, and presented models for designing instruction for them. Well-structured problems are familiar to most instructional designers, and instruction is designed around learning domains (like Gagné's) and objectives. Ill-structured problems, however, are those that do not have clear goals and have multiple solutions. Constructivist learning environments (CLE) best compliment ill-structured problem solving because of the inherent discursiveness present in CLEs and in a problem that has a multitude of possible solutions arrived at through negotiation. Jonassen presented guidelines for developing instruction to facilitate learners' negotiation of ill-structured problems, reminding the reader that his guidelines are still works in progress.

In Jonassen and Rohrer-Murphy (1999), activity theory is presented as a framework for helping instructional designers create CLEs. In brief, activity theory "is a powerful socio-cultural and socio-historical lens through which we can analyze most forms of human activity. It focuses on the interaction of human activity and consciousness (the human mind as whole) within its relevant environmental context" (Jonassen & Rohrer-Murphy, 1999). Because CLEs depend on activity to promote learning, the authors draw clear connecting lines between the two, offering the field of instructional design a six-step process for using activity theory to design CLEs.

Jonassen and Hernandez-Serrano (2002) continued looking at problem solving, this time identifying the benefits of storytelling, or narratives, "as a task analysis tool and as an instructional aid in the form of case-based teaching. Problems are solved by applying the lessons learned from previous problem-solving experiences to the solution of new problems." Specifically, the authors identified stories as a way to perform cognitive task analyses and provide four activities for collecting stories from practitioners. They then outlined three ways in which stories can be used to design instruction—as exemplars, as sample problems, and as sources of advice. Teaching learners how to find instruction in narratives is congruent with the goals of instruction focusing on real-world situations.

The articles presented here are but a slice of the current work being conducted at the intersection of instructional design and cognitive psychology. Because of their different ontological approaches, van Merrienboer suggested that the three worlds he identifies need to be bridged, and he discusses the role of mental models as a potential way of connecting the three worlds (van Merrienboer & Kirschner, 2001). Distinguishing the role of cognitive psychology in the field of instructional design will become both easier and more difficult as the two fields continue to engage in a recursive relationship. Theories and research from each field influence the direction of work in both. Instructional designers should not, however, move away from the foundations of the field; rather, they need to remember to return frequently to the world of knowledge, bringing to it new ways to organize learning and design appropriate instruction—ways garnered from research based in cognitive psychology as well as other leading areas of influence. All three worlds need to grow and nurture each other as the relatively new field of instructional design continues to develop, simultaneously defining and extending its boundaries.

CONCLUSION

Royer (Chapter 3, this volume) identified major changes in the overall field of learning and instruction that also serve well to capture the corresponding changes in instructional design over the years. Changes in theory wrought by the cognitive revolution are apparent in all of the work that we reviewed for this chapter. The early behavioral focus on observable, measurable outcomes of learning was an integral part of Gagné's views and remains an important aspect of traditional instructional design models. However, the cognitive view of the learner as an active processor of information broadened the focus of instructional design to account for and facilitate internal processes through instructional manipulation. This view also allowed for a more complex picture of human performance to emerge. Now, with increased attention to the context in which learning is situated, instructional designers have again broadened their focus to include the influence on learning and instruction of social and cultural variables.

This evolution of focus is reflected as well in beliefs about what instructional designers should do. For Gagné, the field was always about the application of knowledge about human learning, and for the most part, that meant a concern for individual learning and performance. As this knowledge about human learning became codified and used systematically to solve practical performance problems, instructional design models such as Dick and Carey's emerged. Most instructional designers recognized, at least implicitly, that not all learning and performance problems required instructional solutions, but it is the design and development of instruction that traditional models were nonetheless intended to guide. Moreover, although the models could be, and most certainly were, used to design instruction that would be delivered to groups, the focus remained on the individual learner and his or her performance.

Today, with increased understanding of sociocultural variables in learning, instructional designers develop not just instruction, but learning environments in which learners are supported by their interaction with others and with "smart" tools. The focus may not be just on the performance of an individual learner but also on the overall performance of the system.

Finally, what changes in research methodology and practices have occurred in instructional design research? Well-controlled, experimental studies typify the early days of the field and characterize most of the research that Gagné conducted. We have already noted Gagné's awareness that his training strongly influenced the kind of research that he did and his recognition that more qualitative approaches such as protocol analysis add an important dimension to quantitative research. Other qualitative approaches are gaining currency in the field that enable researchers to investigate the interactions and effects going on within complex learning environments. In addition, as instructional design has matured as a field, there is a growing sense that research should help to refine and improve the methods and processes that instructional designers use. As such, scholars have called for more developmental research (Driscoll & Dick, 1999; Richey, Klein, & Nelson, 2004). Sometimes called formative research because of its resemblance to formative evaluation, developmental research has as its purpose the iterative development and refinement of instructional design products, techniques, models, or theories. Richey and colleagues (2004) concluded:

> Such research can identify context-specific findings and determine their relevance for other teaching and learning environments. It can also identify new principles of design, development, and evaluation. Developmental research techniques not only expand the empirical methodologies of the field, but also expand the *substance* of instructional technology research. As such, it can be an important vehicle in our field's efforts to enhance the learning and performance of individuals and organizations alike. (p. 1123, original emphasis)

NOTE

1. We use the term "man" here deliberately. In the studies Gagné conducted at the Air Force Laboratories, the subjects who participated and the ultimate object of the research were men.

REFERENCES

Anderson, J. R. (1976). *Language, memory, and thought.* Hillsdale, NJ: Lawrence Erlbaum Associates.

Anderson, J. R. (1980). *Cognitive psychology and its implications.* San Francisco: Freeman.

Anderson, J. R. (1996). ACT: A simple theory of complex cognition. *American Psychologist, 51,* 355–365.

Armbruster, B. B. (1986). Schema theory and the design of content-area textbooks. *Educational Psychologist, 21,* 253–267.

Asher, J. W. (2003). The rise to prominence: Educational psychology 1920–1960. In B. J. Zimmerman & D. H. Schunk (Eds.), *Educational psychology: A century of contributions* (pp. 189–205). Mahweh, NJ: Erlbaum.

Atkinson, R. C., & Shiffrin, R. M. (1968). Human memory: A proposed system and its control processes. In K. W. Spence & J. T. Spence (Eds.), *The psychology of learning and motivation* (Vol. 2). New York: Academic Press.

Brown, J. S., Collins, A., & Duguid, P. (1989). Situated cognition and the culture of learning. *Educational Researcher, 18,* 32–42.

Chase, W. G., & Simon, H. A. (1973). Perception in chess. *Cognitive Psychology, 4,* 55–81.

Dick, W., & Carey, L. (1978). *The systematic design of instruction.*

Dick, W., & Carey, L. (1985). *The systematic design of instruction* (2nd ed.).

Dick, W., & Carey, L. (1990). *The systematic design of instruction* (3rd ed.).

Dick, W., & Carey, L. (1996). *The systematic design of instruction* (4th ed.).

Dick, W., Carey, L., & Carey, J. (2001). *The systematic design of instruction* (5th ed.).

Driscoll, M. P., & Dick, W. (1999). New research paradigms in instructional technology: An inquiry. *Educational Technology Research & Development, 47*(2), 7–18.

Ertmer, P. A., Driscoll, M. P., & Wager, W. W. (2003). The Legacy of Robert Mills Gagné. In B. J. Zimmerman & D. H. Schunk (Eds.), *Educational psychology: A century of contributions* (pp. 303–330). Mahwah, NJ: Erlbaum.

Fields, D. C. (2000). The impact of Gagné's theories on practice. In R. C. Richey (Ed.), *The legacy of Robert M. Gagné* (pp. 183–209). Syracuse, NY: ERIC Clearinghouse on Information & Technology.

Gagné, R. M. (1941). External inhibition and disinhibition in a conditioned operant response. *Journal of Experimental Psychology, 29,* 104–116.

Gagné, R. M. (1954). Training devices and simulators: Some research issues. *American Psychologist, 9,* 95–107.

Gagné, R. M. (1955). Methods of forecasting maintenance job requirements. In A. P. o. P. a. T. Research (Ed.), *Symposium on electronic maintenance* (pp. 47–52). Washington, DC: U.S. Department of Defense, Assistant Secretary, Research and Development.

Gagné, R. M. (1962a). The acquisition of knowledge. *Psychological Review, 69,* 355–365.

Gagné, R. M. (1962b). Human functions in systems. In *Psychological principles in system development.* New York: Holt, Rinehart & Winston.

Gagné, R. M. (1965). *The conditions of learning.* New York: Holt, Rinehart & Winston.

Gagné, R. M. (1968a). Contributions of learning to human development. *Psychological Review, 75,* 177–191.

Gagné, R. M. (1968b). Learning hierarchies. *Educational Psychologist, 6,* 1–9.

Gagné, R. M. (1970). *The conditions of learning* (2nd ed.). New York: Holt, Rinehart & Winston.

Gagné, R. M. (1972). Domains of learning. *Interchange, 3,* 1–8.

Gagné, R. M. (1973). Observations of school learning. *Educational Psychologist, 10*(3), 112–116.

Gagné, R. M. (1974). Educational technology and the learning process. *Educational Researcher, 3,* 3–8.

Gagné, R. M. (1977). *The conditions of learning* (3rd ed.). New York: Holt, Rinehart & Winston.

Gagné, R. M. (1982). Developments in learning psychology. *Educational Technology, 22*(6), 11–15.

Gagné, R. M. (1984). Learning outcomes and their effects: Useful categories of human performance. *American Psychologist, 39,* 377–385.

Gagné, R. M. (Ed.). (1989). *Studies of learning: 50 years of research.* Tallahassee, FL: Learning Systems Institute.

Gagné, R. M., Mayor, J. R., Garstens, H. L., & Paradise, N. E. (1962). Factors in acquiring knowledge of a mathematical task. *Psychological Monographs, 76*(7).

Gagné, R. M., & Medsker, K. L. (1996). *The conditions of learning: Training applications.* Fort Worth, TX: Harcourt Brace College.

Gagné, R. M., & Merrill, M. D. (1990). Integrative goals for instructional design. *Educational Technology Research and Development, 38*(1), 23–30.

Gagné, R. M., & Merrill, M. D. (1990–1991). In conversation. *Educational Technology, 30*(7), *30*(8), *30*(9), *30*(10), *30*(11), *30*(12), *31*(1).

Gordon, J., & Zemke, R. (2000). The attack on ISD. *Training, 37*(4), 42–45.

Gustafson, K. L., & Branch, R. M. (2002). *Survey of instructional development models* (4th ed.). Syracuse, NY: ERIC Clearinghouse on Information & Technology.

Jonassen, D. H. (1997). Instructional design models for well-structured and ill-structured problem-solving learning outcomes. *Educational Technology Research and Development, 45*(1), 65–94.

Jonassen, D. H. (2003). Instructional design for learning to troubleshoot. *Performance Improvement, 42*(4), 34–38.

Jonassen, D. H., & Henning, P. (1999). Mental models: Knowledge in the head and knowledge in the world. *Educational Technology,* pp. 37–42.

Jonassen, D. H., & Hernandez-Serrano, J. (2002). Case-based reasoning and instructional design: Using stories to support problem solving. *Educational Technology Research and Development, 50*(2), 65–77.

Jonassen, D. H., & Land, S. M. (2000). *Theoretical foundations of learning environments.* Mahwah, NJ: Erlbaum.

Jonassen, D. H., & Rohrer-Murphy, L. (1999). Activity theory as a framework for designing constructivist learning environments. *Educational Technology Research and Development, 47*(1), 61–79.

Kalyuga, S., Ayres, P., Chandler, P., & Sweller, J. (2003). The expertise reversal effect. *Educational Psychologist, 38,* 23–31.

Kalyuga, S., Chandler, P., & Sweller, J. (2000). Incorporating learner experience into the design of multimedia instruction. *Journal of Educational Psychology, 92*(1), 126–136.

Klatzky, R. L. (1980). *Human memory.* San Francisco: Freeman.

Merrill, M. D. (1983). *Component display theory.* Hillsdale, NJ: Erlbaum.
Merrill, M. D. (1994). *Instructional design theory.* Englewood Cliffs, NJ: Educational Technology Publications.
Merrill, M. D. (1999). Instructional transaction theory (ITT): Instructional design based on knowledge objects. In C. M. Reigeluth (Ed.), *Instructional-design theories and models: A new paradigm of instructional theory.* Mahwah, NJ: Erlbaum.
Paas, F., Renkl, A., & Sweller, J. (2003). Cognitive load theory and instructional design: Recent developments. *Educational Psychologist, 38,* 1–4.
Pressley, M., & Roehrig, A. D. (2003). Educational psychology in the modern era: 1960 to the present. In B. J. Zimmerman & D. H. Schunk (Eds.), *Educational psychology: A century of contributions* (pp. 333–366). Mahweh, NJ: Erlbaum.
Reiser, R. A. (2001). A history of instructional design and technology, part II: A history of instructional design. *Educational Technology Research and Development, 49*(2), 57–67.
Richey, R. C. (2000a). The future role of Robert M. Gagné in instructional design. In R. C. Richey (Ed.), *The legacy of Robert M. Gagné* (pp. 255–281). Syracuse, NY: ERIC Clearinghouse on Information & Technology.
Richey, R. C. (Ed.). (2000b). *The legacy of Robert M. Gagné.* Syracuse, NY: ERIC Clearinghouse on Information & Technology.
Richey, R. C., Klein, J. D., & Nelson, W. A. (2004). Developmental research: Studies of instructional design and development. In D. H. Jonassen (Ed.), *Handbook of research on educational communications and technology* (2nd ed.). Mahwah, NJ: Erlbaum.
Scandura, J. (2001). Structural learning theory: Current status and new perspectives. *Instructional Science, 29.*
Sweller, J., van Merrienboer, J. J. G., & Paas, F. G. C. (1998). Cognitive architecture and instructional design. *Educational Psychology Review, 10,* 251–296.
van Merrienboer, J. J. G., & Kirschner, P. A. (2001). Three worlds of instructional design: State of the art and future directions. *Instructional Science, 29,* 429–441.
van Merrienboer, J. J. G., Kirshener, P. A., & Kester, L. (2003). Taking the load off a learner's mind: Instructional design for complex learning. *Educational Psychologist, 38,* 5–13.
Woolfolk, A. E. (1995). *Educational psychology* (6th ed.). Boston: Allyn & Bacon.
Zimmerman, B. J., & Schunk, D. H. (Eds.). (2003). *Educational psychology: A century of contributions.* Mahwah, NJ: Erlbaum.

CHAPTER 9

RESEARCH IN INSTRUCTIONAL TECHNOLOGY

Jennifer Wiley
Christopher A. Sanchez
Tom Moher
University of Illinois at Chicago

ABSTRACT

This chapter traces the changes in instructional technology from the early days of instructional practice based in behaviorist theory, to the social cognitive and constructivist approaches that are currently in vogue. The affordances and uses of different technologies for educational purposes, as well as the theoretical movements that have inspired instructional techniques and new technological developments, are discussed.

The last half of the 20th century saw an impressive array of educational technologies introduced to the classroom. In the 1950s and 1960s, filmstrips and movies brought sound and moving pictures into the classroom. The overhead projector, first developed for military training, allowed

The Impact of the Cognitive Revolution on Educational Psychology, pages 231–248

teachers to face the class instead of the blackboard, and also create reusable teaching materials. The photocopier began to replace the mimeograph machine, enabling teachers to distribute copies of material from newspapers, magazines, and books, and also to design their own units and assessments. Cold war schoolchildren watched the broadcast launches of NASA's first Mercury manned space flights on small televisions in lunchrooms and auditoriums. In the 1970s and 1980s, the invention of calculators promised individual access and mobility, and television found its natural partner in the videocassette player, creating a market for educational video content that was less expensive and easier to keep current than film collections. The introduction of personal computers dominated the 1990s. Apple II computers were replaced with Macintosh and DOS/Windows platforms that were distributed in laboratories and classrooms and a new industry arose: educational software titles. At the turn of the millennium, global, local, and wireless networking dominated the educational technology effort as network devices such as laptop and handheld computers gained commercial popularity.

The introduction of each new technology has led to a great deal of optimism that it would lead to substantial improvements in educational outcomes. Those products that were marketed to the general public, such as the television and the personal computer, were objects of great techno-optimism and were promised to forever change the way we learn. However, no grand improvements in student learning were seen simply due to the introduction of these technologies into the classroom. An important lesson to take from the past few decades of research on instructional technology is that technology in and of itself does nothing to improve learning or instruction. Technology does not eliminate the issues of the classroom, but instead offers a new means of approaching them. Surprisingly, despite the fact that computers have been in classrooms for close to 20 years and that children are highly familiar with computers and use them daily from very young ages, there does seem to be a "novelty" effect for technology even now (see Clark, 1983, 1985; Kozma, 1991, for discussion about the motivating effects of computers). Children seem to like to use technologies in the classroom, and computerized instruction can lead to increased levels of interest and engagement among students. However, interest alone does not guarantee positive learning outcomes. Rather, it is a question of when and how technology can be incorporated and leveraged into the educational process to best support learning.

The goal of this chapter is to trace the theories of instructional technology from the early behaviorist approaches to the present, to highlight the ways that the cognitive revolution has affected the way technology is used in instruction, and to reflect on several shifts in instructional theory and practice that may in part be due to the introduction of the new resources.

THE BEHAVIORIST LEGACY

The earliest approaches to the use of technology for educational purposes were rooted in behaviorist theories of learning. Pressey (1926) realized Thorndike's (1912) vision of a mechanical sequencer 50 years before the onset of the personal computer era, designing a typewriter-like device that presented a frame of text and recorded users' keystroke responses. These "teaching machines" were based on the same principle as simple contingency response boxes: the behaviorist premise that in order for instruction to be effective, learning needed to be student-paced, active, and errorless, and feedback on performance needed to be immediate and personalized (Skinner, 1968).

Skinner's (1968) theory of instruction is based on his ideas of operant conditioning (i.e., that human behavior is shaped by contingencies of reinforcement). If we want a behavior to become more probable, it needs to be reinforced. Thus, in programmed instruction, new and complex patterns of behavior are "learned" by conditioning the student or animal to generate the desired behavior or product, either through extensive coaching, explicit directions, or rewarding responses that approximate the correct response. After the target behavior is produced, supports are gradually removed and only wholly correct responses are reinforced (i.e., shaping or fading). An example of an application of this theory to a learning task offered by Skinner is a vocabulary lesson presented on a teaching machine. A student is taught to spell a new vocabulary word incrementally, first by copying the word, then by filling in a few missing letters, and finally by producing the whole word to fill in a blank in a sentence. Important to Skinner is that the student "has probably learned to spell the word without a mistake" (1968, p. 41). It is also important that the student actively generates their response. Skinner argues that composing an answer (putting letters or words in blanks) is superior to selecting an answer from a multiple-choice set of alternatives, as it not only teaches what it "means" to know that the word is right, but also that it reduces the likelihood that incorrect responses could be strengthened through association. The machine then gives them immediate feedback (the student is shown the correct answer and scores him- or herself). This "feedback" acts as reinforcement and strengthens correct responses and weakens the incorrect responses. Skinner argues that these methods are effective because students can pace themselves and always be busy and alert—whereas lectures and textbook reading can lead to boredom.

When one examines these features of the behaviorist approach to instruction, there are striking similarities to many current approaches in instructional technology. Today's CAI (computer-assisted instruction) systems are somewhat more advanced than their programmed predecessors in their ability to accept multiword responses instead of single letters, numbers, or words. They also can automatically score responses, instead of relying on the student to score themselves. New language analysis programs can also allow for some scoring of open-ended responses. However, these systems provide mostly feedback on answers from prestored sets of messages (explaining why answers are right or wrong in a generic manner). Thus, although these programs provide support to the learner, the feedback is not individualized to the student (Aleven, Stahl, Schworm, Fischer, & Wallace, 2003). Intelligent Tutoring Systems, on the other hand, represent a newer generation of supportive learning environments that utilize cognitive models to provide individualized instruction. This approach will be discussed in more detail later. At the turn of the millennium, however, few contemporary commercial software programs seek to tailor instruction using cognitive modeling, and behaviorist-inspired software technologies continue to dominate the landscape in schools.

THE COGNITIVE REVOLUTION

Behaviorist methods can be successful forms of instruction, but it is important to note the kinds of content that are likely to be successfully acquired with such an approach. The examples that Skinner gives for programmed instruction span such simple tasks as learning spelling, arithmetic, geographical facts, foreign language vocabulary, anatomical terms, and teaching a student to recite a poem. The simple response and feedback chains espoused in the behaviorist tradition are likely to be good ways to teach such simple associations like country names and their capitals, or even country names and their main geographical features. However, it is difficult to envision exactly how such an instructional approach could be used to teach students, for example, why particular geographical features are found in different places around the globe. In a number of places, Skinner refers to his approach as producing skilled behavior. A distinction needs to be made, however, between the learning of simple discriminations, associations, or procedural skills (e.g., vocabulary or arithmetic), and the understanding of complex concepts such as plate tectonics or photosynthesis.

Nevertheless, Skinner maintained that complex topics, such as high school physics, could be taught via fill-in-the-blank exercises. For example, consider a lesson on the emission of light. Fill-in-the-blank exercises are intended to enhance the students' ability to learn what technical vocabulary words fit in sentences of a physics text. In additional exercises, students then use the same terms to talk about familiar objects, such as flashlights and candles, that also emit light. Skinner claims that being able to discriminate when a term should be applied in a particular sentence represents an understanding of the topic. In essence, if students are capable of perceiving under what conditions they should generate a certain answer, they have "learned" what that term means, and not just how to use it in a sentence. What many researchers have questioned is whether the learning of isolated facts can effectively support the understanding of complex concepts. If an approach such as the one outlined above were effective, one could ask whether the learning occurs because of the reinforcement of correct vocabulary use, or whether the examples of everyday objects are critical to providing the student with prior knowledge or a mental model upon which to develop their understanding of the topic.

The notions that prior knowledge and mental representations might be necessary to explain learning were the foundations of many researchers' dissatisfaction with behaviorism, and a main premise behind the cognitive revolution. Even animal behavior researchers such as Tolman (1948) were forced to resort to the use of mental maps as an explanatory construct to explain how rats were able to run a maze correctly when placed at a new entry point. While behaviorism stressed errorless learning of associations, one can see the emergence of schema and mental model theories as conceptualizing learning as the process of constructing, assimilating, adapting, or revising a representation (cf. Alba & Hasher, 1983; Anderson, Reynolds, Schallert, & Goetz, 1977). With the cognitive revolution, the emphasis in instruction shifted from simple associations to helping the student integrate new information with prior knowledge, and prompting the student to elaborate on incoming information to build coherent and connected models of the subject matter (Bransford, Brown, & Cocking, 2000; Gardner, 1985).

From Information Transmission to Active Learning

The cognitive revolution has resulted in a view of knowledge acquisition that seeks to replace the *transmission* model, in which the role of education is to transmit a body of facts to students, with a *constructivist* model, in which the goal is for students to gain an understanding of the subject matter through their own exploration and problem solving. Constructivist approaches emphasize "learning by doing" and active manipulation of information in order to achieve problem-solving goals. They also require elaborate responses, often in the form of external models or authentic learning products. The technology in constructivist learning environments supports the construction of mental representations or models in multiple forms, which help students construct their own meaning from instances, data, or evidence. The use of multiple representations and generalization from concrete examples is also meant to strengthen the likelihood that this knowledge will transfer to new situations. Related to this is the movement toward the use of multiple representations of content in instruction (i.e., instead of memorizing a single verbal description or diagram, students are faced with multiple representations that they need to resolve in order to construct their own coherent understanding of subject matter). Constructivist-based technologies take many forms, but four basic implementations are intelligent tutoring systems, simulation or modeling environments, project- or problem-oriented learning environments, and participatory simulations. These approaches differ in the amount of curricular structure and tools that are offered to the student, and are often combined in specific programs. However, these implementations all share a common emphasis on the development of coherent mental representations of subject matter.

Intelligent Tutoring Systems

Current technology offers the possibility of customizing instruction for each student. In most classrooms, there is but one teacher in a room of 20 to 30 students. Yet, despite this disparity, the teacher is still responsible for several elements of instruction that can be critical to positive learning outcomes: matching students to materials and tasks, prompting and scaffolding the learning process, and giving feedback or assessments of learning outcomes. One way that instructional technology has sought to produce improvements in the classroom is by attempting to fill several of these teacher roles through dynamic, interactive systems that can support learning by structuring or responding to students as they work individually or in groups. Attempts have been made to optimize the match of the learner to the content or structure of a lesson, for instance by prior knowledge or ability assessments. (Some of these systems are more online and dynamic than others, using online tests, progress tracking, or LSA [latent semantic analysis] to derive the student's level of understanding before or during a lesson.) Other approaches rely on teacher or student self-reports of ability level, skills, or learning preferences to determine the version of a tutor or program that a student receives.

At the heart of the benefit of computerized tutoring environments is their ability to provide personalized support and feedback to a student based on their performance. The most successful tutoring environments use a cognitive model of the student to provide context and learner-sensitive feedback, as well as to tailor the curriculum to the student's individual learning needs. Using information gleaned from student responses, including errors and correctly solved problems, the intelligent tutoring system (ITS) can track student understanding in a cognitive model based on an analysis of the skills and procedures needed to perform well in the domain. Because of the reliance on cognitive models of student understanding, ITS learning environments have also been called "cognitive tutors" and have been shown to support learning better than simpler CAI systems (Anderson, Boyle, Corbett, & Lewis, 1990).

The Algebra Cognitive Tutor (ACT) is one example of a PC-based ITS (Koedinger, Anderson, Hadley, & Mark, 1997). Students work through an organized curriculum of problems by reading a text description of the situation and are posed a number of questions about it. The problems students are asked to investigate are real-life situations that require algebraic solutions like figuring out how to make money in a snow shoveling business. They investigate the situation by representing it graphically, in tabular form and in symbolic equations. They use these multiple representations to understand the problem and answer the questions.

Students use the tutor individually, and critical to its design is a cognitive model of student performance that it uses to track students' progress. The tutor is mostly silent, but when help is needed, the tutor can give hints appropriate for the student. Errors are flagged for the student to correct. For common errors, a message is provided explaining what is wrong with an answer. Immediate feedback is based on a particular student's understanding of each problem, and new problems are given based on analysis of weaknesses in the student's set of skills. Additionally, teachers are also supported in their use of this technology. A workbook has been created that accompanies the tutor to give teachers useful ideas for implementing the technology in the classroom. As a package, the Algebra Cognitive Tutor consistently yields learning gains over and above comparison classes who are taught algebra using a traditional curriculum (Koedinger et al., 1997). These gains are seen both on standardized assessments as well as on tests that directly address everyday problem-solving and representation skills.

Simulation Environments

Simulation learning environments allow students to explore multiple representations of content, so that information can be compared, contrasted, or integrated across different sources, configurations, or modalities. Technology is especially useful for visual displays and representations. Computers can easily offer access to animations and visualizations not available through 'traditional' classroom materials (e.g., textbooks), and in the latest VR (virtual reality) systems, learners can actually be immersed in the environment about which they are learning.

Even on its own, the ability to view visual representations seems to offer many affordances for learning (Wiley, 2003). Visualizations can be an effective means of analyzing complex streams of information, like weather patterns or stock market data. Similarly, visual displays may be quite effective for the comprehension of complex parallel systems like traffic patterns. Visual displays are also essential to virtual training environments, which seem to be most effective for content domains that require inherently perceptual or motor skills, like piloting or surgery. This potential for "virtual firsthand experience" through visualizations can not only enhance the saliency of relationships between materials, but can also allow for a larger knowledge base (i.e., verbal and spatial) to be applied to the learning process.

The ability to portray multiple representations can support the mapping of relations between knowledge units, for instance between concrete instances of observed data (like volcanic eruptions or selling prices of houses) and their representations in a graph or other symbolic data display. The available comparison of the two makes both the referents, and important relations between the elements in the system, more apparent, whereas this might not be the case if it were just summarized in a textbook chapter. Similarly, students could be presented with concrete and abstract versions of representations of the circulatory system, for example, which might facilitate the mapping process. Visual representations could also be "layered" in such a way that students can be prompted to see the correspondence between multiple representations at different grain sizes (i.e., the functioning of the circulatory system vs. the role of white blood cells in the blood stream vs. the cell biology of white blood cells). Finally, the classic pairing of text description and image can allow for readers to "see" a system that they are attempting to understand, while the text can make the dynamic relations between the elements of the system explicit.

Computer simulations based on the input of the user also allow students to see what variables may affect phenomena, view different perspectives, and test their own understanding of systems. Here, as in ITS learning environments, feedback is available directly to the student, as they can see what "works" and produces a desired outcome in their simulations. Simulation environments have been especially important for learning in the sciences where visualization is often a key to understanding (White, 1993) and can help students to make important distinctions in concepts such as heat and temperature (Linn, Songer, & Eylon, 1996). Some simulation environments employ an organized curriculum or problem set, while others are used more openly as exploration spaces.

Thinkertools is one example of a PC-based simulation environment (White & Schwartz, 1999), designed to enhance students' understanding of physics, and whose effectiveness has also been empirically demonstrated by improvements on year-end assessments. ThinkerTools approaches physics instruction from a fundamentally different perspective than "traditional" physics instruction. Rather than focusing on traditional, top-down quantitative instruction of formulas and applications, ThinkerTools approaches physics teaching with a focus on "constructing causal models" of understanding through the use of intermediary causal models (computer simulations) or microworlds (White, 1993, 1995; White & Schwartz, 1999).

In ThinkerTools, the units are not solely computer modeling exercises. In addition to instruction on how to run the simulations in the computer modeling program, each unit involves hands-on experimentation, guided inquiry, and hypothesis testing, as well as debate and interaction among individuals regarding the designated topic. To address individual differences in performance and prior knowledge, the teacher can customize the content and parameters of the computer experiments. Both students and teachers are provided with a resource book that provides a detailed road map through each unit.

The design of the ThinkerTools interface may also be important for the positive learning outcomes that have been found. Graphs are simple, as is the animation in the simulated experiments. There appear to be few irrelevant, extraneous "bells and whistles." Students are also involved in inputting data, and constructing the graphs and notes from their experiments, so the correspondence between the simulations and graphs is reinforced.

ThinkerTools represents the fusion of the traditional classroom with technology. Students conduct simulation experiments on the computer, the computer allows them to save their work, and further provides them the chance to test multiple hypotheses easily and quickly. Students then must apply their gained knowledge to subsequent "real-world" demonstrations. This integration of real-world experience and computer-aided instruction ensures that the information students are exposed to is connected to their existing knowledge, is supported through interaction with peers, fits the global goal of promoting scientific inquiry, and ultimately leads to an increase in students' understanding of physics (White & Schwartz, 1999).

Project-Based Learning Environments

Project-based learning environments generally employ more complex, realistic, and less-directive discovery spaces, inspired by situated learning approaches. One example of such an approach is the Jasper Woodbury Problem Solving Series (Cognitive and Technology Group at Vanderbilt, 1997) in which students are presented with 12 interactive video environments, and are challenged to solve problems that require the application of mathematical concepts (like designing a playground). Students work together to build models, compute measurements, and design solutions. These programs have demonstrated gains in student learning of mathematical concepts, as well as changes in communication skills and attitudes toward math. However, these experiences have also highlighted some of the attendant complexities of the introduction of such activities in the classroom. Perhaps most importantly, the lack of explicit structure and sequence in the learning activity forces a heavy reliance on the metacognitive ability and background knowledge of students. In recent years, researchers have begun to focus on the importance of embedding scaffolding or help systems as an integral component of learning applications (e.g., Aleven et al., 2003; Reiser et al., 2001).

Collaborative Simulations with Embedded Technology

The fourth approach within recent uses of technology may be best thought of as a hybrid between collaborative problem solving and simulation environments. One example of this new approach are participatory simulations (Resnick & Wilensky, 1998). Participatory simulations engage groups of learners in the enactment of a dynamic model, with students controlling model parameters, often by interacting with common physical devices embedded with computational capabilities. The critical design feature in this approach is that students take part in collaborative activities, the result of which is an emergent representation of the phenomena they are trying to understand. A more specific example of this approach is Geney, a collaborative problem-solving application to help children explore genetic concepts using multiple wireless PDAs and a central personal computer (Danesh, Inkpen, Lau, Shu, & Booth, 2001). Here, the application is used to implement a game played by the entire class, in order to teach basic genetic concepts of dominant and recessive genes. Each PDA is shared by a pair of students, but each member has their own stylus for input. In this game, each PDA represents a pond of fish. Within each pond, fish age and die. What students can do is exchange fish with other pairs through their infrared ports, and then mate their fish and observe the characteristics of the offspring. (A limited set of traits is used to reduce complexity.) The goal of the game is for the students to produce a fish with a particular set of characteristics. To do so, the class must work cooperatively, gathering genetic material from multiple ponds. Each pond only represents a portion of the gene pool. However, students can upload their information to a central PC to see the complete family tree of fish. The collaborative activity seems to be very engaging and exciting. Even quieter children who were usually less inclined to work with others were brought into discussions. This use of technology seems promising, but research on it is still in its early stages, and learning outcomes have yet to be examined.

In a pilot experiment of our own using an analogous kind of activity (Moher et al., 2003), we have investigated how collaboration with handhelds can support the acquisition of scientific reasoning strategies. Similar to the Geney application, students must engage in cooperative problem solving in order to turn a field of circles on a large plasma screen all one color. Each student has a wireless PDA that controls a single dot. In other words, the students act as the variables in this experiment. Successful attainment of the goal requires that the class, as a whole, recognizes that in order to achieve a desired pattern only one person can click their PDA at a time. **In other words, students must adopt a strategy of manipulating only one variable at a time (a control of variables strategy) in order to successfully** complete the exercise. We found that after a class of third-grade students engaged in this activity, the majority of them were able to generalize the control of variables strategy to a new transfer problem (determining which dog food makes dogs sick). Performance of individuals on the transfer problem showed a significant advance versus a similar pretest question, as well as in relation to normative levels of understanding of this concept reported in the literature for this age group. These data suggest that collaborative problem solving or simulation activities are a promising use of handheld and other mobile computing technologies, and may be an important future direction for educational technology research.

Although the above examples utilize technology in very different ways, these approaches also share a number of common elements, as do many examples in this generation of instructional technology. A number of learning environments or contexts that have produced positive learning gains use inquiry/problem-solving tasks to help focus students on important concepts during their exploration. Providing students with a specific inquiry task to guide their activity, like "What caused the eruption of Mt. St. Helens?" or "Why are all the California tree frogs dying?" seems to be important for keeping students on task and working constructively (Wiley & Voss, 1999). Such problem-solving tasks can also help to "anchor" cognition in a realistic, meaningful learning activity.

Many current examples of instructional technology also involve learning by active manipulation of data. This often involves the use of multiple representations and viewing data from different perspectives, or in different forms. All these examples also have a method of feedback to the student. The feedback in the ACT is most tailored, but the feedback that students receive from seeing their simulations in ThinkerTools, and the results of their collaborations on a larger screen in the activities using handhelds, also seem important to the learning process. Many examples of current instructional technology also involve some sort of hands-on manipulation or activity interweaved with the technology. In most cases, the benefits of both the hands-on or exploratory activities, and the technology-supported activities, will be most beneficial when they are reflected upon or integrated conceptually through discussions. And, it is important to note that both ACT and Thinker Tools provided support to teachers so they could accomplish this integration of technology into the classroom discourse. This leads to another set of common elements that cross many current learning contexts: the use of collaboration, discussion, and/or debate in the learning process. The social processes of learning, and the role that technology can play in supporting collaborative learning, is receiving more and more attention in educational research, and is an important part of many learning environments.

TRENDS IN INSTRUCTIONAL TECHNOLOGY

The above examples illustrate several trends that seem to have emerged in educational research in the past decade. The first trend we note is from individual to group learning contexts. The second is from an emphasis on cognitive activity to a recognition of the role of metacognitive activity in the learning process. The third trend we note is a movement from creating learning activities to fit around technology, to a movement toward embedding technology into meaningful learning activities.

From Individual to Collaborative Learning Contexts

As a number of other authors in this volume have remarked, the role of social processes in learning has received increasing attention in the past few years. Emphasis within the cognitive psychology tradition has been overwhelmingly focused on individuals acting and learning alone. This can be an excellent approach to effective instruction, especially in the paragon case of the Algebra Cognitive Tutor, which gains much of its power from the detailed analysis of the cognitive skills needed for an individual to understand the domain. If this method could be followed for all content areas it would clearly lead to powerful learning gains. However, since all subject matters are not as easily reduced to a set of production rules, we must investigate other effective forms of instruction. One promising area is the examination of learning that occurs through social interaction.

While the original reasons for having small groups of children share a computer may have been pragmatic (due to scarce resources), there are reasons to believe that collaboration may be important for positive learning outcomes (Webb & Palincsar, 1996). Collaboration is essential for real scientific inquiry, which happens in laboratories and among communities of researchers on a regular basis in the real world. Although there have been numerous findings that groups do not outperform individuals in the social psychology literature, a few studies have found learning benefits versus individual controls (Barron, 2000, 2003; O'Donnell & Dansereau, 1993; O'Donnell, Dansereau, Hall, & Skaggs, 1990). The recent work of Anderson and his colleagues has also demonstrated that peer interaction may be a critical mechanism for the development of reasoning skills among schoolchildren (Anderson et al., 2001).

Many new technologies have become available to support collaboration, which may be responsible for this growing trend and interest in collaboration as an effective means of instruction. Although collaborative learning seems to be an important new direction for research, any new technology needs to be investigated, specifically in terms of how it can support collaborative interaction and when such interaction actually results in better learning, as results are still decidedly mixed on this topic.

From Cognitive to Metacognitive

Intuitively, many educational researchers believe that collaboration can benefit learning because students can provide each other with prompts for reflection and explanation, contribute multiple sources of knowledge, and explicitly generate more planning and monitoring statements of goals. All of these are elements of successful metacognitive activity, and collaboration may be especially important for supporting metacognitive skills and awareness. In addition, the role of supports and prompts for metacognition as students are engaged in learning activities is an important direction for future research in interactive learning environments.

Technology has enabled new possibilities in terms of placing students in rich multimedia problem spaces or virtual environments that allows for learning questions to be posed in such a way that they "feel" more authentic and situated to the learner. Also, the lack of explicit curricular structure and sequence in some learning environments may allow for the excitement of "discovery" or "exploration." This freedom to explore and discover, however, is a double-edged sword. While enthusiasm and interest may be enhanced by the discovery process, discovery may only work with students who have adequate metacognitive ability and background knowledge. Appropriate scaffolds are necessary for students who might be deficient in these areas, and are realizable through technology. In recent years, researchers have begun to focus on the importance of individualizing scaffolding or help systems as an integral component of learning applications (e.g., Aleven et al., 2003; Reiser et al., 2001).

From Technology-in-a-Box to Immersive Technology

The vast majority of new technologies were not developed specifically for education. Most of our current educational technologies have been developed by asking the question "How can we use this tool for education?", and then adapting existing resources to the educational task at hand. The tool exists first and instruction is fitted around it. This leads to many awkward, less than ideal situations. As a prime example, desktop computing is a problematic fit for school classrooms, especially elementary school classrooms where children would rather be moving around the room. Even among older children, there is often a mismatch between the number of available computers and the number of students. When a mismatch occurs between resources and students, the teacher is forced to establish a scheduling regimen around the scarce resource (further burdening teachers, rather than assisting them). For example, in a classroom with one microscope, if each of 28 kids gets 15 minutes at the scope to observe and record their observations, the teacher must plan a full day's worth of "parallel" activities—activities that won't be impacted by the comings and goings of kids to and from the microscope—for the rest of the class to do while waiting for their turn for the microscope. In a classroom with one computer, certain activities, such as the use of the Internet as a resource for open-ended research questions, becomes almost impossible to coordinate. At the elementary school level in particular, the physical needs and abilities of younger children are not well suited to extended interaction with desktop computers. Over the past several years, motivated by the active and collaborative learning exigencies outlined above, researchers have begun to explore alternative means for students to interact with computers and each other in meaningful learning activities.

One approach to dealing with these problems has been, essentially, the miniaturization of the computer. Palm- and tablet-size computers provide a less expensive and considerably more portable platform that can be utilized at the physical point of phenomena. Highly portable computing devices are being used to support students' collection of environmental probe data in the real world (e.g., the BioKids project; Parr, Jones, & Songer, 2002) as well as in simulated settings in the classroom (Moher, Johnson, Cho, & Lin, 2000). Handhelds are also being used for activities such as personal reflection and planning (Soloway et al., 1999).

A complementary approach has been the enlargement of the display. As screen technologies improve and prices drop, larger tubes, panels, and projectors, capable of supporting small-group as well as whole-class activities, have begun to fall within the price points of schools. A new class of applications called single display groupware has begun to emerge (Druin,

2002; Inkpen, 2000), combining multiple private-user affordances with a large common display for collaboration.

New computing devices and embedded technologies are being developed and implemented in participatory simulations. An interesting example is Thinking Tags (Colella, Borovoy, & Resnick, 1998), small necklaces with digital and LED displays that sense the presence of proximate tags and exchange information, which have been used by groups of students for simulating disease vectors and inheritance of genes. Wilensky and Stroup's HubNet (1999) architecture allows students to network their local devices (computers or calculators), and project the simulation onto a large common display. This provides students with a view of the aggregate state of the simulation, which may be critical for reflection and understanding. Finally, new technologies have been developed that allow for the creation of virtual worlds. A completely different sense of "immersion" arises from the use of technologies intended to situate users within virtual reality environments—in the depths of Washington's Puget Sound (Windschitl & Winn, 2000) or riding along on an electron (Dede, Salzman, Loftin, & Ash, 2000).

With all these changes and developments in technologies and their affordances, at present, we are still largely engaged in the arduous process of adapting business technologies to schools. The one-person, one-keyboard, one-screen physical configuration of computer systems is well suited to the deskbound, continuous-access world of the office worker, but does not translate easily to schools where there is a great deal more physical movement, and where the availability of computers is brief, infrequent, and synchronized to a daily schedule. We need to ensure that sufficient, appropriate resources are accessible to teachers that will allow them to use available technologies to an optimal level. At the same time, we need to explore emerging technologies that can more economically overcome the "availability gap" and support effective learning. New handheld and embedded-chip technologies that enable mobile and active learning seem quite promising in this regard.

THE FINAL WORD: REASONS FOR PESSIMISM AND OPTIMISM

Although there might still exist some important questions regarding the educational effectiveness of the latest technological learning environments, there are several reasons for continued optimism about educational uses of technology. Technology continues to provide new and motivating learning contexts for students. Technology is becoming more available and students are increasingly comfortable with technology. Finally, perhaps the most promising reason on the horizon, is that new technologies are emerging that will allow for teachers to bring technology into the classroom without having to mold course planning around the computer. This being said, empirical demonstrations of how and when educational technologies actually result in positive learning outcomes are sorely needed. Technology, when based in cognitive models of student learning, and used appropriately, can raise the level of performance. However, it is clear that simply introducing technological gadgets into the classroom will not improve learning. The lesson as we moved from behaviorist to cognitive principles of instruction was that it was important to consider and support the learner's mental model of the subject matter. Current educational technologies that focus students' attention on important concepts—either through structured instruction, prompts and feedback, through simulations, or through authentic problem solving—have produced positive learning gains. Recent trends suggest that the next movement in educational technology is considering the learning activity in terms of its social context, supporting activity and interaction among students, and perhaps even discovering new kinds of learning activities.

REFERENCES

Alba, J.W. & Hasher, L. (1983). Is memory schematic? *Psychological Bulletin, 93,* 203–231.

Aleven, V., Stahl, E., Schworm, S., Fischer, F., & Wallace, R. (2003). Help seeking and help design in interactive learning environments. *Review of Educational Research, 73,* 277–320.

Anderson, J. R., Boyle, C. F., Corbett, A. T., & Lewis, M. W. (1990). Cognitive modeling and intelligent tutoring. *Artificial Intelligence, 42,* 7–49.

Anderson, R. C., Reynolds, R. E., Schallert, D. L., & Goetz, E. T. (1977). Frameworks for comprehending discourse. *American Educational Research Journal, 14,* 367–381.

Anderson, R. C., Nguyen-Jahiel, K., McNurlen, B., Archodidou, A., Kim, S., Reznitskaya, A.,Tillmanns, M., & Gilbert, L. (2001). The snowball phenomenon: Spread of ways of talking and ways of thinking across groups of children. *Cognition and Instruction, 19*(1), 1–46.

Barron, B. (2000). Achieving coordination in collaborative problem-solving groups. *Journal of the Learning Sciences, 9,* 403–436.

Barron, B. (2003). When smart groups fail. *Journal of the Learning Sciences, 12*(3), 307–359.

Bransford, J. D., Brown, A., & Cocking, R. (Eds.). (2000). *How people learn: Mind, brain, experience and school* (Expanded ed.). Washington, DC: National Academy Press.

Clark, R. E. (1983). Reconsidering research on learning from media. *Review of Educational Research, 53,* 445–460.

Clark, R. E. (1985). Confounding in educational computing research. *Journal of Educational Computing Research, 1*, 137–148.

Cognition and Technology Group at Vanderbilt. (1997). *The Jasper project: Lessons in curriculum, instruction, assessment, and professional development.* Mahwah, NJ: Erlbaum.

Colella, V., Borovoy, R., & Resnick, M. (1998). Participatory simulations: Using computational objects to learn about dynamic systems. *Proceedings of the CHI '98 Conference*, Los Angeles.

Danesh, A., Inkpen, K., Lau, F., Shu, K., & Booth, K. (2001). Geney™: Designing a collaborative activity for the Palm™ handheld computer. *CHI 2001, 3*(1), 388–395.

Dede, C., Salzman, M., Loftin, R. B., & Ash, K. (2000). The design of immersive virtual learning environments: Fostering deep understandings of complex scientific knowledge. In M. J. Jacobsen and R. B. Kozma (Eds.), *Innovations in science and mathematics education: Advanced designs for technologies of learning.* Mahwah, NJ: Erlbaum.

Druin, A. (2002). The role of children in the design of new technology. *Behaviour and Information Technology, 21*(1), 1–253.

Gardner, H. (1985). *The mind's new science: A history of the cognitive revolution.* New York: Basic Books.

Inkpen, K. M. (2000). Designing handheld technologies for kids. *Personal Technologies Journal, 3*(1&2), 81–89.

Koedinger, K., Anderson, J., Hadley, W., & Mark, M., (1997). Intelligent tutoring goes to school in the big city. *International Journal of Artificial Intelligence in Education, 8*, 30–43.

Kozma, R. B. (1991). Learning with media. *Review of Educational Research, 61*(2), 179–211.

Linn, M., Songer, N. B., & Eylon, B. S. (1996). Shifts and convergences in science education and instruction. In R. C. Calfee & D. C. Berliner (Eds.), *Handbook of educational psychology.* Riverside, NJ: Macmillan.

Moher, T., Ding, X., Wiley, J., Conmy, D., Hussain, S., Singh, P., & Srinivasan, V. (2003). Combining handhelds with a whole-class display to support the learning of scientific control. *ACM Conference on Human Factors in Computing Systems Extended Abstracts*, pp. 882–883.

Moher, T., Johnson, A., Cho, Y., & Lin, Y. (2000). Observation-based inquiry in a virtual ambient environment. *Proceedings of the Fourth International Conference of the Learning Sciences*, 238–245.

O'Donnell, A., & Dansereau, D. F. (1993). Learning from lectures: Effects of cooperative review. *Journal of Experimental Education, 61*(2), 116–125.

O'Donnell, A. M., Dansereau, D. F., Hall, R. H., & Skaggs, L. P. (1990). Learning concrete procedures: Effects of processing strategies and cooperative learning. *Journal of Educational Psychology, 82*(1), 171–177.

Parr, C. S., Jones, T., & Songer, N. B. (2002). CyberTracker in BioKIDS: Customization of a PDA-based scientific data collection application for inquiry learning. In P. Bell, R. Stevens, & T. Satwicz (Eds.), *Keeping learning complex: The Proceedings of the Fifth International Conference of Learning Sciences*, pp. 474–475.

Pressey, S. L. (1926). A simple apparatus which gives tests and scores - and teaches. *School and Society, 23*(586), 373–376.

Reiser, B. J., Tabak, I., Sandoval, W. A., Smith, B., Steinmuller, F., & Leone, T. J. (2001). BGuILE: Stategic and conceptual scaffolds for scientific inquiry in biology classrooms. In S.M. Carver & D. Klahr (Eds.), *Cognition and Instruction: Twenty five years of progress.* Mahwah, NJ: Erlbaum.

Resnick, M., & Wilensky, U. (1998). Diving into complexity: developing probabilistic decentralized thinking through role-playing activities. *Journal of Learning Sciences, 7*(2), 153–172.

Skinner, B. F. (1968). *The technology of teaching.* New York: Appleton-Century-Crofts.

Soloway, E., Grant, W., Tinker, R., Roschelle, J., Mills, M., Resnick, M., Berg, R., & Eisenberg, M. (1999). Science in the palms of their hands. *Communications of the ACM, 42*(8), 21–26.

Thorndike, E. L. (1912). *Education: A first book.* New York: MacMillan.

Tolman, E. C. (1948). Cognitive maps in rats and men. *Psychological Review, 55,* 189–208.

Webb, N. M., & Palincsar, A. S. (1996). Group processes in the classroom. In D. C. Berliner & R. C. Calfee (Eds.), *Handbook of educational psychology.* New York: Macmillan.

White, B. (1993). Causal models and intermediate abstractions: A missing link for successful science education? In R. Glaser (Ed.), *Advances in instructional psychology* (Vol. 4). Mahwah, NJ: Erlbaum.

White, B. (1995). The ThinkerTools project: Computer microworlds as conceptual tools for facilitating scientific inquiry. In S. Glynn & R. Duit (Eds.), *Learning science in the schools: Research reforming practice.* Mahwah, NJ: Erlbaum.

White, B. Y., & Schwartz, C. V. (1999). Alternative approaches to using modeling and simulation tools for teaching science. In W. Feurzeig & N. Roberts (Eds.), *Modeling and simulation in science and mathematics education.* New York: Springer-Verlag.

Wilensky, U., & Stroup, W. (1999). Learning through participatory simulations: Network-based design for systems learning in classrooms. *Proceedings of the Computer Support for Collaborative Learning.*

Wiley, J. (2003). Cognitive and educational implications of visually-rich media: Images and imagination. In M. Hocks & M. Kendrick (Eds.) *Eloquent images: Word and image in the age of new media.* Cambridge, MA: MIT Press.

Wiley, J., & Voss, J. F. (1999). Constructing arguments from multiple sources: Tasks that promote understanding and not just memory for text. *Journal of Educational Psychology, 91,* 301–311.

Windschitl, M., & Winn, W. (2000). A Virtual environment designed to help students understand science. In B. Fishman & S. O'Connor-Divelbiss (Eds.), *Proceedings of the Fourth International Conference of the Learning Sciences.* Mahwah, NJ: Erlbaum.

CHAPTER 10

TRANSFER AND PROBLEM SOLVING

A Psychological Integration of Models, Metaphors, and Methods

Gary D. Phye
Iowa State University

ABSTRACT

The following three themes are the storyline around which this chapter is developed. The first theme is the identification of a philosophy of science that guides the study of adaptive human behavior where human learning is a primary agent of change, and transfer and problem solving are but special cases. The second theme is a selective historical review of transfer and problem-solving research in the United States that identifies: (1) major theoretical models and metaphors and (2) the methods and materials used to test respective theories. This chronology includes (1) initial efforts (circa 1900–1920), (2) verbal learning research (circa 1920–1970), and (3) the cognitive revolution (circa 1970–present). The third theme is a subtheme that involves identifying similarities and differences in the study of transfer and problem solving between 1900 and 1970. This subtheme is carried over into

The Impact of the Cognitive Revolution on Educational Psychology, pages 249–292

the early decades of the third era (cognitive revolution). By the decade of the 1990s, we see attempts at integrating the study of transfer and problem solving with the introduction of problem-solving transfer. This integrative effort relies on a reintegration of (1) a philosophy of science of human behavior, (2) pragmatic conceptual models and metaphors of transfer and problem solving, coupled with (3) a reconsideration of materials and experimental research methods.

INTRODUCTION

Transfer and problem solving have long played a central role in experimental and educational psychology. This interest can be traced to research conducted at the beginning of the 20th century by such psychologists as Thorndike (training for transfer and problem solving/methodology), Woodworth (training for transfer/methodology), Angell (training for transfer/methodology), Coover (training for transfer/methodology), and Judd (problem-solving transfer/methodology). This research was conducted at a time in the United States when educators were seeking to answer questions about learning in the classroom. In many respects, this interest in the application of psychological principles to educational practice was the legacy of William James at Harvard. James, trained as a philosopher, and commonly viewed as the father of American psychology. At the end of the 19th century, James (1899/1958) published a small volume entitled *Talking to Teachers* that provided the impetus for empirical research efforts designed to test the philosophical ideas that provided the basis for 19th-century pedagogical practices.

A theme that extends through this chapter is one based on a philosophy of science approach identified as pragmatism. Pragmatism as a philosophy of science approach to the investigation and explanation of human behavior is best exemplified by the approach commonly called experimental psychology. In the United States, many of the first generation of educational psychologists were experimental psychologists. While experimental-educational psychology in the United States may have a short life span that extends back approximately 100 years, it has a much longer philosophical history (Boring, 1957). Furthermore, while philosophy has traditionally addressed issues involving the "nature of the human condition," psychology evolved as a discipline that sought to address issues of "human behavior." This distinction between human nature and human behavior should not be ignored, and the ramifications of the distinction will be explored both in terms of educational-psychological theory and educational-psychological methodology.

Psychology's focus on behavior is highlighted by the fact that psychology sought causal explanations of behavior (how do children learn, does this mode of instruction influence problem solving, etc.). Arguably, it is this seeking of causal explanations for human behavior that provides the functionalist tradition that characterizes psychology today. "Despite the fact that early German psychologists took physics as their model rather than biology, a kind of psychology known as functionalism soon flourished in this country. The adaptability of behavior was emphasized; both comparative and applied psychology were established; and especially, from Thorndike on, the topic of learning was a focus of theoretical development" (Gibson, 1994). Historically, much of the learning research at the turn of the 20th century was being conducted in educational settings and involved methodological issues related to transfer and the need for the use of an independent control (Coover & Angell, 1907; Thorndike & Woodworth, 1901a, 1901b). These early research efforts to empirically address pedagogical questions helped establish the experimental research tradition employed today by both experimental psychologists and educational psychologists. A case in point is an interesting account of the initial development of control groups in psychology recently published by Dehue (2000). Ironically, the hue and cry about the current emphasis on the use of comparison groups in the development of credible educational research data has lead some educators to believe that this is a new and unreasonable practice.

ESTABLISHING A PRAGMATIC PHILOSOPHY OF SCIENCE FOR EDUCATIONAL PSYCHOLOGY

Functionalism

As experimental psychology at the beginning of the 20th century was moving away from a philosophical level of analysis and establishing itself as a separate discipline, two distinctive attributes of psychology can be noted. The pragmatism of William James (1890) set the functional tone for psychology that has been maintained into the 21st century. Generally speaking, "functionalism is the view that mental processes and behaviors of living organisms help them adapt to their environments" (Heidbreder, 1933, cited in Schunk, 2000, p. 20). This school of thought also flourished at the University of Chicago with the works of John Dewey, James Angell, and Harvey Carr. It should not be overlooked that the eminent sociologist George Herbert Mead was also at Chicago during this time, as was John B. Watson (the initial promoter of behaviorism in the United States). Psychological functionalism's emphasis on the adaptability of human behavior is not an isolated event. Rather, the perspective is global in the sense that adaptability must be viewed as an integration of both macro (societal) and micro (individual) influences and issues. In many respects, functionalism emerged as a reaction to an elements approach to understanding higher cognitive processing identified as structuralism. Structuralism in psychology at the end of the 19th century was an outgrowth of the influence of Wilhelm Wundt's experimental approach to the study of mental functioning (Boring, 1957). As has been pointed out by Bredo (1997), the historical roots of today's mainstream social constructivism can be traced back to the Chicago School and their approach to psychological functionalism. In terms of theory development and methodology, functionalism's emphasis on integrating both individual and societal influences necessitated the development of assumptions about how to describe the integrative effort across varying levels of analysis. This led to the insistence that integrative efforts focus on the *correspondence* between or among levels of behavior being analyzed.

The use of "levels of analysis" as an epistemological tool by psychologists is frequently employed in cognitive science and has been well documented by Stanovich (2001). When using levels of analysis as an epistemological tool, it is critical to insist on the assumption that there need not be a "cause–effect" relationship inferred between or among levels of analysis (Gibson, 1994). Rather, the assumption of a corresponding relationship is assumed. Thus, qualitative differences (differences in kind) as well as quantitative differences (differences in degree) may exist between or among corresponding levels of analysis. This theoretical assumption is basic to functionalism and keeps theory building from taking the simplistic route of assuming cause–effect relationships between levels of analysis that leads to the charges of reductionism and a mechanical model of human behavior as espoused by radical behaviorism of the 1960s.

At the turn of the 20th century, transfer of training and problem solving were receiving attention from experimental psychologists that studied human learning. The zeitgeist was one of addressing research questions that related to the adaptability of the human organism. Broad questions about the etiology of behavior (heredity or environmental), the nature of human learning (reflexive or acquired), and how these questions should be addressed (level of analysis) dominated much of experimental psychology. More relevant to education, early experimental-educational psychologists like E. L Thorndike, C. H. Judd, and John Dewey (while still at the University of Chicago) studied school-age children in academic environments.

While Thorndike, Judd, and Dewey studied learning in educational environments, the level of analysis they employed differed greatly. In fact, upon leaving the University of Chicago for Columbia University during the first decade of the 20th century, John Dewey gravitated more to the philosophical level of analysis than the psychological when dealing with educational issues. However, what makes Dewey's educational philosophy so appealing was the influence of his early work in psychology. A seminal paper, *The Reflex Arc Concept in Psychology* (Dewey, 1896), set the tone for experimental psychology's pragmatic approach to the study of human behavior. This paper addresses the issue of "how" to study human behavior without having to resort to a dualistic position on the "mind–body" issue. Dewey's position on this issue was consistent with that taken by Charles Sanders Peirce in reaction to the Kantian dualistic position on the mind–body issue.

> This experimentalist theory of the meaning of concepts asserted an intimate connection between knowledge and action. Peirce argued that in order to attribute meaning to concepts one must be able to apply them to existence. Kant had referred to the situation in which knowledge and action are strictly separated as "practical" and the situation in which knowledge and action are intimately connected as "pragmatic." It was for this reason that Peirce (initially) decided to name his philosophy pragmatism—and not practicism or practicalism. (Biesta & Burbules, 2003, p. 6)

Thus, Dewey's pragmatic approach to the experimental study of human behavior set the course for a functionalist approach in psychological research that continued through the verbal learning era to this day. "The basic idea of psychological functionalism is that it does not conceive of psychological phenomena such as *perception, consciousness,* or *mind* as entities on their own, but accounts for these phenomena by making it clear how they emerge from more general processes to perform a specific *function*" (Biesta & Burbules, 2003, p. 31). This is a central tenet of Dewey's theory of action that also characterizes what Dewey meant by learning.

It is the pragmatist's approach to dealing with the mind–body issue that seems to have been lost on many educational researchers today. This is currently reflected in endless debates among educational philosophers and some educational researchers about the nature of knowledge (reality) and what constitutes a "curriculum" (read standards and benchmarks and acceptable data). These debates frequently arise within the larger context of what constitutes the "discipline" of education (or on a smaller scale, a department of curriculum and instruction). In contrast, Dewey's pragmatic approach to educational research as it pertains to the body–mind issue reflects the idea that educational research is not so much research *about* education as it is research *for* education (Biesta & Burbules, 2003).

Dewey and Learning

For those like myself who do not claim to be philosophers, the most readable summary of Dewey's pragmatism and the implications for educational research has been written by Biesta and Burbules (2003). In acknowledgment to the tremendous contribution this monograph has made to this section, I am again borrowing their words:

> One way to summarize Dewey's theory of action is to say that it amounts to a *theory of experimental learning*. Dewey characterized living organisms—including the human organism—as capable of establishing and maintaining a dynamic, coordinated transaction with its environment. Through this process the predispositions of the organism become more focused and specific, which is another way of saying that through the tentative, experimental way of establishing coordinated transaction, the organism has *learned*. This learning is, however, not the acquisition of information about how the world "out there" really is. It is learning in the sense of the acquisition of a complex set of predispositions to act. In this process the world becomes more differentiated. It becomes, in other words, infused with meaning. This theory of learning also lies at the basis of Dewey's understanding of inquiry and research. Dewey suggested, so we could say, that we understand inquiry and research as process of experiential and experimental learning, which result in more diverse ways of action, reflection, and understanding. (pp. 37–38)

For Dewey, experimental learning involved the action of cognitively formulating *potential* alternative courses of action. This cognitive activity must be followed by a testing of the potential alternative courses of actions in order to acquire knowledge. These are two salient points that have influenced my scholarship. First, experimental learning is a process akin to problem solving where the distinction is made between habit and learning. Nonetheless, habits (including prior knowledge) are predispositions that influence the generation of alternative cognitive actions (ideas in the form of hypotheses) that serve a specific function. These alternative cognitive actions (hypotheses) are tested to determine which course of action best serves the function at hand. In this regard, learning is not a "causal attribution." Rather, learning is a process of constructing knowledge where knowledge is viewed as meeting two criteria: (1) it is information that has been organized conceptually (declaratively, procedurally, and strategically), and (2) has been tested and serves one or more functions.

My second point is that learning as a constructive process is predicated on memory (habits for Dewey). Therefore, memory is a necessary but not a sufficient condition for learning. From this perspective, memory and learning share a reciprocal relationship. Again there must be a correspondence between actions that serve a common function in order for successful adaptation to any situation (including classrooms).

As will be discussed in the sections devoted to theory development and research methodology, extensive advancements in both education and psychology were being made during the first half of the 20th century, most of which has been directly influenced by pragmatism. In educational psychology journals, a great deal of the published literature was devoted to instructional issues, learner ability issues, and issues involving the classroom environment. In experimental psychology journals, the focus was on the determination of principles of learning. This involved research focusing on identifying and defining basic "laws" or principles of learning that were constant or occurred across situations, learner abilities, and environments This research was conducted in laboratories under controlled conditions and was thought to personify "basic" research while the educational psychologists' efforts were viewed as "applied" research. The point to remember is that the philosophical foundation for both groups of researchers was Dewey's pragmatism as expressed by functionalism as a means of studying human behavior.

The Demise of Grand Theories

Functionalism's heyday was during the verbal learning area (circa 1920–1960). Unfortunately, a paradigm shift was occurring in psychology and education that greatly reduced the influence of functionalism. This was the impact of behaviorism during the middle of the 20th century. Interestingly, radical behaviorism as a philosophical movement also had its inception at the University of Chicago with the writings and research of John B. Watson who was a graduate student when Dewey was at Chicago. However, it was the work of B. F. Skinner that is today associated with the radical behaviorism that grew in popularity during the 1950s and 1960s. In an effort to deal with the dualism of the mind–body issue, radical behaviorists simply eliminated "mind" or "cognition" from the equation and focused strictly on observed behavior. Operant conditioning as formulated by radical behaviorists was very functional in nature and provided a simplistic analysis of human behavior in terms of antecedent and consequent behaviors. In many ways, it was the pendulum swing to radical behaviorism as an explanation for human behavior that provided an impetus for the cognitive revolution in psychological and educational research.

By the middle of the 20th century in both psychology and education, grand theories were losing favor to models. Behavior modification as an operant conditioning model of learning was in vogue. Paired-associate and serial learning models were being studied by verbal learners. In education circles, models of instruction, conditions of learning, and taxonomies of educational objectives were developed. Of importance to this discussion is the observation that by the middle of the century, psychology and education were moving further and further apart both theoretically and methodologically.

This paradigm shifting signified the emergence of specific models and metaphors that served to focus both educational research and psychological research. This increase in specificity (micro-theorizing) was positive in terms of added precision. However, the down side was a bifurcation between psychology and education theory. Furthermore, this loss of contact with grand theory seems to have created a situation in some educational research circles where models and metaphors developed to guide educational research and practice did not correspond with advances in our understanding of adaptive human behavior and advances in psychological research focusing on learning, memory, and cognition.

My suggested solution to this problem is a move "back to the future" where grand theory in the form of Dewey's pragmatism and its derivative "functionalism" is reunited with current models and metaphors currently employed in educational research and practice. While it is relatively easy to deconstruct general theory, the practical problem is that one loses sight of the forest for the trees. Reconstruction is the logical corollary to deconstruction if we are to advance both theory and practice in education. In essence, this is a call for a correspondence (not a cause–effect relationship) between and among various levels of the educational enterprise (theory development, research activities, educational practice). Or as noted by Biesta and Burbules, "Dewey saw his work as a response to a crisis in modern culture. The central element in this crisis, according to Dewey, is the disintegrating influence of modern science on everyday life—or, with the phrase Dewey preferred, *common sense*" (p 16).

BACK TO THE FUTURE: A CHRONOLOGICAL JOURNEY

The following brief review of the initial studies of transfer and problem solving conducted in the United States is organized around the operational definitions employed at the time the research was conducted. The storyline will focus on three factors or variables found in most educational settings: the *learner, materials to be learned,* and *learning conditions* (Edwards & Scannell, 1968).

MODELS (CIRCA 1900–1920)

Transfer of Training

For a brief glimpse of the initial research conducted in transfer of training and problem solving, our focus is on the research conducted by Edward L. Thorndike and Charles H. Judd. The basic research paradigm for the initial studies of transfer was the transfer of training method. "Any method for measuring the influence of practice at one activity upon the rate or other characteristics of the learning of a second activity is a transfer method" (McGeoch & Irion, 1952, p. 24). During this era, variations of this paradigm were being employed to address the nature of this influence from differing levels of analysis. Faculty psychology as a school of thought believed that transfer was broadly defined when it came to verbal learning and addressed questions like "Does the study of Latin broadly influence the subsequent learning of English?" Today, we recognize this as a curriculum question and the answer is, not much.

Faculty psychology was a Scottish school of philosophy that arose during the 18th century in opposition to the British empiricism of J. Locke, J. Mills, and J. S. Mills. The primary proponent of faculty psychology was Thomas Reid (1710–1796). For our purposes, it is Reid's focus on faculty theory that sets the stage for our discussion of transfer of training at the turn of the 20th century. Reid postulated various mental powers such as comparison, calculation, language, and causality. Faculty theory postulated an active mind rather than the *tabul rosa* assumption postulated by the British empiricist John Locke. Faculty psychology was viewed as a nativist or hereditarian theory. According to Sahakian (1975), faculty theory had significant impact on psychological and instructional thought in the American colonies from 1776 to 1876 and was widely taught in American colleges from about 1770 to 1830. This influence did, however, have a longer-lasting impact.

> At the turn of the century, educational psychology often included a study of the "intellectual faculties"—the doctrine of formal discipline, for example, being based on faculty psychology. The theory led to the requirement of courses in Greek, Latin, and Mathematics, on the grounds that these trained the intellectual facilities, which could then deal with any novel information. Empirical research, not surprisingly, demonstrated that this assumption was incorrect (see, e.g. Thorndike and Woodworth, 1901a, b). Thus, practicing memorizing poetry did not improve a person's ability to memorize nonsense syllables, and there was no support for the idea that training improved the "faculty" in general. *Later and more sophisticated research has shown that training does transfer, but not in the ways postulated by faculty theory.* (Stagner, 1988, p. 72, emphasis added)

The work of both Judd (1908) and Thorndike (Thorndike, 1913; Thorndike & Woodworth, 1901a, 1901b) was instrumental in debunking the faculty psychology approach to the study of transfer. However, from a theoretical level of analysis there was a distinct difference in the type of research question being addressed by Thorndike and Judd and the way the experimental methods were defined operationally. Judd approached the study of transfer from what today would be called a "top-down" approach. For Judd, the basis for positive transfer was studied at the level of general principles that were shared by the training and the transfer task. Judd's approach to the study of transfer would fit nicely into a cognitive approach as we understand the concept today. For Thorndike (Thorndike & Woodworth, 1901a, 1901b), the basis for transfer was the connecting of elements (through practice with knowledge of results) between the training and transfer tasks. The level of analysis was one we would recognize today as a "bottom-up" approach and would fit nicely with today's cognitive science approach to instructional psychology. It is the operational definitions of task variables, and the level of analysis of independent and dependent variables, that are the most apparent differences in the approaches these two researchers took when approaching the study of transfer of training.

Problem Solving

It is infrequent that one finds the term "problem solving" in books devoted to learning that were published during this period in the United States. This is no doubt due in part to the overwhelming influence of Thorndike's prolific writing during this era.

For example, the only index entry related to problem solving in the 1913 edition of *The Psychology of Learning* is the term "problem attitude." This term was used by Thorndike within a discussion of the elements of improvement that contribute to improvement during practice. When considering the impact of attention, he listed five commonly accepted aids to improvement. These commonly accepted aids were (1) interest in the work, (2) interest in improvement, (3) significance, (4) problem–attitude, and (5) attentiveness.

At this time, research efforts in the United States focused on trial and error learning with the issue of problem solving being implicit in the nature of the learning task, such as: (1) type of learner; (2) type of learning materials (letters or words vs. higher-order mental activities such as principles of rules); (3) making connections versus the role of insight. Problem-solving research revolved around the question "When confronted with a novel problem, how does the learner reach a solution?" (Hilgard & Bower, 1966, p. 10).

Summary

As educational psychology emerged at the beginning of the 20th century, within the context of experimental psychology, the beginnings of what we think of as behavioral and cognitive perspectives can be observed in the way common research questions were being addressed. This is especially the case in the study of transfer of training and problem solving. The aforementioned concern about the relative roles of "prior habits" and "insight" in the transfer process are an example. However, both approaches fit comfortably within the early functionalist movement (or school of thought) that focused on the overall adaptability of the human organism (Haskell, 2000). The point is that in the United States, both of these approaches to the study of learning (transfer and problem solving) evolved in opposition to faculty psychology that had made a major impact on pedagogical thought and practice during the 19th century.

Thus, early work by Thorndike and Judd in training for transfer set the stage for much of the experimental research on learning conducted in the United States during the first two decades of the 20th century. Consequently, the following summary comparison of their respective experimental approaches to a large extent reflects the mainstream study of educational psychology research as conducted during this period in the United States.

The cited work of Thorndike and Judd that follows in the next section will be compared and contrasted in terms of three common research issues: (1) the classroom element of focus, (2) the complexity or abstractness of the materials used in the experiments, and (3) the dependent variables that served as the data for analysis. Both researchers had a common interest in identifying the positive impact of training on performance in subsequent tasks. However, they differed in terms of the classroom elements that were the focus in (1) environmental setting versus learner background, (2) the nature of the materials participants manipulated, and (3) the target behaviors (dependent variables) observed.

For example, Thorndike's emphasis was placed on the manipulation of environmental variables to improve subsequent behavior. This is consistent with a behavioral focus where the unit of analysis is on basic elements and change is thought to occur as the result of a "bottom-up" process (from elements to principles).

For Judd, the major focus was the learner. Judd manipulated the learner's prior knowledge during training by teaching the principle of refraction. The subsequent task was a novel one that would be characterized as a problem task with a global rule or principle solution with the focus being an attempt to identify what the learner brought to the task. While not ignoring the importance of "bottom-up" processing, Judd contended that meaning and understanding were facilitated by insight (a "top-down" perspective).

One last point of difference should also be emphasized because this difference sets the stage for a great deal of the research efforts undertaken by "verbal learning" researchers during the decades between 1920 and 1970. While Thorndike focused almost exclusively on the positive impact of training (positive transfer) on subsequent performance, Judd also identified instances of no impact of training on subsequent tasks (zero transfer) and instances of a negative impact of training on subsequent performance (negative transfer).

METHODS AND MATERIALS (CIRCA 1900–1920)

Transfer of Training: Thorndike and Woodworth

The types of tasks and materials used by Thorndike and Woodworth (1901a, 1901b) to test their view of transfer as a process of connecting specific elements helps highlight the level of investigation that characterized Thorndike's (1913) theoretical approach to transfer. Today, we would characterize Thorndike's theoretical perspective as behavioral or cognitive science. He addressed learning issues including transfer and problem solving with the most specific experimental research methods available at the time. Both of the 1901 studies coauthored with Woodworth were published in *The Psychological Review.* Interestingly, both carried the same title: *The Influence of Improvement in One Mental Function upon the Efficiency of Other Functions.* In the first article (1901a), the results of several experiments are reported where participants were trained to estimate the area of geometric forms. The method employed was a form of the "method of savings" where accuracy of magnitude judgment (measured in terms of area) was the data collected (dependent variable). The method of savings as a measure of learning and retention is discussed in detail later in this section. Basically, the training was initiated with judgments of the area of various triangles of varying size. The second task involved judging the magnitude of rectangles.

The idea behind the study was to determine the impact of training a mental function (estimate of magnitude) using triangles on the improvement in estimating magnitude when the task was estimating area of objects that have a different physical structure (other geometric forms). The authors concluded that while there was ample evidence to support intertrial transfer (during training) as the result of knowledge of results, transfer across tasks was far less dramatic. The authors concluded that there might have been some training influence on an allied mental function (discrimination or attention), but the series of studies reported were not able to discern such influence. However, they did conclude that when the functions were similar (estimating magnitude of triangles vs. estimating magnitude of rectangles) small positive influences were observed under the conditions employed.

> For functions so similar and for cases so favorable for getting better standards and better habits of judging, the amount of improvement gotten by training in an allied function is small. Studies of the influence of the training of similar functions in school and in the ordinary course of life, so far as we have made such, show a similar failure to bring large increases of efficiency in allied functions. (Thorndike & Woodworth, p. 395, 1901a)

In the second series of experiments reported by Thorndike and Woodworth (1901b), the focus was on the training of the functions of attention, observation, and discrimination and the influence on subsequent tasks. The task in this series of studies was the training of participants to accurately observe detailed features of words on the printed page. Participants would practice marking every word containing the two letters *e* and *s* until they demonstrated an increase in speed of responding and in accuracy. Before and after this task, participants were tested on their accuracy and/or speed of marking other combinations of letters, misspelled words, inserted letters, words of a certain length, different parts of speech, and so on. The authors believed that these experiments reflected the functions (today, we would use the term "processes") "comparable to the training in school life which pretends to improve our general habits and powers of attention, discrimination, and accurate work" (p. 544). Generally speaking, the results replicated those in the first study. Transfer within a series of practice trials was evident and quite apparent. However, when attempting to measure the influence of one function (attention) on another (observation or discrimination), little positive influence was observed.

Levels of analysis employed in the above series of transfer studies is characteristic of the level of analysis that defined an "element" for Thorndike in his theory of transfer (1913). Also, this is the series of studies for which a "no-treatment control group" was originally employed. In this series of studies, Thorndike and Woodworth appear to operationally define the functions of attention, observation, and discrimination as speed and accuracy of performance when attempting to determine the influence of practicing the marking of words with *e* and *s* on subsequent performance of such tasks as marking words with *i* and *t*, marking verbs, marking other parts of speech, misspelled words, and so on. Thus, elements were specifically defined as single letter combinations, parts of speech, misspelled words, and so on. In their conclusions, the authors believed that their findings provided a replication of the conclusions arrived at in the earlier study

(1901a). They suggest that future research efforts focusing on the interdependence of mental functions be more carefully devised and that the precision of the methodology be improved "to detect the exact elements of any function that were changed by training, to measure such changes and to find which of such changes brought about increased efficiency in other functions and how" (Thorndike & Woodworth, 1901b, p. 564).

Learning as a Method of Saving. "The transfer method, although often treated as an independent measure, also measures retention (McGeoch, 1942). Any methods for measuring the influence of practice at one activity on the rate or other characteristic of the learning of the second activity is a transfer method" (p. 17). Historically, this was the approach employed in Ebbinghaus's study of memory retention. In terms of the earlier discussion of operational definitions, data collected to provide evidence of *learning* took the form of time, number of trials, or errors. Usually the percentage of saving was computed as a ratio of the learning rate in training divided by the learning rate during the subsequent transfer task. If training to criterion required 20 practice trials and the subsequent task required only 10 trials to reach criterion, the savings in effort was 20 trials – 10 trials = 10 trials, or 50% of the original trials. If the number of practice trials to criterion was the same in both training and the subsequent task, no transfer was observed.

The method of savings approach produces reasonably credible data when the learning tasks are nonsense syllables and the subsequent task is a new list of nonsense syllables; or, motor learning involving training on a pursuit rotor task is followed by a maze tracing task. In these cases, the tasks are ones where prior knowledge, meaning, or experiences are almost nonexistent. In a sense, these activities that define the learning tasks serve as a control for individual differences in the past history of the learner.

When the learning tasks move to more familiar materials (math, reading, and science), the prior history of the participants can have very strong impacts on not only original learning rates during training but during the transfer task as well. In addition to the issues involving learner differences, in a transfer study, a "method of savings" approach is difficult to employ with meaningful learning materials. This is due to the problem of ensuring that the difficulty level or meaningfulness of the learning material in both tasks are comparable. If the comparability of task difficulty cannot be ensured, the method of savings metric as a means of determining the impact of task A on task B does not provide credible data.

Transfer of Training: Judd

The types of tasks and the materials used by Thorndike and Woodworth can be contrasted with the types of tasks and materials Judd employed to focus on the investigation of "higher-order" cognitive elements such as principles or generalizations. Today we would probably refer to this as an exercise in training students to learn and use strategies. The title of the article in which Judd describes the principle taught (refraction) and the task (throwing darts at underwater targets) is *The Relation of Special Training to General Intelligence* (1908). This article included a discussion of several training for transfer studies ranging from motor tasks to the teaching of abstract principles. While the level of analysis was cognitive in nature, these studies were conducted as a means of disputing the ideology that the training of mental functions is always specific.

Two groups of fifth- and sixth-grade pupils participated and were given the task of throwing a small dart and hitting a small target submerged under water. The problem presented by this task was the "deflection" that the light suffers through refraction. The target is not where it seems to be. Consequently, participants must adjust prior experience with throwing darts at targets to meet the new situation.

> In this experiment one group of boys was given a full theoretical explanation of refraction (prior to practice). The other group of boys was left to work out experience without theoretical training. These two groups began practice with the target under twelve inches of water. It is a very striking fact that in the first series of trials the boys who knew the theory of refraction and those that did not, gave about the same results. That is, theory seemed to be of no value in the first tests. All the boys had to learn how to use the dart, and theory proved to be no substitute for practise. At this point, the conditions were changed. The twelve inches of water were reduced to four. The difference between the two groups of boys now came out very strikingly. The boys without theory were very much confused. The practise gained with twelve inches of water did not help them with four inches. Their errors were large and persistent. On the other hand, the boys who had the theory, fitted themselves to four inches very rapidly. That theory evidently helped them to see the reason why they must not apply the twelve-inch habit to four inches of water. Note that theory was not of value until it was backed by practice, but when practice and theory were both present the adjustment was rapidly worked out. (Judd, 1908, p. 37)

An additional issue that has frequently been overlooked is that Judd's reaction to the work of Thorndike was based on more than simply a difference in the level at which the study of transfer should be approached (level of analysis) and the nature of the tasks and materials. Based on earlier work with learning and knowledge of results, Judd was convinced that "prior knowledge" could also have a debilitating effect on transfer performance. Thus, for Judd transfer could be either facilitating (positive transfer) or debilitating (negative transfer). In contrast, Thorndike (1913) was still viewing transfer as the study of transfer under the general heading of the "influence of improvement" and was not very receptive to the idea of negative transfer. In my reading of these pioneering works on transfer of training, I see this as a larger theoretical issue than the level of analysis issue. I base this assertion on the gist of Judd's 1908 article and the gist of Chapter 12, Volume II, of Thorndike's classic series *Educational Psychology: The Psychology of Learning.* Both Judd and Thorndike tacitly acknowledge at several points that both a "bottom-up" and "top-down" approach to the study of the transfer of training has merit. Furthermore, when discussing the transfer of training, Bayles (1936) made the following point: "The similarity has been noted by Sandiford, a pupil of Thorndike, who claims that both theories amount to one and the same thing" (p. 427).

> The major difference between the two (Thorndike and Judd) was more a matter of emphasis than a basic difference in theoretical orientation. The major difference was the focus on "elements" by Thorndike and "generalizations" by Judd. This point was explicitly made by Judd when discussing the relatedness of mental functions: "Our problem is one of analysis and not of classification" (Judd, 1908, pp. 35–36). Both approaches to studying transfer of training were in opposition to faculty theory. As pointed out by Bayles (1936), at the turn of the 20th century there were three theories of transfer of training: (1) elements theory, (2) generalization theory, and (3) faculty theory.

> Between 1915 and 1920 the argument waxed hot and heavy between Thorndike and Judd as to whether transfer was to be obtained through the agency of identical elements or of generalizations. Judd's dart-throwing experiment seemed to show that identical elements in the environment were not enough. It appeared to show the advantage of a "generalized" training; that is, an individual must have been shown during his training that situations seem to work according to discoverable *rules,* and that one must become able to apply the rules (or generalization) before one can take advantage of environmental recurrences and use one's training profitably. (Bayles, 1936, p. 426)

Problem Solving: Thorndike and Judd

During the first part of the 20th century, much of Thorndike's work in problem solving focused on animal learning and the puzzle box. Commonly employed animals were cats, dogs, monkeys, and even fish. Learning was operationally defined as a reduction in error rate and elapsed time to solution. This comparative approach to addressing problem solving was extensively employed by Thorndike and his followers. As we have already seen, Judd, among others, were restricting themselves to the study of human learning with the learning materials being abstract and cognitive in nature. This difference in their approach to problem solving appears to reflect their differing views about the role of "insight" in attaining a solution that circumvents constrains of the problem situation.

The trial-and-error versus insight issue in problem solving is one of relative emphasis. The trial-and-error approach of Thorndike gives priority to prior knowledge or habits as the explanation for the successful solution to a problem situation. The insight approach to problem solving that is more characteristic of Judd's approach gives priority to the process of restructuring the problem situation. As we know today, both prior knowledge and insight are necessary for successful problem-solving performance.

VERBAL LEARNING MODELS: ANTECEDENTS OF THE COGNITIVE REVOLUTION (CIRCA 1920–1970)

A great deal of educational research addressing both transfer and problem solving occurred between 1920 and 1970 in the United States. During this period, the prior influence of Thorndike and Judd can be seen in terms of the classroom elements of focus, the types of materials learned, and the performance measures that served as the data for analysis. *Conditioning* and *learning* as theoretical explanations for behavioral change were moving further and further apart in terms of operational definitions for materials, settings, and behaviors being investigated. More and more research was focusing on verbal behavior rather than motor behavior. In the case of verbal learning there was a dramatic shift away from the exclusive use of nonsense syllables as stimulus materials. In part, this reflected the waning influence of Ebbinghouse's approach to the study of memory and a growing cognitive zeitgeist witnessed by the study of reading, mathematics, and so on, by educational psychologists.

This is not to suggest that there was an integration of the cognitive and behavioral approaches of learning research to any great extent. Rather, researchers focusing on a finer level of analysis were rigorously investigating serial learning and paired associate learning and still employing nonsense syllables and simple word lists. Many researchers were approaching the study of *meaning* and *understanding* from an experimental psychology perspective called *verbal learning*. The research focus was still on understanding the basic conditions that fostered the acquisition of new verbal behavior and the factors that negatively affected memory retention (forgetting) for the acquired verbal behavior. A perusal of the table of contents of Thorndike's *The Psychology of Learning* (1913), and McGeogh and Irion's classic (1952), *The Psychology of Human Learning*, indicate very similar topics being covered in the two texts.

A great deal of the transfer research conducted during this time frame (circa 1920–1970), was designed to address the questions raised at the turn of the century by Judd concerning the type of impact training has on the subsequent transfer task. These questions basically investigated the manipulation of task variables related to training (practice intervals, massed or distributed practice, knowledge of results, type of learning materials, etc.) and the determination of (1) positive, (2) zero, or (3) negative transfer affects. As mentioned above, the predominate level of analysis by experimental psychologists studying verbal learning was more akin to the behavioral elements approach taken by Thorndike than the more cognitive approaches that characterized Judd's work.

This is not to suggest that cognitive ideas were not having an impact in educational psychology. Educational psychologists were asking many of the same research questions concerning the influence of training on subsequent tasks, but were working with school-age participants and subject matter like mathematics, reading, and science as the knowledge to be learned. Obviously, these factors reflect a more cognitive orientation to the question of learning materials in academic environments. For an excellent introduction to these studies, Edwards and Scannell's (1968) chapter on generalization and transfer of training is a place to start.

For example, two studies by Muehl (1960, 1961) illustrate the type of transfer research being conducted late in the era under analysis. These studies were designed to address questions about successful approaches to teaching early reading. In the first study, the idea was that the type of visual discrimination training kindergarten students received prior to vocabulary learning would have differential impact on the acquisition-learning of vocabulary words. Three groups of students were formed. Each group received discrimination training with three different types of material to be learned. One group was given discrimination training on the vocabulary words to later be learned. The second group was given visual discrimination training on words that were different than those that would later be learned. The third group received visual discrimination training with geometric forms. All three groups received the same list of vocabulary words to be learned following visual discrimination training. This experimental design basically consists of two treatment groups and a control group. The geometric form visual discrimination group served as a control for the influence of visual discrimination training with words. "Muehl concluded that positive transfer to the learning task by the pupils in group one was due to their acquiring a "set" for observing responses to distinctive letters in the stimulus words. This kind of response then helped to make words in the vocabulary list more distinctive for this group" (Edwards & Scannell, 1968, pp. 377–378).

In the second study, Muehl (1961) tested for differences in "whole-word" visual discrimination training and "single-letter" visual discrimination training. Two different groups of kindergarten students were participants in this study. Differences were observed in training that favored the performance of the letter discrimination group. However, no differences were observed in performance on the subsequent reading task used to test for learning differences. These studies are informative because they reflect the use of experimental research designs in addressing questions pertinent to the teaching of early reading. As Edwards and Scannell (1968, p. 378) also point out, "they illustrate the effective use of positive transfer from a relatively discrete, isolated activity to a more generalized state."

The studies described above illustrate research conducted along lines similar to those initiated to Judd, but there was also comparable research from an approach more similar to Thorndike's approach. An example of research in this tradition that also investigated early reading is a study reported by Levin, Watson, and Feldman (1964) that used writing as a pretraining activity with 6-year-old children. Results indicated that pretraining was more effective when pretraining materials emphasized high relevance for later learning.

Phrased differently, the greater the similarity between the pretraining materials and the transfer task, the greater the amount of positive transfer on the subsequent task. This study, among others of a similar ilk, gave rise to the distinction of what came to be designated as "near" and "far" transfer. In other words, the scale for transferability of prior knowledge was operationally defined in terms of the properties of the stimulus materials employed in the training and transfer tasks. If the stimulus properties of the tasks were very similar, they illustrated near transfer, whereas if they were very different, they illustrated far transfer.

Transfer Models

During this era, transfer came to be operationally defined as any impact (positive or negative) on subsequent behavior when compared with an appropriate control group. In terms of a descriptive transfer model, this is fine. Also, in terms of a more credible predictive transfer model, focusing on the prediction of positive, zero, or negative transfer based on similarity or dissimilarity of learning materials served an outstanding purpose. Research in this era provided educational psychology with the *interference theory of forgetting* with two factors (proactive and retroactive inhibition models of forgetting). Basically, interference theory identifies the conditions (e.g., task order and similarity of learning materials) that produce negative transfer. As illustrated in the next paragraph, transfer (positive or negative) in a classroom learning situation need not be restricted to an investigation of the immediately preceding activity. Rather, it can also be understood in terms of the stimulus characteristics of the transfer and training tasks.

The focus on stimulus and response surface elements shared by training and transfer tasks is best summarized in the Osgood (1949) transfer surface model that comes from experimental psychology research. Osgood's transfer and retroaction surface is a three-dimensional surface used to graphically represent similarity and dissimilarity of stimulus elements and response elements. "What Osgood's surface states is that the amount of transfer in positive or negative directions is a function of shifts in similarity between learning and retention of both the stimulus conditions and the responses required" (Hilgard & Bower, 1966, p. 314). For our purposes, Osgood's work provides a glimpse of the experimental zeitgeist influencing the study of transfer of learning during this era.

Thus, by the end of the first half of the 20th century, descriptive and predictive models of transfer are prevalent in the research literature. However, there is little explanation value concerning "how" the transfer process occurred, or more importantly, "who" will engage in the transfer process when it is appropriate to do so. Still lacking in the 1960s was an explanatory model for the transfer process. This was not a concern new to the 1960s. This point had been raised years earlier by Bayles.

In terms of an explanation for "how" transfer occurs, Ernest Bayles (1936) raised a critical point about all three theories of transfer (faculty theory, Thorndike's elements theory, and Judd's generalization theory). All three theories assume *automatic transfer.* "As we see it now, the difficulty with the faculty theory was that it assumed transfer to be automatic; dependent entirely upon conditions posed by the environment. Whenever the environment established a condition conducive to transfer, the properly trained individual, according to the theory, was supposed to take advantage of it and act accordingly" (p. 425). Interestingly, Bayles contended that there was little difference between Thorndike's identical elements and faculty theory with regards to the idea of automatic transfer. "But here again is a notion of automatic transfer; a notion that the individual will transfer his training whenever the environment establishes a favorable set of conditions" (p. 425). While Bayles contended that Judd's generalization theory had a better record on the issue of automatic transfer than either faculty theory or elements theory, it still falls short of an adequate explanation. "The difficulty is that the theory of transfer by generalization still leaves the emphasis on the environment as the dominate motivating agency; it still interprets transfer as being automatic when the environment sets the stage" (p. 426). Interestingly, the assumption that transfer is an automatic process triggered by environmental variables (affordance) is still being championed today in some theories of instructional design, situated learning, and higher-order cognitive processing.

By the end of this era, educational psychology was making attempts to integrate what had been learned from 50 years of research about transfer conducted from both the behavioral and cognitive theoretical perspectives in order to influence educational practices. For example, in 1968, Edwards and Scannell provide the following suggestions about *teaching for transfer*. The focus is on the instructional activities that promote positive transfer.

> Before positive transfer can be expected, learning of a process, or fact or idea must occur. If the student has learned only partially the elements to be transferred, the possibilities will be very limited. This is a necessary condition to positive transfer, but hardly a sufficient one. If the learning is neglected at this point, retroactive inhibition will exercise its effect on retention. Review becomes a further necessary condition, and during review the steps below, designed to facilitate positive transfer may be utilized: 1) designate the principle(s) to be derived from the learning; 2) apply the principle(s) to other highly similar situations; 3) apply the principle(s) to other situations, varying in degree of relation to the original learning; and 4) apply the principle(s) to many widely varying situations, including some in which the principle is not relevant. The final step represents both a learning and a test situation. (p. 386)

An emphasis on moving from a radical behavioral position on transfer and learning can also be observed in the work of Gagne and his colleagues. Robert Gagne had as much impact on instructional design and instructional psychology as any psychologist writing during the last half of the 20th century. In *Conditions of Learning* (1965), Gagne provided educational psychology with one of its first systematic approaches to identifying eight types or conditions of learning, ranging from simple to complex, in his hierarchical model. "A strong emphasis within Gagne's analysis is upon the *structure of knowledge*, an important supplement to principles of learning whenever a practical instructional task is under consideration" (Hilgard & Bower, 1966, p. 571). The emphasis on cognition from an instructional focus can be recognized in the highest types of learning discussed in the hierarchy (principle learning and problem solving).

In summary, the study of transfer of training was changing in terms of the operations used to define the phenomenon. Theoretically, the zeitgeist was one of seeking the finest-grained level of analysis possible given the technology for measuring learning and transfer. Technology in this case must also be broadly defined to include the advances in experimental research design, method of measurement, and statistical methodology.

VERBAL LEARNING: METHODS AND MATERIALS (CIRCA 1920–1970)

Transfer

A close inspection of the methodology and materials employed by the end of the first half of the century reflects a dramatic change in the way studies of transfer were designed, analyzed, and interpreted. These changes influenced the types of research questions being addressed and the manner in which the questions were being investigated. Examples include general "warmup effects" during acquisition learning or more specific "learning to learn" effects that arise from individual differences in prior knowledge that precede learners' participation in the study. There were also statistical analysis advances that replaced the "method of savings" analysis.

The aforementioned Muehl studies (1960, 1961) and the Levin and colleagues study (1964) employed research designs that took advantages of developments in methodology and statistical analysis that reflect changes in the approach to the study of transfer that were occurring during this period. The Muehl studies employed comparison groups and a mixed factorial design with a between-groups factor and a within-subjects repeated measures (trials) factor. The Watson and colleagues study employed a control and experimental group in a between-groups factorial design. In both cases, specific research questions were being addressed that required greater precision in terms of controlling for sources of influence that would decrease the internal validity of the design and hence the credibility of findings

In terms of statistical issues, Muehl (1960) cites Linquist's (1953) type I mixed design analysis of variance as the statistical analysis employed to assess the impact of prior knowledge with a repeated measure factor. The Muehl and Levin and colleagues studies were focusing on the study of positive transfer within the context of training for transfer.

The second example comes from experimental psychology where a major focus during this period was on the study of *negative transfer* measured in the form of errors and the study of forgetting. The basic research paradigm employed in the study of forgetting was one of acquisition learning and retention. In this case, task A (acquisition learning) and task B (acquisition learning) was followed by a measure of retention (recall or recognition) for either task A or task B. Research in this area provided educational psychology with the *interference theory of forgetting* with two factors (proactive and retroactive inhibition models of forgetting). Basically, interference theory identifies the conditions (e.g., task order and similarity of learning materials) that produce negative transfer. As mentioned above, this focus on stimulus and response surface elements shared by tasks A and B is best summarized in the Osgood (1949) transfer surface model, which comes from experimental psychology research.

Problem Solving in the United States

With the exception of transfer of training studies where the subsequent task was abstract and required the use of a principle or rule, problem-solving research as we know it today was not extensively conducted (1920–1970). Rather, the focus was on elements of reasoning and the study of thinking. The classic *A Study of Thinking*, published in 1956 by Bruner, Goodnow, and Austin, is prototypical of the experimental study on thinking that was being conducted near the middle of the century. The focus was on the manipulation of task variables (e.g., type of knowledge of results, etc.) that influenced thinking in the form of decision making during deductive reasoning. During this era, a group of experimental psychologists in the United States was investigating the experimental psychology of original thinking. An excellent review of this research signaled a resurgence in research and theoretical work in problem solving and thinking (Ray, 1967). However, many of the studies described in this review (Bugelski & Huff, 1962; Dunker, 1945; Gibson, 1941; Guilford, 1950; Kendler & Kendler, 1962; Luchins, 1942; Maltzman, 1955; Mednick, 1962; Van de Geer, 1957) tended to be ignored as behaviorism was broadening its base among psychologists and educators during the first half of the 20th century. Several of these researchers (Dunker, 1945; Gibson, 1941; Luchins & Luchins, 1950) were influenced by Gestalt psychology and focused their efforts on the investigation of *Einstellung* or set and functional fixedness. These cognitive-processing phenomena (set and functional fixedness) are mental activities that are species specific and an inherent (unlearned) aspect of central nervous system processing. I consider this work to be a precursor of the process dissociation technique that has become a central independent variable manipulation employed in current memory and cognition research focusing on the functional identification and description of cognitive processing engaged in during problem-solving transfer. This historical thread will be picked up again in the next section.

Problem Solving: Western Europe

Although Gagne and Bruner were focusing on thinking and problem solving from a cognitive perspective that emphasized instructional factors influencing learning, transfer, and problem solving, a different cognitive influence was coming from Western Europe during this period. Again because of space issues, I am restricting the current review to a few prominent examples of this perspective. During the 1920s and 1930s, the work of F. C. Bartlett (1932) on remembering and the sharpening and leveling of memory that occurs across time was influencing many experimental psychologists, particularly those in Europe. Bartlett's approach to the study of remembering was quite different from the study of forgetting that was the focus in the United States. Obviously, remembering and forgetting are opposite memory functions. However, the methods employed were quite dissimilar. Bartlett took a very naturalistic approach to the study of memory while psychologists in the United States were working in their laboratories collecting data under highly controlled conditions using nonsense syllables as learning materials and paired-associate or serial learning tasks. In contrast, Bartlett had his participants read a passage, about a mythical tribe of warriors and then twice asked them to write down what they remembered about the passage. These participants were then followed up at varying intervals and asked to again write down what they remembered from the passage. In some cases, this went on for 7 years. Rather than viewing his stories as parts or elements that could be forgotten or remembered in chunks, Bartlett viewed them as organized wholes that were subject to "sharpening" or "leveling" with the passage of time. Bartlett proposed that these memories consisted of a more general, abstract organizational entity that he called a "schema." Higher-order cognitive operations were organized as a single schema or an organized collection of schemata. This level of analysis and the use of schemata as a construct in the development of academic learning theory will see its greatest influence during the latter third of the 20th century as illustrated prominently in an article by Richard Anderson (1977) and his notion of schemata and the educational enterprise.

Gestalt psychology was also coming of age in Germany during the first half of the 20th century. In a manner similar to what was observed in the United States, it was the issue of "level of analysis" that fermented a split with the elements approach of Wundt. Thus, Gestalt psychology is another variety of rebellion against the structuralism of Wundt that brought an atomistic or elements approach to the study of learning. "Gestalt theorists accepted a major emphasis on consciousness but stressed the presence of organized wholes that did not depend on the formation of associative links between elementary sensations" (Stanger, 1988, p. 240). During the second era under consideration (1920–1970), the work of Max Wertheimer, Wolfgang Kohler, and Kurt Koffka was being introduced in the United States on a large scale.

The Gestalt school focused on the investigation of principles that organize our perceptions. These organizational principles of perception were developed through the study of visual illusions and complex problem solving. The cognitive influences on the study of problem solving initiated by Gestaltists is profound and continue to influence the study of problem solving in the United States. In educational psychology this influence is most easily recognized in the classic problem-solving research conducted by Richard E. Mayer and his students.

The aforementioned principles reflect the Gestalt view that the study of behavior (level of analysis) is molar, not molecular. Without going into detail, just a few of the perceptual organizing principles are: (1) figure and ground, (2) pragnanz, (3) proximity and similarity, (4) attitude, and (5) closure. The Gestalt school does not preclude the influence of learning by association. Rather, the emphasis was relative and best summed up in the shibboleth "the whole is greater than the sum of its parts." Gestalt influences can be found in a number of "real-life" settings. Many of these can be found in education. For example, Wertheimer's major book, *Productive Thinking* (1945), urged less reliance on rote learning and more emphasis on problem solving.

In summary, the study of problem solving and transfer as we know it today was greatly influenced by diverse cognitive influences emanating from Western Europe. By the 1960s, the stage was being set for what we have come to call the cognitive revolution in the study of transfer and problem solving in education by researchers in the United States.

Epilogue

During the first two-thirds of the 20th century, several trends are notable given the story line of this chapter. First, the models of transfer and problem solving were progressing from descriptive models to predictive models. This is reflected in the teaching for transfer model of instruction, which predicts that transfer will be demonstrated by the learner when the instructional setting is manipulated in a particular manner by the teacher. What is still missing, however, is a strong explanatory model of the transfer process engaged in by the learner. Second, problem-solving tasks have expanded in complexity, requiring mental operations that came to be called mediators or cognitive processes rather than simple trial-and-error approaches. These mediators or processes involved reasoning, strategy development, concept development, and so on. In part, due to the influence of Gestalt psychology on experimental psychology, the study of problem solving was progressing from the early descriptive and predictive models of learning to attempts at explaining the mediators or processes engaged when a learner is confronted with a new or novel situation.

THE COGNITIVE REVOLUTION: THEORETICAL ISSUES (CIRCA 1970–PRESENT)

By the beginning of the cognitive revolution, educational psychologists had also abandoned "grand philosophical theories" and were defining their theoretical models in terms of research methodology. While there was general agreement as to what constituted empirical research models, there was a proliferation of models that focused on specific aspects of the aforementioned learning process (e.g., acquisition learning, retention of learning, impact of prior knowledge on acquisition, individual differences, etc.). While common research designs provided replication of results, the interpretation of these results might vary to a considerable degree. This difference in interpretation of common findings is in part the result of the use of differing metaphors for the development of theoretical models of various aspects of the learning process. Simply put, the interpretation of experimentally derived findings is never a simple matter of "going with the data." The metaphor that guides the development of research hypotheses inevitably grounds the interpretation of results and consequent advancements in both educational theory and educational practice.

Metaphors

It was at the beginning of the cognitive revolution when the working relationship between experimental psychology and educational psychology involving the sharing of a common focus pertaining to theory, research method, and research design took separate paths. In part, this parting of the ways can be traced to the pursuit of metaphors to better understand the learning process and the cognitive processing that served as the basis for observable behavior.

Verbal learning researchers, who had provided the basis for a better understanding of fundamental learning processes, approached the cognitive revolution by adopting the "functional memory metaphor." This metaphor provides the context for an organic model (the human biological system) when pursuing explanations of transfer and problem solving. This shifting of focus reflects a paradigm shift from the investigation of settings (including instruction) to the identification of cognitive processes engaged in by the learner. Phrased differently, the primary research focus was a better understanding of the learning, transfer, and problem-solving processes through an intensive investigation of the learner's information-processing abilities related to memory function.

In contrast to experimental psychologists, educational psychologists were influenced by the issues involved in instructional design and by the development of instructional technology. Both of these historical factors appear to have focused educational research efforts more on (1) environmental setting and (2) instructional variables that impact the study of transfer and problem solving. The "computer metaphor" reflects a paradigm shift in educational research that prompted the development of an inorganic model (the computer) as a means of explaining how the learner functioned cognitively. Phrased differently, the primary research focus was on an attempt to develop a better understanding of learning, transfer, and problem-solving processes through an intensive investigation of instructional variables and practice settings (individual study, cooperative study, computer simulations, artificial intelligence, intellectual tutoring systems, etc.).

This selected overview of the cognitive revolution is developed around the following storyline. There is a brief introduction to the "computer metaphor" and the research questions addressed in the context of this metaphor. This will be followed by an equally brief overview of the "memory metaphor" and the research questions this metaphor elicited. I will contend that as educational researchers, we must understand the way our metaphors hinder and help our progress in moving from descriptive and predictive models of transfer and problem solving to explanatory models of problem-solving transfer that includes a consideration of learner characteristics.

The computer metaphor has been extremely valuable, has provided conceptual tools for theory development, and has strongly influenced the way we currently study transfer and problem solving in education. The computer metaphor also provides the basis for an information-processing level of analysis of the transfer and problem-solving process. This level of analysis has been very valuable in the development of research questions involving instruction or teaching issues and issues involving classroom settings.

However, an information-processing level of analysis based on the computer metaphor has its shortcomings. For instance, because it is an inorganic (mechanical) model, the computer metaphor has little to offer in terms of either motivation issues or a consideration of individual differences. Both of these are extremely important contributors to learning activities in the classroom. However, as mentioned above, the computer metaphor has limited our thinking about transfer and problem solving when considering learner characteristics. For example, the computer metaphor contributes to a view of memory processing as basically a "storage or warehouse" phenomenon. With the computer metaphor, there can be an overreliance on the input–storage–automatic retrieval sequence of cognitive processing when attempting to explain problem solving and transfer performance (Phye, 2001).

It is at this point where the work in learning, memory, and cognition during the cognitive revolution goes far beyond that observed in educational psychology. Most interestingly, it is the human memory metaphor that puts the learner at the center of the research process when attempting to address issues involving transfer and problem-solving performance. In contrast, much of what we currently see in articles published in educational psychology journals are manipulations of instructional variables and settings variables. This instructional research is quite good and provides us with a better understanding of two (e.g., materials to be learned and learning conditions) of the three learning variables found in any classroom. Unfortunately, unless we do a better job of putting the learner back into the equation, our ability to describe, predict, and explain transfer and problem solving in terms of human abilities and characteristics are limited.

The *functional memory metaphor* emerged during the last part of the 1960s (Neisser, 1967) and the decade of the 1970s. The cognitive evolution in the theoretical study of human learning proliferated as a result of new data collection instruments and advances in research practices. In experimental psychology, learning was approached as "learning, memory, and cognition." The "functional memory" metaphor was guiding much of the research efforts in learning and the level of analysis was an information-processing approach. The functional memory metaphor is an organic model and research efforts focused on the identification of functional memory components such as short-term memory, working memory, long-term memory storage, and various memory processes (memory search, memory retrieval, implicit memory, explicit memory, consciousness, etc.). In other words, this was a new approach and a new attempt to describe cognitive learner characteristics and activities that support schooling efforts.

In addition to a shifting in the way learning was being analyzed, this was also a time when there was an integration of experimental methodology that went largely unnoticed by many educational researchers studying transfer and problem solving. In contrast to the radical behaviorism position, cognitive psychologists made a basic distinction between learning and performance. This distinction between learning and performance became a basic point of departure for the study of transfer, especially when different types of tasks are used to assess what has been learned as a result of practice under a given set of conditions.

The general point is that any transfer task has its own characteristic response patterns. If a retention task and a transfer task are identical, you have only an immediate and a delayed retention task. In this case, cognitive processing may be automatic. However, if the task used to test for transfer has its own response characteristics, by definition it is a problem-solving task. This transfer task (test) may take the form of a well-defined or ill-defined problem. It is the response characteristics of the transfer task that will determine whether or not mediation is involved in the transfer activities. Thus, we are back to the question of automaticity raised decades ago by Bayles (1936).

This search for cognitive mediators or the alternative, simply relying on an assumption of automaticity of transfer, may be an artifact of the level of analysis employed in the investigation of transfer. If a researcher is willing to accept "general transfer" as an acceptable criterion for defining transfer, then a general *description* of the impact of practice (positive, zero, or negative) on the transfer test is adequate. However, if the desire is to *explain* the basis for the transfer effect in terms of cognitive processes employed, the criterion is one of "specific transfer." Furthermore, by employing both criteria in the investigation of transfer, a determination of the automaticity of processing can also be made. In other words, automaticity of processing need not be a theoretical assumption when studying transfer. Rather, automaticity is a testable hypothesis in any task involving problem-solving transfer that requires some form of a mindful strategy other than trial and error.

Research Model: Transfer

In terms of methodology and research design, advances were being made in experimental psychology during the early stages of the cognitive revolution to address issues involving general and specific transfer issues. Also, the nature of the transfer tasks participants were being asked to undertake were becoming more and more complex in terms of cognitive activities. This increase in the complexity of tasks investigated and the attempts to move to theoretically stronger models (e.g., from description to explanation) necessitated the development of more intricate research designs. For example, in the *Fundamentals of Human Learning and Cognition*, Ellis (1972) identifies two basic research designs employed in the study of transfer during the middle of the 20th century. The research design in Table 10.1 is employed to study *gross (general) transfer.* In this design, a comparison of the experimental group and the no-treatment control group performance on task B defines the presence or absence of transfer. If the

Table 10.1. Research Design for the Study of Gross Transfer

	Initial Task	*Transfer Task*
Experimental Group	Learn Task A (German Vocabulary)	Learn Task B (Russian Vocabulary)
Control Group (No Treatment)		Learn Task B (Russian Vocabulary)

performance of the experimental group (assessed as fewer errors, more correct responses, or faster rate) on task B is superior to control group performance on task B, then positive transfer is observed. If superior performance on task B is observed in the control group, then negative transfer is observed. If no advantage is observed for either the experimental or control group on task B, then zero transfer is observed. This design only tells us if transfer occurred. It does not provide any description of the cognitive processes involved in the transfer process. Thus, while the design in Table 10.1 provides credible evidence that transfer occurred, we are unable to specify what factors are responsible for transfer.

Table 10.2 portrays an extended research design where two control groups are employed in order to obtain information about *specific transfer.* This design provides the ability to control for (partial out) the effects of *general transfer.* The same post-test comparisons are made (e.g., task B). However, we now have more information. The second control group engages in task X during the initial task. Task X represents a task that is minimally related to task B but does provide for practice in learning how to learn tasks of the same class (Ellis, 1972, p. 840).

Table 10.2. Research Design for the Study of Specific Transfer

	Initial Task	*Transfer Task*
Experimental Group	Learn Task A (German Vocabulary)	Learn Task B (Russian Vocabulary)
Control Group 2	Learn Task X (French Vocabulary)	Learn Task B (Russian Vocabulary)
Control Group 1 (No Treatment)		Learn Task B (Russian Vocabulary)

By adding this second control group, one can determine if the observed transfer is the result of gross transfer or transfer specific to the similarities between tasks A and B. For example, transfer performance (task B) for the three groups is compared. If the experimental group and the no-treatment control group 1 differ, but there is no difference between the experimental group and control group 2 that was exposed to the initial task X, then transfer is specific to similarities between tasks A and B. However, if both the experimental group and the control group 2 outperform the no-treatment control group 1, then whatever is transferring from task A to task B is not specific to the two tasks. Rather, performance on the transfer task (task B) is not specific to the two tasks but due to a general kind of activity. These research design innovations served as precursors to the study of problem-solving transfer that reflects an integration of transfer theory and problem-solving theory that emerges during the 1990s.

Academic Learning: Theoretical Model

Since transfer and problem solving are being treated as special cases of the academic learning process, prior to moving to a discussion of problem-solving transfer, a closer look at a current definition of academic learning is warranted. Consistent with the correspondence principle, a theoretical conceptualization of problem-solving transfer should correspond to some degree with our theoretical understanding of academic learning.

Learning has so many meanings that it is necessary to first add a descriptive adjective. For example, in the classroom, there are different types of learning experiences. Some experiences involve motor learning that serves as the basis for motor skill development. In some cases, social learning activities focus on behavior that serves as the basis for social skills development. On the other hand, academic learning involves the processing and communication of cognitive information about the subject matter we teach in the classroom. Obviously, these types of learning are interdependent and do not occur in isolation. However, each has its own model and metaphor.

Academic learning as a specific "type of learning activity" has been defined in the following manner (Schunk, 2000; Shuell, 1986). "Learning is an enduring change in behavior, or in the capacity to behave in a given fashion, which results from practice or other forms of experience" (Schunk, 2000, p. 2). This definition of learning is consistent with a cognitive focus and captures the criteria most educational professionals consider central to learning.

Three elements of Shuell's (1986) definition of academic learning (behavioral change, endurance, and practice) deserve further attention. Together these three elements stress the idea that academic learning is not typically a one-shot instructional experience. It is a process of (1) acquiring new information, (2) the refinement and organization of what is already known with new information through practice, and (3) the successful use of that knowledge. Academic learning is the product of practice that provides the basis for relatively long-term change in one's personal knowledge. Ideally, this change will increase students' ability to be successful as they move from grade to grade where the curriculum requires more complex and specialized forms of personal knowledge.

Within this context of academic learning, we now turn to problem-solving transfer. This discussion of the academic learning process foreshadows the logical argument that current models of problem-solving transfer reflect efforts to integrate transfer of training and problem solving when the metaphor guiding theoretical integrative efforts is the human memory metaphor.

PROBLEM-SOLVING TRANSFER: METHODS AND METAPHORS (THE WHAT AND HOW OF IT)

Many exciting publications pertaining to the study of transfer and problem solving appeared during the last third of the 20th century. Consequently, it is impossible to acknowledge all of the important contributions to the cognitive revolution that have been made by individual researchers in educational psychology and instructional psychology. However, I do want to acknowledge a seminal theoretical article by Mayer and Wittrock (1996) that appeared in the *Handbook of Educational Psychology*. The chapter entitled "Problem-Solving Transfer" is an excellent blueprint for an integration of the study of problem-solving and transfer. I view this chapter as the theoretical basis for integrating research involving problem solving and transfer within the context of academic learning. Furthermore, I see the employment of the aforementioned research methods employed by experimental psychologists using the functional memory metaphor to guide research efforts as the means by which this integration of instruction, setting, and learner variables are accomplished.

Theoretically, to understand problem-solving transfer, it is necessary to define what Mayer and Wittrock (1996) meant by the concept. Problem solving is cognitive processing directed at achieving a goal when no solution method is obvious to the problem solver (Mayer, 1992). According to this definition, problem solving has four main characteristics. First, problem solving is *cognitive*—it occurs within the problem-solver's cognitive system and can be inferred indirectly from changes in the problem-solver's behavior. Second, problem solving is a *process*—it involves representing and manipulating knowledge in the problem-solver's cognitive system. Third, problem solving is *directed*—the problem-solver's thoughts are motivated by goals. Fourth, problem solving is *personal*—the individual knowledge and skills of the problem-solver help determine the difficulty or ease with which obstacles to solutions can be overcome (Mayer & Wittrock, 1966).

"A problem occurs when a problem solver wants to transform a problem situation from a given state into a goal state but lacks the obvious method for accomplishing the transformation" (Mayer & Wittrock, 1996, p. 47). This definition is consistent with the classic definition proposed by Dunker (1945) and exemplifies the aforementioned contribution of Gestalt psychology to the current study of problem-solving transfer.

Mayer and Wittrock (1996, p. 48) define the transfer process as occurring "when a person's prior experience and knowledge affects learning or problem solving in a new situation. Thus transfer refers to the effect of knowledge that was learned in a previous situation (task A) on learning or performance in a new situation (task B)." According to cognitive theories of transfer, transfer is mediated by the problem-solver's cognitive and metacognitive strategies. Within this theoretical context, a *strong claim* for specific evidence of problem-solving transfer would exist when the specific transfer of a cognitive or meta-cognitive strategy is observed.

Mayer and Wittrock (1996) also distinguish between knowledge transfer and problem-solving transfer. The basic distinction is comparable to the distinction between general knowledge and specific knowledge. Knowledge transfer is general and is addressed in a number of different ways. For further discussion of this point, the review by Dochy, Segers, and Buehl (1999) of knowledge transfer studies and the varied methodologies employed by researchers is very informative. Generally, Mayer and Wittrock's examples of problem-solving transfer tend to be examples of specific transfer with the use of a well defined problem as the transfer task.

Identification of Meta-Cognitive, General, and Specific Strategies during Problem-Solving Transfer

While between-group comparisons may provide good measures of gross and specific problem-solving instructional practices (see Tables 10.1 and 10.2), it is also possible to identify general, specific, or meta-cognitive strategies employed by a problem-solver. This is accomplished by extending the basic research design shown in Table 10.2. It is at this point where the design methodology of process dissociation can be introduced. Process dissociation techniques are commonly used by experimental psychologists employing the functional memory metaphor to identify differing cognitive processing or the use of differing cognitive strategies.

Process dissociation techniques involve the manipulations of materials to be studied and the instructions employed when designing a problem-solving transfer study. Process dissociation techniques are constructed on an understanding of how people process information. The following research design example (Table 10.3) is a brief description of a problem-solving instructional episode (task A) employed when using a process dissociation technique in order to determine if strategies that have been acquired and retained during instruction (practice) impact problem-solving transfer performance. This is followed by a second process dissociation technique employed during the problem-solving task (task B) in order to identify the problem strategy retrieved from memory by a student during successful problem solution.

In the research design example that follows, I have built on Robert Sternberg's (1985) theoretical work involving analogical thinking and reasoning. As noted in Figure 10.1, two very different types of problems, each of which has formal conventionalized characteristics, are inductive and deductive logic problems. These two types of problems were selected

Table 10.3. Cognitive Process Instruction and Delayed Problem-Solving Transfer Paradigm

Monday	*Wednesday*
Cognitive Process Instruction (Task A) *Analogy or Syllogism Instruction*	*Problem-Solving Transfer (Task B)*
Analogy Instruction Groups	*Common Transfer Task*
A set of 20 verbal analogy problems **Study Conditions** 1. Advice 2. Feedback **Practice Sequence** Test trial—Assess prior knowledge Test trial—Assess acquisition-learning Study trial—Study conditions (see above) Test trial—Assess acquisition-learning Study trial—Study conditions (see above) Test trial—Assess acquisition-learning	1. Twenty new verbal analogy problems all having cause–effect solution strategies. 2. Ten with cause–effect solution strategies and 10 with association solution strategies. 3. No mention made of previous instruction. 4. Problem-solving task consisted of a single test trial.
Syllogism Instruction Groups	*Common Transfer Task*
A set of 20 categorical syllogism reasoning problems. **Study Conditions** 1. Advice 2. Feedback **Practice Sequence** Same as analogy instruction groups.	Same as analogy practice group.
No-Treatment Control Group	Same as Analogy and Syllogism Groups.

because they are distinctive and require different strategies for problem solution. Consequently, problem-solving transfer between types (domains?) would not be anticipated (e.g., zero transfer). However, problem-solving transfer within a problem type would be anticipated (e.g., positive transfer). By manipulating the types of problems and the instructions encountered in the instructional task (task A) and the types of problems encountered in the transfer task (task B), one can identify strategies that problem-solvers retrieve from working memory and long-term memory to facilitate problem solving in the transfer task.

The research design (for task A), shown in Table 10.3, is a mixed factorial (two instructional groups by four practice trials) design with repeated

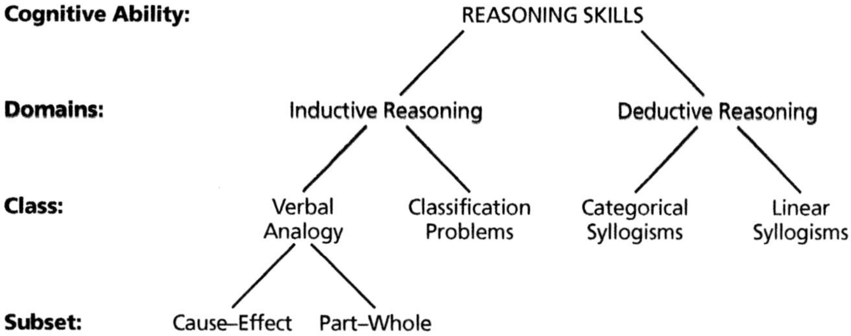

Figure 10.1. Partial hierarchical representation of inductive domain and deductive domain reasoning strategies.

measures. This design provides the basis for identifying the transfer of a meta-cognitive problem-solving strategy (inductive reasoning strategy or deductive reasoning strategy) during an *instructional episode* (task A). As mentioned earlier, the gestalt concept of "set" would allow us to predict no particular benefit from practice with deductive reasoning problems when the delayed problem-solving task (task B) consists of solving new analogy problems.

The assumption of no meta-cognitive strategy transfer would be confirmed by first comparing the problem-solving transfer performance (task B) of the deductive reasoning practice group with the no-treatment control group. If the deductive strategy training group performed significantly better than the no-treatment control group, then one could claim that prior knowledge about reasoning in general (deductive reasoning) facilitates problem-solving performance when the problems task involves inductive reasoning.

However, prior knowledge in the form of inductive reasoning is expected to facilitate problem-solving performance when the transfer task (task B) involves the solving of new analogy problems of the type encountered during instruction. This assumption about the transfer of a general strategy for solving inductive reasoning problems can be tested by comparing the performance of the inductive and deductive training groups. If the inductive training group performs significantly better than the deductive training group, prior knowledge in the form of a general strategy for solving inductive reasoning problems is demonstrated.

Finally, a test for the transfer of a specific analogy problem-solving strategy (cause–effect solution strategy) can be made. This is accomplished by testing for differences in performance by the inductive reasoning instruction group on the two different types of analogy problems encountered on the delayed problem-solving task. This would be performance on the 10 problems requiring a cause–effect strategy for successful solution in contrast with the 10 problems requiring an association strategy for solution. By inference, superior performance on problems requiring a cause–effect strategy is based on the retrieval of prior knowledge from long-term memory about "how" this type of problem is successfully solved.

Essentially, the example given above of "how" to use dissociation techniques to isolate cognitive processing is predicated on knowledge of research findings involving the conditions that produce positive, zero, and negative transfer as identified and defined during the verbal learning era. When this knowledge is integrated into the basic transfer of training research paradigms used for data collection by Thorndike and Judd at the turn of the 20th century, one has a working model for the investigation of problem-solving transfer.

However, this is not the full picture. As Dewey the pragmatist and functionalist reminds us, we must also consider tasks A and B within the larger context of adapting to the schooling environment over time. By analogy, tasks A and B are but snapshots in the DVD of life. Phrased differently, each problem-solver brought to the study a prior history or knowledge of problem solving and the use of problem-solving strategies. Thus, tasks A and B are but an isolated situation. Thus the bigger issue is one of attempting to control for individual differences in knowledge prior to tasks A and B. However, this is a secondary issue relative to this storyline; those interested in one such effort can pursue the Phye (2001) article looking at instructional issues in problem-solving transfer.

As indicated at the beginning of this discussion of problem-solving transfer, an experimental approach to identifying the "what" and "how" of problem-solving transfer can be accomplished by manipulating academic tasks and materials. This can provide credible empirical evidence for making strong inferences about cognitive processing involving a meta-cognitive monitoring strategy, a general problem-solving strategy (inductive reasoning), and a specific inductive reasoning strategy (cause–effect problem solution). However, we have not addressed the issue of "who" is successful in the process of problem-solving transfer. How do individual differences in memory abilities influence problem-solving transfer performance?

PROBLEM-SOLVING TRANSFER: LEARNER CHARACTERISTICS (THE WHO AND WHAT OF IT)

This combining of experimental and psychometric methods has a long history in psychology. A call for the analysis of individuals in groups is not new (Cronbach, 1957, 1975; Underwood, 1975). Both of these immanent psychologists called for the need to combine a group level of analysis and an individual level of analysis in the study of learning and achievement. This call has recently been reissued (Kosslyn et al., 2002). It is this approach to which we now turn.

By modifying the research design discussed above (Table 10.3), a pilot study can be developed that will provide insight into the issue of individual differences in working-memory monitoring ability. This is accomplished by employing only the inductive reasoning instructional task (task A) and a delayed problem-solving task (task B) where problems have only one solution strategy, the one practiced during instruction (task A).

The additional data collected in the pilot study would be obtained on the last practice trial during instruction (task A) by asking participants to do two things. As usual, indicate the correct solution and also identify the solution selected during the first test trial. In other words, we now have a measure of working memory monitoring during task A. Operationally, working memory monitoring ability during practice is defined as the number of correct identifications of the solution employed during initial exposure to the set of 20 problems in task A. Whether or not the initial solution was appropriate is immaterial. This measure simple reflects the ability to remember prior performance from working memory that occurred earlier in the instructional episode.

This design now provides an opportunity to address a research hypothesis about the role that working memory monitoring ability plays in successful problem-solving transfer performance (task B). However, this requires a paradigm shift in terms of statistical analysis logic. As suggested elsewhere (Phye & Sanders, 1992), causal modeling provides a means for addressing this question. The causal model in Figure 10.2 reflects a pilot data analysis where prior knowledge is defined as a person's first trial performance during practice (task A—Table 10.3). This is assumed to be an accurate reflection of prior knowledge that participants brought to the task. The working memory monitoring ability measure was described above. The delayed transfer task consisted of 20 new analogy problems. Thus we have three data points and the question to be addressed is two-fold. First, what do individual differences in appropriate prior knowledge brought to the instructional task contribute to delayed problem-solving transfer performance? Second, what role does individual differences in working memory monitoring ability (at work during practice) have on problem-solving transfer performance? Is working memory monitoring ability a variable that moderates the influence of practice on problem-solving transfer performance?

The causal model with standardized path coefficients shown in Figure 10.2 is simply pilot data that provides insight for further research. From a causal modeling perspective that takes a path analysis approach, individual differences in appropriate prior knowledge brought to instruction does di-

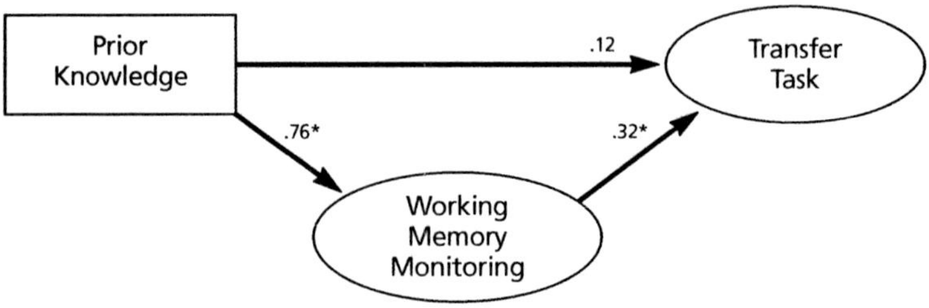

Figure 10.2. Trimmed causal model demonstrating the centrality of working memory.

rectly impact delayed problem-solving performance to a small degree. Furthermore, the relationship between amount of appropriate prior knowledge and amount of successful problem-solving transfer is a positive relationship (albeit small). However, when instruction involves practice (as in task A), working memory monitoring ability is a critical mediating variable during practice. In other words, effective problem-solving transfer performance is mediated by the ability to monitor one's working memory performance during instruction. Here, the relationship between individual differences in appropriate prior knowledge is positively related to individual differences in monitoring performance that is positively related to successful problem-solving transfer performance. Note that the introduction of working memory monitoring ability as a learner characteristic provides an effective means of identifying "who" will benefit most from problem-solving instruction. This also raises the research question, Can we teach students to improve their ability to monitor their own cognitive activities during problem-solving instruction?

In summary, the integration of experimental and psychometric approaches further refines the discussion of problem-solving transfer to include a discussion of individual differences that address the question of "who" is a successful problem-solver. In a real sense, this integration of methods also provides one way of addressing the "automaticity" issues that have been raised. In other words, successful problem-solving transfer is not simply the product of an "affordance" created by a combination of task variables and the environmental situation.

Rather, active memory processing is critical (Dewey's action principal). However, task variables and environmental situation have a critical role to play. Without a doubt, problem identification is the initial processing stage in the problem-solving transfer process. Here, familiarity plays a crucial role and automaticity in the form of Dewey's habits is important. However, problem representation, strategy development, and solution verification are corollary processes. It is during these stages in the problem-solving transfer process where active cognitive processing is involved.

Within the context of the method and design elements to be considered when investigating problem-solving transfer, Figures 10.3 and 10.4 represent an information-processing approach to laying out a number of variables that need further investigation. Figure 10.3 reflects a focus on instruction (task A) and includes the manipulation or identification of task variables, learning materials, and subject variables related to motivation and focuses on the acquisition process. Figure 10.4 reflects a focus on transfer and is an attempt to determine the impact of source memory (task A).

The basic problem-solving transfer paradigm depicted in Figures 10.3 and 10.4 can also be thought of as a way of addressing research questions pertaining to the general topic of academic learning. Consistent with the earlier discussion of transfer and problem solving as special cases of the learning process, research questions pertaining to instructional issues that facilitate acquisition and promote retention involve task A. By following students from task A to task B, the ability to use what was acquired in task A under the constraints imposed in task B can be determined. Within the

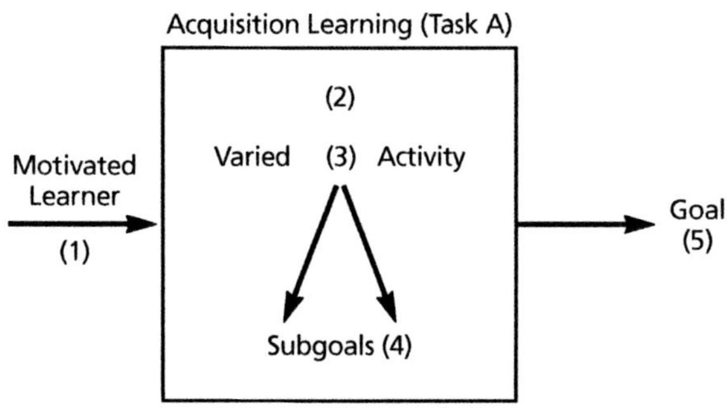

(1) **Motivated Learner**: Each participant indicated her or his confidence in the solution constructed for each of the 20 problems encountered during the first practice trial.

(2) **Problem Situation (training)**: Two groups of participants learned how to solve 20 verbal analogy problems. Two groups of participants learned how to solve 20 categorical syllogism problems.

(3) **Varied Activities**: All participants (regardless of task) experienced the following training sequence (a) test, (b) study, (c) test, (d) study, (e) test, (f) study, and (g) retention measures. During the retention trial, participants were asked to provide for each problem, both the correct solution and the solution they constructed on the initial practice trial.

(4) **Subgoals**: Assumably, these were (a) the confirmation of appropriate solution, (b) the correction of errors, (c) proceduralization of schema for memory representation through effective search and retrieval processing.

(5) **Goal**: Improve performance or master task.

Figure 10.3. Hypothesized cognitive processing during acquisition learning (training) for task A.

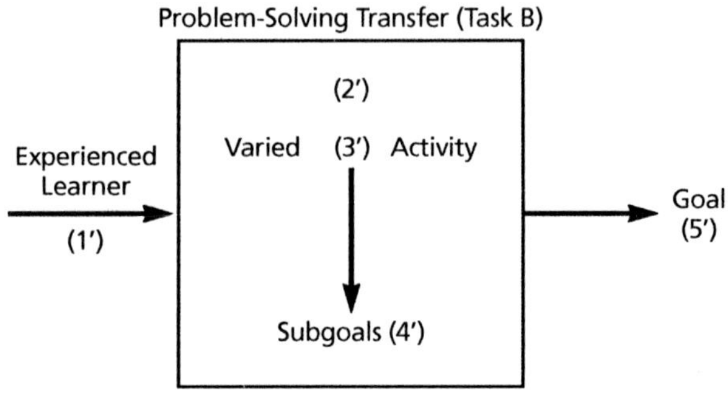

(1′) ***Experienced Learner**: Participants have had recent (2 days) prior training with either verbal analogy or categorical syllogism problems.

(2′) **Problem Situation (delayed transfer)**: Single exposure to 20 new problems. Half required a cause–effect solution strategy (comparable to analogy problems in training) and half an association solution strategy.

(3′) **Varied Activities**: Recall from long-term memory of appropriate prior knowledge. This involves the subprocesses of memory search and schemata retrieval. This may be thought of as accessing prior procedural knowledge.

(4′) **Subgoals**: Use prior experience to facilitate (a) problem identification, (b) problem representation, (c) solution construction, and (d) solution execution.

(5′) **Goal**: Solve as many problems as possible.

Figure 10.4. Hypothesized cognitive processing during the delayed problem-solving task that served as the transfer task B.

context of the memory metaphor, the basic definition of academic learning has been met and one has a research design that provides data collection reflecting a true integration of the instruction and learning process. From a "common sense" perspective, while tasks A and B are but different phases of the same "learning process," determinates of change within each phases and between phases can be identified.

For example, Jonassen (2002) has published an excellent review of a number of problem-solving designs that could be employed by instructional designers when studying problem-solving acquisition. Any of these designs could be employed as task A of the problem-solving transfer paradigm described in Figure 10.3. The addition of an appropriate delayed transfer task (Figure 10.4) would also provide evidence of permanency (retention) and use of prior knowledge in a transfer situation (memory retrieval of task A knowledge); in other words, problem-solving transfer.

Back to the Future

My use of the phrase "back to the future" is predicated in part on the idea that progress is neither a straight line nor a pendulum swing. Rather, change experienced by human beings is more likely to approximate an epigenetic model. In education, the analogy is the spiral curriculum proposed by Jerome Bruner. An epigenetic model of change involves forward progress combined with a recycling (or cycling back) process. Hopefully the recycling process involves an integration of the successful old and the promising new ideas involving educational research and provides a synergy for educational practice.

From a philosophical perspective, Dewey's pragmatic research approach to the study of learning still has much to offer as a guiding force. The cognitive nature of Dewey's grand theory is consistent with, and has demonstrated fidelity with, current mainstream psychological theory, models, and metaphors. This awareness is critical for educational researchers who will inevitably involve some aspect of adaptive behavior either on the part of teachers or students in their research design. The teaching–learning equation is difficult to consider without one or both of these human actors.

In terms of models and metaphors, a pluralistic approach is not only warranted but necessary given the nature of the research questions being addressed. The point being made is that it is the research question that dictates the model or method, not the other way around. In the educational arena today, it is not a "chicken or the egg" issue. Rather, educational researchers are being asked to address the very practical question, "How do educational practices impact student learning and achievement?"

In closing, I urge educational psychologists to remember Dewey's philosophical position that educational research is not a matter of trying to identify what education is, but what education is for. More to the point, there must be a correspondence between philosophy, psychological models, metaphors, and methods. I hope that this integrative effort that focuses on an extremely significant educational issue, *problem-solving transfer,* has been of some value in this regard.

REFERENCES

Bartlett, F. C. (1932). *Remembering: A study in experimental and social psychology.* Cambridge, UK: Cambridge University Press.

Bayles, E. E. (1936). An unemphasized factor in current theories regarding the transfer of training. *Journal of Educational Psychology, 27,* 425–430.

Biesta, G. J. J., & Burbules, N. C. (2003). *Pragmatism and educational research.* Lanham, MD: Rowman & Littlefield.

Boring, E. G. (1957). *A history of experimental psychology.* New York: Appleton-Century-Crofts.

Bredo, E. (1997). The social construction of learning. In G. D. Phye (Ed.), *Handbook of academic learning: Construction of knowledge* (pp. 3–45). San Diego, CA: Academic Press.

Bruner, J. S., Goodnow, J., & Austin, G. (1956). *A study of thinking.* New York: Wiley.

Bugelski, B. R., & Huff, E. M. (1962). A note on increasing the efficiency of Luchins' mental sets. *American Journal of Psychology, 75,* 665–667.

Coover, J. E., & Angell, F. (1907). General practice effect of special exercise. *American Journal of Psychology, 18,* 328–340.

Cronbach, L. J. (1957). The two disciplines of scientific psychology. *American Psychologist, 12,* 671–684.

Cronbach, L. J. (1975). Beyond the two disciplines of scientific psychology. *American Psychologist, 30,* 116–127.

Dehue, T. (2000). From deception trials to control reagents: The introduction of the control group about a century ago. *American Psychologist, 55,* 264–268.

Dewey, J. (1896). The reflex arc concept in psychology. *Psychological Review, 3,* 357–370.

Dochy, F., Segers, M., & Buehl, M. M. (1999). The relation between assessment practices and outcome studies: The case for research on prior knowledge. *Review of Educational Research,* (69), 145–186.

Dunker, K. (1945). On problem solving. *Psychological Monograph, 58.*

Ebbinghaus, H. (1964). *On memory.* New York: Dover Publications. (Original work published 1885)

Edwards, A. J., & Scannell, D. P. (1968). *Educational psychology: The teaching – learning process.* Scranton, PA: International Textbook Company.

Ellis, H. C. (1972). *Fundamentals of human learning and cognition.* Dubuque, IA: Wm. C. Brown.

Gagne, R. M. (1965). *The conditions of learning.* New York: Holt, Rinehart & Winston.

Gibson, E. J. (1994). Has psychology a future? *Psychological Science. 5,* 69–76.

Gibson, J. J. (1941). A critical review of the concept of set in contemporary experimental psychology. *Psychological Bulletin, 38,* 781–817.

Guilford, J. P. (1950). Creativity. *American Psychologist, 5,* 444–454.

Haskell, R. E. (2000). *Transfer of learning: Cognition, instruction, and reasoning.* San Diego, CA: Academic Press.

Heidbreder, E. (1933). *Seven psychologies.* New York: Appleton-Century-Crofts.

Hilgard, E. R., & Bower, G. H. (1966). *Theories of learning* (3rd ed.). New York:Appleton-Century-Crofts.

James, W. (1890). *Principles of psychology.* New York: Holt Reinhart & Winston.

James, W. (1958). *Talks to teachers.* New York: Norton. (Original work published 1899)

Jonassen, D. H. (2002). Integrating problem solving into instructional design. In R. A. Reiser & J. Dempsey (Eds.), *Trends and issues in instructional design and technology.* Upper Saddle River, NJ: Prentice-Hall.

Judd, C. H. (1908). The relation of special training to general intelligence. *Educational Review, 36,* 28–42.

Kendler, H. H., & Kendler, T. S. (1962). Vertical and horizontal processes in problem solving. *Psychological Review, 69,* 1–16.

Kosslyn, S. M., Cacioppo, J. T., Davidson, R. J., Hugdahl, K., Lovallo, W. R., Spiegel, D., & Rose, R. (2002). Bridging psychology and biology: The analysis of individuals in groups. *American Psychologist, 57,* 341–351.

Levin, H., Watson, J. S., & Feldman, M. (1964). Writing as pretraining for association learning. *Journal of Educational Psychology, 55,* 181–184.

Lindquist, E. F. (1953). *Design and Analysis of experiments in psychology and education.* Boston: Houghton Mifflin.

Luchins, A. S. (1942). Mechanization in problem solving: The effect of Einstellung. *Psychological Monograph, 54.*

Luchins, A. S., & Luchins, E. H. (1950). New experimental attempts at preventing mechanization in problem solving. *Journal of Genetic Psychology, 42,* 279–297.

Maltzman, I. (1955). Thinking: From a behavioral point of view. *Psychological Review, 62,* 275–286.

Mayer, R. E. (1992). Thinking, problem solving, cognition (2nd ed.). New York: Freeman.

Mayer, R. E., & Wittrock, M. C. (1996). Problem-solving transfer. In D. C. Berliner & R. C. Calfee (Eds.), *Handbook of educational psychology* (pp. 47–62). New York: Macmillan.

McGeoch, J. A. (1942). *The psychology of human learning.* New York: Longmans, Green and Company.

McGeoch, J. A., & Irion, A. L. (1952). *The psychology of human learning* (2nd ed.). New York: David McKay Company.

Mednick, S. A. (1962). *The associative basis of the creative process. Psychological Review, 69,* 220–232.

Muehl, S. (1960). The effects of visual discrimination pretraining on learning to read a vocabulary list in kindergarten children. *Journal of Educational Psychology, 51,* 217–221.

Muehl, S. (1961). The effects of visual discrimination pretraining with word and letter stimuli on learning to read a word list in kindergarten children. *Journal of Educational Psychology, 52,* 215–221.

Neisser, U. (1967). *Cognitive psychology.* New York: Appleton-Century-Crofts.

Osgood, C. H. (1949). The similarity paradox in human learning: A resolution. *Psychological Review, 56,* 132–143.

Phye, G. D. (2001). Problem-solving instruction and problem-solving transfer: The correspondence issue. *Journal of Educational Psychology,* 93, 571–578.

Phye, G. D., & Sanders, C. E. (1992). Accessing strategic knowledge: Individual differences in procedural and strategy transfer. *Contemporary Educational Psychology, 17,* 211–223.

Sahakian, W. S. (1975). *History and systems of psychology.* New York: Wiley.

Schunk, D. H. (2000). *Learning theories: An educational perspective* (3rd ed.). Upper Saddle River, NJ: Merrill.

Shuell, T. J. (1986). Cognitive conceptions of learning. *Review of Educational Research, 56,* 411–436.

Stanger, R. (1988). *A history of psychological theories.* New York: Macmillian.

Stanovich, K. E. (2001). The rationality of educating for wisdom. *Educational Psychologist, 36,* 247–251.

Sternberg, R. J. (1985). *Human abilities: An information processing approach.* New York: Freeman.

Thorndike, E. L. (1913). *Educational psychology, Volume II: The psychology of learning.* New York: Teachers College, Columbia University.

Thorndike, E. L., & Woodworth, R. S. (1901a). The influence of improvement in one mental function upon the efficiency of other functions: The estimation of magnitudes. *The Psychological Review, 8,* 384–395.

Thorndike, E. L., & Woodworth, R. S. (1901b). The influence of improvement in one mental function upon the efficiency of other functions: Functions involving attention, observation, and discrimination. *The Psychological Review, 8,* 553–564.

Underwood, B. J. (1975). Individual differences as a crucible in theory construction. *American Psychologist, 30,* 128–134.

Van de Geer, J. P. (1957). *A psychological study of problem solving.* Haarlem: Utigeverij de Toorts.

Wertheimer, M. (1959). *Productive thinking.* New York: Harper & Row. (Original work published 1945.

CHAPTER 11

SOCIAL PERSPECTIVES ON THE COGNITIVE REVOLUTION AND EDUCATION

From Alien Beings to Robust Trustees

Peter Freebody
Griffith University

ABSTRACT

In this chapter I take a selective look at aspects of the conjoined histories of psychology, education, and cognitive science in an effort to make three arguments. The first is that, while the so-called cognitive revolution was not simply a revival of pre-behaviorist mentalism, it did nonetheless signal a revival of ideas either discarded or left unresolved by key psychologists working around the turn of the 20th century. I offer the cases of Binet and Huey as brief illustrations of the point that these ideas largely revolved around the question of how to theorize relationships among mind, behavior, and social experience.

Second, I explore moderate and strong arguments for the place of the Social in cognitive theorizing as they might be productively brought to bear on the

The Impact of the Cognitive Revolution on Educational Psychology, pages 293–323

study of educational practices. I draw on Bransford, Brown, and Cocking's compilation *How People Learn: Brain, Mind, Experience and School* (1999), considering it as an attempt to use a moderate form for the place of the Social. I conclude that, despite its other substantial strengths as a statement about the relationship between research and educational practice and policy, it is finally unsatisfying in its characterization of the significance of social experience. I then draw out, with a few examples of classroom interaction, observations made or at least hinted at by Binet and Huey, and the ongoing concerns of Jerome Bruner, to make the stronger case for the place of the Social in cognitive theorizing. I argue that it is no longer productive to persist with the undertheorized "bucket" theory of context evident in the moderate case (where things happen "inside" a context, and where what is "outside" individuals is taken as a "factor" in accounting for their behavior). Rather, I endorse a more profound appreciation of social context, at least as profound as members of a society appreciate it as evidenced ways in their daily social practices, that is, as (1) transformable, (2) both the project and the product of their coordinated actions, and (3) simultaneously and continuously shaping and renewed by their actions. This leads to a view that the study of education is enhanced by documentations of naturalistic uses of language as it is used in everyday educational practices, as these in turn are organized by natural language and common sense, and through a perception that educational events are in a dialectic relationship to commonsensical ways of reasoning about the world—drawing on and critique them simultaneously.

Finally, I suggest that the influences among educational psychology, cognitive science, and education have not all been from the sciences to the professional practice. Rather, an abiding concern with educational and acculturational matters has persistently stretched psychologists of various persuasions to more and more comprehensively "social" takes on human activity, such that we can arrive at the equally intriguing question of influence of educational practice and research on the development of cognitive science.

INTRODUCTION

It had to happen that we would look back and say "it had to happen." Cognitivists are quick to point out that our retrospective accounts are unreliable, smoothed out into linear stories with unambiguous judgments, fanciful chains of causes and effects, and all of it apparently heading toward the now-obvious inevitabilities—pulled along by what "had to happen." In our retrospections, they warn, we may even recall revolutions.

This Proustian dilemma becomes more troubling when we turn our attention to gauging the nature and extent of the impact of the cognitive revolution on educational psychology, and in particular the possible role that social perspectives might have had in that impact. The 40 or so years since the revolution that this volume celebrates have encompassed the professional lives of many of its contributors. The ebb and flow of the intervening debates have helped shape our professional understandings and investments. As educational cognitive psychology matured, the smoothed-out causes became effects, which became in turn new causes, and so on. A shared narrative retrospection developed to house and nurture a subculture.

In this chapter, I spend some time in retrospect, and then project what the moderate and strong versions of the "sociocognitive" position may mean for the practice of educational research from now on. The chapter is not intended to review the now-extraordinary corpus of research and theory developed from within sociocognitive perspectives. Rather, its primary aim is to make plausible two claims. The first is that the contribution of cognitive approaches to educational psychology would be enhanced by an even more intensive theorization of the Social as a comprehensive organizational feature not only of education, but of cognition.

The secondary claim is that it is salutary to turn on its head this central question of the cognitive revolution's impact on education: For over 100 years a prime source of dynamism in psychology has in fact come from the preoccupation of many of its most influential practitioners with matters to do with the acculturation, development, and education of the young, with how novices actually become people, and even sometimes experts, and, more generally, with how communities, and societies can fashion the kinds of people they think they want or need (Wenger, 1998). Even the briefest glance at how individuals, groups, communities and societies accomplish this acculturation, development, and education of the young almost makes a mockery of the conventional drive for parsimonious explanation: Almost a century ago John Dewey remarked on the astonishing feat of education: "by various agencies, unintentional and designed, a society transforms uninitiated and seemingly alien beings into robust trustees of its own resources and ideals."

The argument here is that the imperatives and aspirations of education have provided motivation for pushing psychology to focus on the "messiness" of educational experience and to find in that messiness glimpses of complexity, orderliness, and even artfulness. So I aim to show in part that it is worth reversing the question: How have the realities of everyday educational experience impacted on educational psychology? Is it educational realities that have driven (or at least enticed) psychology toward the waystation of cognitivism? And are they now luring it toward investing even more heavily in a reinvigorated appreciation of social order, as both the source and accomplishment of cognitive processes? How can education even more forcefully inform psychology?

In its most general terms, the open-textured collection of orientations to human learning, thinking, and acting that we might label "sociocognitive" takes it that learners are involved directly in social contexts, and that any explanation worth its air-time must first and foremost avoid any simplistic separation of the learner from the Social. This involvement is evident in the give and take of interactions in important relationships that themselves embody the propositional, procedural, and (in some accounts) the moral contents of the reciprocal attributes of the "expert" and the "novice." This reciprocation is activated when these categories of people are engaged, and knowingly engaged, in purposeful, effortful interactions—parents aren't teachers all the time, but when they are, the child had better know that it's "student time."

Conventionally, sociocognitive orientations take it that these relationships are situated in, mediated through, and constituted around cognition (Resnick, Saljo, Pontecorvo, & Berge, 1997; Rogoff & Lave, 1984), viewed generally as ways of marshaling, directing, and generally managing the interaction between what is known and what is not known, building the epistemological bridges between experience, belief, and knowledge as they are being crossed. Meaning and knowledge are always in the process of construction, never fixed statically within an environment, always dynamic. People working with meaning and knowledge are always seeing and moving through, being offered different kinds of understanding and agency.

To open those observations out into a case for the role of the Social in cognitive perspectives and, in turn, for the impact of this on educational psychology, we can start by discussing how sciences change and how this relates to changes in neighboring areas of inquiry and professional practice. We can draw here on the work of Alain Desrosieres, the Director of the French National Bureau of Statistics, who recently provided a thoughtful account of the history of development of statistics as, literally, "the measure of the state." Desrosieres (1998) showed how individuals and professional subcultures invest in particular theoretical, empirical, and methodological practices and how particular concepts become objectified so that they can be inserted into debates as objects that are both known and known-in-common:

> The student, research worker, or statistical data-user receives compact concepts, encapsulated into concise and economical formulas—even though these tools are the result of a historical gestation punctuated by hesitations, retranslations, and conflicting interpretations.... In science-in-the-making (or "hot" science), truth is still a wager, a subject of debate; only gradually, when science cools down again, are certain results encapsulated, becoming "recognized facts." ... (pp. 2, 5)

Within one collection of conceptual frameworks, social phenomena such as the "processes of learning" are taken to exist independently of their methods, sites, or conditions of production through measurement.

They are taken to be discovered. From this perspective, the important issues are the reliability, validity, and utility of the theoretical or statistical representations. In contrast, we find other frameworks that emphasize the local, variable, contingent, and multiple nature of cultural, social, and institutional practices. This latter view insists that the very existence of phenomena such as the "processes of learning" is at least in significant part a matter of the conventions that have developed to name, transmit, and assess them.

Desrosieres's point is that the natural and social sciences drive toward the abstraction of categories of equivalence (e.g., "the *processes of learning* over here are the same as they are over there"). The aim is to make these categorical objects "grip" and transcend the singularity of individual situations and occurrences of, in our case, teaching and learning. The potential cost is always their empirical adequacy as a description of any particular single occurrence of teaching and learning. A central observation for Desrosieres's is that the production of durable objects through science is not just for the management of the work of scientists; it is as well a key move in the construction, unification, and administration of the state's key formations (such as public education through schooling). Versions of the "processes of learning," for example, will grip partly because they afford practical administrative procedures in classrooms, schools, regional educational offices, central authorities, and educational publishing houses.

Desrosieres observed that the primordial dialectic driving the conduct of science is this: The "durably solid forms" of the natural and human worlds (whether they be constructs such as immunity levels, unemployment rates, cognitive processes) must remain always undebated so that human action can proceed at the same time as they must remain always debatable so that human action can change. We can think of education as one such category of human action: Educators need to act as if they work in a space occupied by durably solid forms—learning, tests, grades, syllabus, motivation, abilities, good and less-good outcomes, and all the rest, including constructs drawn form cognitive psychology. At the same time, we must behave as if such forms are always debatable, always amenable to more refined or productive understandings and practical implementations.

The view that a science will oscillate between "hot" and "cool" phases alerts us to the possibility of continual, cyclic changes, especially in an enterprise as complex as psychology, and even more especially as it strives for descriptive and explanatory adequacy of an open-textured collection of social practices such as "education." It is the dialectic of "solid forms"—at the one time durable and debatable—that indexes these cycles, and that indicates the fundamentally social nature of educational psychology as itself a set of social practices around key puzzles.

THE SETTING FOR A REVOLUTION?

So what and when was the cognitive revolution? In Anglophone countries in at least the early part of the 20th century saw comprehensive and generally successful attacks on introspection-based Wundtian mentalism. These attacks came to be consolidated around the approach termed "behaviorism," and chapters in this volume document aspects of the lines of attack and their various consequences. It seemed, by the middle of that century that psychologists had so thoroughly conducted this theoretic cleansing that, in many circles, there was not, nor could there foreseeably be, any place for concepts of "mind" in the disciplined study of human behavior. The strong forms of behaviorism claimed that minds did not exist; the softer forms conceded that they might, but that they didn't matter because nothing that might or might not go on in them could count as an explanation of anything anyway.

It is important to recognize, however, that while the behaviorist position was universally acknowledged, not all of its precepts were universally accepted. Concepts of mind continued to be debated and developed, generally in university departments other than those named psychology. The work of John Dewey, George Herbert Mead, and their followers illustrate how social theoreticians and philosophers of knowledge and human experience felt the need to acknowledge explicitly the victories enjoyed by behaviorism in psychology at the same time as they reworked notions of mind, admittedly tiptoeing around some of the more threatening minefields. (Mead's collected works, for example, published by University of Chicago Press in 1934, were entitled *Mind, Self, and Society from the Standpoint of a Social Behaviorist.*)

Any account of the cognitive revolution and of the various domains it influenced presupposes among other things a theory of historical change in intellectual and institutional life. Dynamic movements in and around the social sciences, for instance, may have led some psychologists to develop a roving eye. In one account, social scientists, even experimental psychologists, simply lost interest in the restraints that behaviorism had applied to their work and their thinking for half a century. They got bored with the decor and the company, checked out, and looked for more interesting accommodation. Other accounts look to the agency of strong and innovative individual social scientists, reading off this revolution as a consequence of their initiatives and insights. The latter approach seems to characterize George Miller's description of the cognitive revolution, in which Miller himself had a significant role to play. Miller's account (2003) is instructive in a number of ways: essentially, he describes behaviorism as, in its day, "an exciting adventure for experimental psychology," as only really adhered to by experimental psychologists in the United States, and as publicly and evidently "failing." Miller dates the commencement of the cognitive revolution to another September 11, this one in 1956, the second day of a symposium organized at the Massachusetts Institute of Technology. This day's presentations included a variety of new approaches to the ways in which humans and machines process information, and included a presentation by Chomsky on syntax, a preview of his *Syntactic Structures*. In the years that followed, a number of key texts appeared that indicated that this shift in orientation entailed more than simply the granting of permission for ideas about the mind to be included in scientific accounts of human behavior. Among these texts were Miller, Galanter, and Pribram's *Plans and the Structure of Behavior* in 1960 and John Flavell's introduction to the work of Piaget in 1963 (much of that work having been left untranslated into English for about 30 years).

But even now, with the benefits of inevitability, it seems remarkable that hard-nosed educational psychologists, tempered in the crucibles of experimental design and inferential statistics, chastened by Campbell and Stanley's ominous warnings about "threats to validity" (the social sciences' equivalent of the domino theory), and the ruling imperatives of operationalization, observability, and associationism, would apparently devolve into speculators on the "mental models" market.

What became clear in the decades that followed is that psychologists' central investments were in the category "science" as centrally defined in terms of method and methodology—rather than in a collection of topics or propositions about its key puzzle, human behavior. "Mind" was admitted, but the methodological strictures that grew within the natural sciences and that had been recruited in much behaviorist work remained in force. It was these that were taken to guarantee the status of scientific work.

The time certainly seemed right for a revolution in something: Since this is an essay concerning social perspectives, I should stress the important developments that were reshaping ideas about society, education, and the production of a particular kind of effective, intelligent, and moral citizen in the United States and elsewhere. It is commonplace to attribute many of the changes that occurred in the social sciences between the early 1960s and the mid-1970s to the accumulation and interaction of pressures such as the credit squeeze in Western countries in the early 1960s, the influence of the Keynesian economics of growth and redistributive policies, civil rights and other populist movements in the United States and elsewhere, the mobilization of political and moral debate along generational, class, and racial lines, around the Vietnam War, the development of computers and other communicational and data-storage technologies, the growth of mass domestic media and the rise of new youth cultures and economies, and so on. The simultaneous and interactive impact of these seems more than sufficient to explain a constellation of cultural, intellectual, ideological, and institutional revolutions.

But there were also more local if less dramatic factors at work within the social sciences themselves, in particular in the ways in which they were and were not having an impact on educational practice and policy. One of the key publications that both catalyzed and capitalized on this sentiment was Bruner's (1960) text on the "process of education." It is significant that one of the first major applications of these new ways of orienting to human behavior and thought should be in the field of education, and notable that, in that book, Bruner pushed ideas about "mind," challenging the conduct of psychology to account more faithfully and cogently for educational phenomena and concerns. In reading Bruner's account today you will still sense its urgency—its sense of impatience, almost indignation, about how seriously educational practice had lagged behind what it could and should have been, even in a society as materially and intellectually resourceful as the United States. Thirty years later, Bruner was to reissue these complaints and challenges (Bruner, 1990; see Bakhurst & Shanker, 2001) in terms that I pick up later in this chapter.

What, in hindsight at least, seemed loomingly evident by the early to mid-1960s was that the generations of gifted and committed students, scholars, practitioners, and policymakers who had, over 50 years, invested in the Great Digression—behaviorism—had harvested precious little by way of answers to the big educational questions. Many of us had our first teaching assistant jobs delivering courses written or overseen by professors still comfortably accommodated by the Great Digression. We were required to give lectures and tutorials to teacher education undergraduates on topics such as "learning and negative reinforcement: Implications for teaching." I can't speak for my colleagues, but I could stare at those words as long as I liked, and, rather than epiphanies, all that would come to me were "implications" of the most breathtaking banality. These I offered with a seriousness bordering on whimsy to young teachers-in-training. I knew that they were about to work in minority, disadvantaged, or isolated schools; that they were trying to act on their commitment to educating children whose biographies, cultural backgrounds, and even languages, they would come to discover, were effectively mysteries to them, using curriculum materials and assessments that might as well have been dropped by parachute into the principal's office. Where was the generative repertoire of techniques, the paradigm with purchase, the powerful idea—anything?

Something had to happen, and somebody had to do it, and there was nothing else for it, it had to be the educational psychologists; there simply was not much good help to be had elsewhere in the neighborhood. The other supposedly "informing, foundational" disciplines of educational sociology, philosophy, history, and economics seemed generally to prefer to remain unburdened by mundane educational side issues such as the actual practices of teaching and learning (there were a few exceptions, such as Dreeben, 1969). Yes, teaching and learning were the ostensible topics of much discourse, but, as actual, concrete daily practices, they made only cameo appearances in semantic struggles that were waged in what progressivists might now describe as "higher-order thinking," talismans from a far-off place or time. The details of the mundane practices of teachers and students were the province of psychologists. The draft was on. Whether or not a cognitively informed educational psychology has indeed come to grapple with the details of educational practice is another question, but by the mid- to late 1960s, the breach had at least been named.

BEHAVIORISM'S UNFINISHED BUSINESS

The cognitive revolution was not simply a renaissance of introspection-based Wundtian mentalism. It capitalized on new developments in linguistics, and engaged with a rapidly growing adolescent computer science, and the new twinkle in its eye, artificial intelligence, but at a deep level, it reconnected with ongoing philosophic deliberations on the central topic and resource of human reflection, the mind. The cognitive revolution connected psychology with the significant contemporary themes of language, knowledge, communication, and their multiple and increasingly sophisticated technologies. Specifically, it represented an engagement with Chomskyan formulations about language development and use that explicitly called on, and indeed installed, a particular concept of "mind," and with post-Turing speculations and machine-experiments that aimed to emulate either the products and/or processes of mentation.

To see the extent to which the Great Digression had dissociated theory and research from experience and practice, we can consider the state of educational research at the advent of the behaviorist critique and the motivations that behaviorism left unattended. The argument here is that one of the reasons the cognitive revolution redefined educational research so thoroughly in the early to mid-1960s was precisely its equally warm embrace of an earlier set of vestigial concerns of psychological scientists, from the behaviorism of 40–50 years before. This latter turn is itself partly explained by the poorly developed analytic and explanatory concepts and systems within educational research itself. What seems to be the case is that educational researchers simply did not have at their disposal ways of readily addressing the factors that become so compelling to us immediately we turn our attention to the practices and settings of actual teaching and learning. To illustrate this view I take an unfairly brief look at two educational researchers who were prominent about 100 years ago, immediately before the serious onset of behaviorist thinking in anglophone psychology—Alfred Binet and Edmund Huey.

Alfred Binet was founder and director of the Independent Society for the Psychological Study of the Child. In 1904 he was commissioned by the Paris Schools Authority to develop assessment instruments that would enable the Authority to distinguish those children who could benefit from public schooling from those who needed special remediation. Binet and his associates developed the New Methods for Diagnosing the Intellectual Level of Abnormal People, the precursor of what we now recognize as intelligence tests. Binet's tests, although spawning many offspring around the world, were never actually used in France. They were also put to purposes different from those initially intended by Binet, and purposes that Binet, in all probability, would oppose (Pollack, 1995). In particular, Binet's position was that "the best way to predict success in school was to measure success in school" (Rogers, 1995, p. 653), in ways strongly grounded in the contours of everyday school life, including the curriculum, the social organization of classroom and school life, and the relationships made possible therein.

As the tests were applied more broadly across diverse cultural settings, the concept of intelligence developed a functional autonomy, gradually assuming a position in theoretical, professional, and policy discourses that hovered, like a "variable," above everyday activity. One of the moves this afforded was the "covarying out" of IQ scores to examine the residual effects of a particular kind of training on performance, effectively, to ask the question: "How much effect did this intervention have on increasing the intelligent practice of these learners after we statistically remove considerations of intelligent practice?"—as wild a form of dissociative thinking as is possible from any vaguely situated educational perspective.

One of the victories of the cognitive revolution was to move intelligence away from center-stage educational theorizing and policy. Binet-style tests are now typically taken to reflect "achievement," a label that does not really avoid the criticisms already leveled at the notion of intelligence (such as its singularity, stability, and generalizability), but which at least grounds the variable in the contours of daily practice.

Binet's story enables us to see the connection between four developments:

1. The institutional, policy-driven imperative of using funds for public schooling in apparently productive ways. Public education was expanding in the early years of the 20th century, particularly in the cities of North America and Europe. To be simultaneously efficient and legitimate, managing classrooms full of students of all varieties became important.

2. The drive toward the production of an assessable (quantifiable), free-range variable, in this case, intelligence.
3. The retheorization of a human's various practices into stable, assessable, individual attributes, such that the new "social fact" of intelligence could be inserted as an object into contemporary debates and thereby provide a criterion that could redefine and reorganize practice and serve policy.
4. The ways in which the unintended consequences of research can have important and lasting effects on the nature of research and on the conduct of educational practice and debate.

In 1908, Edmund Burke Huey was the first U.S. psychologist to produce a systematic and scholarly study of the teaching and learning of reading in school. It is useful to recover some of Huey's (1908) own words, written a century ago, to realize how some of his central interests are directly paralleled in much contemporary common sense and debate about reading and writing and about learning in general. The ongoing dialectic named by Desrosieres gives shape to Huey's discussions of reading. He was sharply aware that the task of theorizing reading and writing was also the task of understanding the culture, the individual's engagement with the culture, and the place of "mind" in that account:

> To completely analyze what we do when we read would almost be the acme of a psychologist's achievement, for it would be to describe very many of the most intricate workings of the human mind, as well as to unravel the tangled story of the most remarkable specific performance that civilization has learned in all its history. (p. 6)

As well, Huey did not regard learning to read as independent of the materials that were to be read, or thereby of the culturally dominant ways of thinking and acting. That is, Huey stressed reading as a form of moral formation and regulation, retaining a belief that reading was and should be tied up with learning the values and ways of a culture. To read was a way to become a "robust trustee" of a society's resources and ideals:

> the pupil should be practiced in grasping the essential meanings, in selecting and gathering from books and papers what they have for his [sic] purposes, in ignoring the irrelevant, and in feeling values always ... to impregnate the souls with the race's highest ideals and tastes. (p. 382)

In many ways, Huey's *Psychology and Pedagogy of Reading* opened a Pandora's box. For its time, Huey's work was a rich encyclopedia of ideas and research about reading and writing. In fact, it was perhaps too rich, allowing many selective, not to say self-serving interpretations. It was to be mis/

interpreted and mis/applied by behaviorists, such as Thorndike and later, Gray, Gates, and others, to build a psychological model of reading rendered compatible, like the products of Binet's work, with industrial-era U.S. educational policy (Luke, 1988). The residual traditions of those models—the construction of reading as behavior, skill, and, now, information processing—still form the heart of many approaches to reading instruction, both in schools and in adult programs.

What is notable, from the perspective of our interest here, is Huey's curiosity about the everyday features (what might now be called "ethnographic" or situated characteristics) of how people actually read and write and learn to read and write, and his frustration with the assumptions that dominated much contemporary educational practice. Huey didn't mince words when it came to the need to see literacy as a collection of socially relevant practices and the irrelevance of many teaching routines:

> As a school subject, reading is an old curiosity shop of absurd practices. ... We have made a fetish of our doctrine of formal discipline, and formal reading has kept its artificial place in our curriculum supported in part by this now fast-decaying prop. (pp. 9–10)

We can almost hear him pounding the table as he described what he saw as the features of effective reading programs:

> Reading and writing and drawing are learned in the service of what the children are doing as a social community. Reading is not made an end in itself, and does not gather the mannerisms and the debris of technique that accompany reading done for its own sake and by [school subject] "Reading's" own special methods. (p. 300)

At second glance, from a century on, Huey's seemingly portentous vocabulary—curiosity shops, fetishes, mannerisms, and debris—come to assume considerable force because they reflect an insight of his whose implications are still not well worked into theories about teaching and learning. This insight can be put like this: Schools reshape real-world practices and bodies of knowledge into curricular programs, with their attendant materials, pedagogies, and assessments (so, for Huey, learning to read becomes "the A-B-C method"). These programs retain their durability partly because they seem to serve well the managerial needs of teachers in contemporary classrooms, a criterion never to be trivialized. The management of the bodies and attention of a large number of institutionally inexperienced learners by one adult/expert presents fundamental managerial problems. Whatever the knowledge-bearing capacity of the interactional formats put into place, if they do not satisfy the basic job specification presented by this management problem, they are not usable. This from a Year

One "shared big-book reading" lesson, using a text about what to do if you encounter a dragon (the Appendix lists the transcription conventions):

T	IF YOU MEET A DRAGON. This is actually a guide for what you will do if you will meet a dragon. If you meet a dragon. What's it tell us to do? Tickle his^?
S	Back.
T	Back, tickle his back\/. Tickle his^?
S	Nose.
T	Nose. What is the ending to that word?
S	'e'
T	'e' . What sort of an 'e' is it?
Ss	Bossy.
T	Bossy 'e' or magic 'e'. In 'nose' a bossy 'e' says^?
T&S(s)	"Say your name."
T	That's right, so it's no:::ose. No:::ose, Tickle his nose. I know that you've been sick, Jamie, but you need to concentrate, thanks. Tickle his^?
S	Leg.
T	Leg\/. Tickle his^?
S	Toes.
T	Toes\/. It's a rhyming book. So it's TICKLE HIS BACK, TICKLE HIS NOOOSE, TICKLE HIS LEG, TICKLE HIS TOES. TICKLE HIS TAIL, TICKLE HIS^?
S&T	Chin.
T	CHIN, we've got three full-stops, that's telling us what will come next. I wonder. TICKLE HIS TAIL, TICKLE HIS CHEEK, AND THAT WILL BE THE END OF HIM. So if you tickle him in all those places, you're going to calm him down. This time I'll read one page and you read one page, and we'll see who gets mucked up first. I'll do the first one. IF YOU MEET A DRA::AGON. Your turn… ((lesson continues))

This is reading and management. But the "mannerism and debris" are not ^…? that's right, random\/; rather, they comprise the ritualized production of monologue-via-multilogue, a set of coordinating procedures that appear to unite not only the collected students as one speaking party, but also the sentences of the students and teachers into one grammatical line. The epistemological activities of both parties are thereby collected into a singular pedagogical point—a single line as if from a single voice (not coincidentally, just like a written text), with the students speaking as a cohort and further as a cohort in full tune with the teacher's reading of the text, tracking as perfectly as possible the flow of her thought. If a child does not appreciate or at least act on the significance of *that* by 6 years of age, he or she is in for a long, hard haul, just as is any theory of learning.

This realization that pedagogies were at one time knowledge-delivery and learner-management systems is what led Huey to distinguish real reading from "Reading." One of his many insights, hinted at in Binet's "contours of school life," was that particular reductionist forms of teaching reading and writing serve schooling well—that reading can indeed be turned into a rich resource for institutionalizing the "mannerisms and debris" of schooling in general. Resisting or failing at this form of "discipline" leads to the version of the Matthew effect that our reading professors never told us about. From a systemic perspective, the more efficiently the teacher can turn propositions and procedures into standard, portable, trainable routines, the more thoroughly those routines will be enshrined as appropriate, valued curricular material and pedagogy.

The contributions of both Binet and Huey were put to work in two senses that neither might have appreciated: In the growing educational bureaucracies of the early 20th century, their ideas, and in Binet's case, analytic techniques, played a significant part in the administrative management of human capacities and practices through public schooling. In the growing institutionalized discipline of psychology, their observations about the Social were dropped from the story of their contribution. Explanatory sources that were for them close to the heart of the matter—local, institutionalized, and cultural social practices—were deemed by the psychology of the day to be someone else's business. This was a stark reflection of the relatively impoverished repertoire of the social sciences of the day for addressing cognitive and social processes, and, perhaps more importantly, for wondering about the relationship between them, and about the ways in which agency and interaction supplement and extend one another in ways that are particular to social settings such as classrooms.

THE MODERATE CASE FOR THE SOCIAL

Recently the U.S. National Research Council (2000) commissioned a synthesis of research in cognitive sciences, broadly construed, and a systematic application of that knowledge to educational practice. That piece (edited by Bransford, Brown, & Cocking and entitled *How People Learn:*

Brain, Mind, Experience and School) represents a rigorous and wide-ranging distillation of contemporary knowledge. It serves as an illustration of what it means to make the moderate case for the Social. I draw briefly on that document to characterize this case. Among the premises of *How People Learn* is that:

> [a] scientific understanding of learning includes understanding about learning processes, learning environments, teaching, socio-cultural processes, and the many other factors that contribute to learning. Research on all of these topics, both in the field and in laboratories, provides the fundamental knowledge base for understanding and implementing changes in education. (p. 233)

Clearly, processes, practices, and environments are collected as important "factors that contribute to learning"; the distinctive environments of the "field" and the "laboratory" are both acknowledged as sites for the development of a "fundamental knowledge base." What is critical in appreciating the views that distinguish the moderate and strong forms of the sociocognitive case lies in definitions of learning:

> The environment supplies information, and equally important, provides a structure to the information, as when parents draw an infant's attention to the sounds of her or his native language. (p. 234)

A crucial ambiguity, almost a double-entendre, centers on the opening expression "the environment supplies information." Later in the point we find that "parents' drawing attention" counts as "the environment," indicating that the social organization of knowledge and the social organization of how *that* organization is passed onto learners are now crucial structural concerns for cognitive theorizing. The issue of how parent–child interaction can be seen as "information supplying" is a point that calls into question the nature of the raw data that "cognitive processes" are intended to "process." As Bruner (1986, p. 13) put it, the critical question that cognitive approaches need to address is this: What are the primary phenomena to be accounted for? By reducing thought to "information," understanding human mental processes only in terms of reduced logico-scientific modes of thought, stripping it of its primordially social/moral/experiential character, "most contemporary forms of cognitive science can only seek to transcend the particular by higher and higher reaching for abstraction, and in the end disclaim in principle any explanatory value at all where the particular is concerned."

Bruner argued that in an important respect, the cognitive revolution had missed the "key puzzle"; like behaviorism before it, cognitive psychology had delivered least where it had promised most. Challenging the equation of knowledge with "information" forms a central plank of critiques of contemporary versions of cognitive psychology (e.g., John Searle's comment [1994, p. 222] that it is "the worst mistake in cognitive science"; see also Scheffler & Howard, 1995). But the reduction of knowledge to information, the leeching out of the ambiguity, relational, moral, and affective aspects of the ways in which humans come to know about and acting knowingly in their environments, is not new. It is a crucial part of the evolution of the enterprises of educational practice and research.

The growing currency of the term "cognitive" in psychological circles in the middle part of the 20th century did not relate to a reinstatement of mentalist concerns. Rather, its meaning was specific, drawing attention not to the "psychological" in general, but rather to the truth-evaluate-able, or algorithmic aspects of human functioning. This was partly to protect the growing interest in mental activity from the criticisms that behaviorists had successfully leveled at Wundtian-style mentalism earlier in the 20th century. But this protection, Bruner and others have argued, has come at too high a cost, in particular to those who considered the cognitive movement's most promising domain of activity to be in education. Cobb and Bowers (1999), among others, have argued that the state of development in this field can be characterized as on the brink of a second or even third wave of revolution (see Shotter, 2001), whereby the lines between strictly cognitive and social concerns are "blurring" and new forms of analysis are called upon to document, with more empirical adequacy, the social complexities in which learning occurs. The case is put that these forms that may entail more ethnographically oriented methodologies and interpretive approaches to human exchange, texts, and discourses (see, e.g., the work of Wortham, 1994, 2001). For several years now these complaints have had a hint of "revolution" about them. The guests may not yet be bored, but they seem at least restless, whispering among themselves about the decor and the company.

Some have gone so far as to start urging that we revisit core aspects of the cognitive program. Sheppard (1997), for example, has made the case that the cognitive revolution brought about two significant changes in the ways in which human activity was conceptualized. The first, she argued, was to shift the explanation of human action into the scientific domain of psychology, subject to rules of truth criteria and information gathering and testing, and so on, thus opening the black box that had remained shut, on strictly scientific grounds, for the duration of the Great Digression. Second, cognitive psychology aimed to provide a new and better metaphor for the mind, a metaphor based loosely on analogies to computer science, and characterizable as an "information-processing" metaphor. In this respect, the analogy continued, social experience was regarded as the software, the brain as the hardware, and their interface as "cognition." Neat, Henry-

Ford-ist, and comprehensively inadequate in the face of the emphasis on reflexivity that became the distinctive contribution of the next generation of social scientists.

A STRONGER CASE FOR THE SOCIAL

> Social interaction is the primordial means through which the business of the social world is transacted, the identities of the participants are affirmed or denied, and its cultures are transmitted, renewed, and modified. (Goodwin & Duranti, 1990, p. 283)

One initial move in strengthening the case for the Social in cognitive and other psychological accounts of education is to aggressively rework our understandings of "context" as an analytic device (Goodwin & Durant, 1992, provide an introduction to this issue). The immediate operational "context" of a learning event can be thought of as something that participants co-accomplish, rather than, as Drew and Heritage (1992) put it, a "bucket" in which certain activities both occur and become more understandable to the participants and analysts alike—simply because "this is the kind of bucket in which this occurs."

The claim here is that it is not productive to persist with this undertheorized "bucket" theory of context, where things happen "inside" a context, or what is "outside" individuals is taken to be a "factor" in any account of their behavior. Rather, the theoretical aim is to appreciate context in the same ways that the participants in social events evidently appreciate it, as (1) transformable, (2) both the project and the product of their coordinated actions, and (3) simultaneously and continuously shaping and renewed by their actions. Why demand that our theories be less perspicacious and perceptive than the people we claim to be studying? This view has it that the fundamental dialectic to be accounted for in sciences dealing with human behavior is this: People's activities cannot be explained solely in terms of how they act out roles predetermined by natural, community, or institutional contexts; they can and continually do specify, reinflect, and transform those contexts through their talk and their other actions. Neither can people's activities be explained solely in terms of how they act out and on their internally developed and determined programs of individual action; they can and continually do closely attend to the activities of others involved with them in daily events (such as learning in a classroom) and coordinate their activities accordingly. So a key question for our purposes here is: Have sociocognitively oriented researchers paid as much attention to the details of everyday educational events as the participants in those events necessarily do? To study these phenomena systematically is to study, at least as a significant part of the program, naturally occurring events in settings such as classrooms.

Over the last 20 years, there has been a growing awareness (but generally not among educational psychologists) that the specific features of the interactions that characterize learning events need themselves to be significant components of any account we may wish to give of educational practice. The argument proceeds that, in the interests of building formal mental or social models, traditional ways of locating the "problems" of teaching and learning have typically bypassed the phenomena that are the most important source of explanation for the kinds and levels of learning that we observe. Early attempts to examine talk in settings such as classrooms and homes tended to regard the context or site of the talk either as not particularly relevant to the purposes of the talk or to the ways in which it was conducted, or as, somewhat mechanistically, determinant of the nature and structure of that talk. Thus, some studies characterized classroom talk, for instance, as the transmission of information, without taking account of the peculiarly "pedagogical" nature and patterning of the interactions, or as rigidly and comprehensively explained as simply "classroom talk" (Edwards & Westgate, 1994, critique examples of both of these approaches).

Most of these early studies explored the implications for the use, valuing, and development of students' knowledge in such interactive settings, such that Edwards and Westgate (1994) concluded from their encyclopedic review of classroom talk that

> [classroom talk] is certainly not conducted normally on a basis of shared knowledge. Its outstanding characteristic ... is one participant's claim to all the knowledge relevant to the business at hand. (p. 124)

Importantly, many of the classrooms that have been studied have not appeared, at least on the surface, to be as traditionally teacher-dominated as the above conclusion might imply (see Barnes, Britton, & Rosen, 1969). However durably Edwards and Westgate's conclusion holds up over time, it points not so much to the embeddedness of "facts" in "knowledge" but, more relevantly to our purposes, to the embeddedness of knowledge in interactions that organize authority relations around tasks and materials. So here is an empirical starting point for psychological accounts of formal educational practice, a distinctive category of events configured by distinctive social constraints and options, with a notion of "learning" as its central topic and resource—the focus of the interactions and the means by which they are recognizably brought off, somehow, by the participants:

T I want you to get your Maths Facts stuff. No:o get your <u>te:est</u> books out please. We're going to have our times tables test.

S Do you have to number them?

T No, you don't have to number them, because I'm not going to tell you how many there are. Anyway you should know how many there are, because, how many times tables are there?=

S =12

T Well then you should put down 12. That's why I was wondering when everyone kept on saying, 'Well how many, how many numbers do I have to write down?' I was waiting for them to wake up. Have a good little think. Put up your hand if I have tricked you this time. You asked a silly question of 'how many do I have to write down'. Yes?

S Do I have to write down the sum?

T No, you just write the answer. Will you keep your work very covered please? Right 3 × 4, 3 × 4, <u>and</u> I'm not going to go extra slow either. Number //[2]

S //[O:oh] ((groan))

T 3 × 6, 3 × 6, if you don't know it in about 4 seconds you probably don't know it, full stop. Number 3 – 3 × 9 – 3 × 9, Number 4 – 3 × 2, Number 4 – 3 × 2. ((lesson continues))

What can be thought about, when questions can and should be asked, when and why not going slow and covering your work matters, what should be known, what probably won't be, how quickly you'll know the difference, and the rest, are all ways in which a publicly produced working notion of mentation is itself part of the accomplishment of the everyday coordinated work of teachers and students. These third graders are shown when you need a "little think," what the consequences are of a "good" one, and how you can be tricked when yours has been not quite good enough. We see a shared notion of "mind" built through continuous public use (see, e.g., Saloman, 1993, on early approaches to ideas about socially distributed cognition) as one of the major contributions of education to individual achievement and social order—and more importantly, to a clear and shared view of how these are legitimately connected. It is through the systematic patterning of these interactions, over 10 to 12 years of schooling, that people learn, in addition to mental arithmetic and all the rest, how it is that their minds can be said to function, how their thoughts are publicly accountable. More significantly, they can learn how subjectivities are made publicly available for scrutiny and assessment.

Along with all the rest, schools produce a shared sense of how the public is rendered private and vice versa, and how an institution can sanctionably evaluate those renderings.

Heap (e.g., 1982, 1985) has spelled out the implications of such a view of educational practice. His focus has been on reading education, but his argument applies more generally: Unless we attend to procedural definitions and how teaching and learning activities are organized to produce them, we can never know whether our uses of theories about teaching and learning are appropriate to any specific interactional context of its application (see also Freebody 2003). Those theories, "in their flawed ways," tell us what learning is, and how it is done, supposedly in terms that are portable and context-free. But students' knowledge of the "what and how" of learning is socially mediated through interactions with other persons. Those interactions do not display "essence-of-learning" (Heap, 1982, p. 24); rather, procedural understandings are made available within activities managed and coordinated socially. That is, they are not put on offer somehow independently of the "mannerisms and debris" of relationship management or knowledge management. Specific learning and teaching activities both draw on and put in place certain kinds of relationships, all being continually rebuilt *only* around and via the use of shared notions of "mind" and "knowledge." In accomplishing procedural knowledge such as "times-tables," they accomplish the sharing of what "mind" and "knowledge" are as warrantable objects of thought and talk.

We can now reintroduce the distinctive phenomena of "teaching and learning"—as open-textured collections of coordinated, practically paired, sociocultural practices—and assert that these practices reflexively reflect and rebuild anew distinctive cultural, moral, and relational orders. It is these orders in turn that allow us to delineate, name, and evaluate teaching and learning activities, inserting these new phenomena into research, debate, and policy on matters of teaching and learning.

But where did this category of social events come from as a historical formation? In his classic study of the role of the printing press in the corporatization of the 16th century academy, Ong (1958) commented on the consequences of the "pedagogical juggernaut" that led to the "decay of dialogue" as Western educational systems came to rely on "set texts," and the new organizational forms of knowledge whose spread they afforded:

> The teacher was ... part of a corporation which was uncalculatingly but relentlessly reducing the personalist, dialoguing element in knowledge to a minimum in favor of an element which made something a corporation could traffic in, a-personal and abstract, almost as though it were something which existed outside a mind, as though one could have knowledge without anybody to do the knowing. (p. 152)

We can see here an analysis of radical change in educational organization and practice, a change that we may term the "corporatization" of education, whereby individual achievement, documented in the representation of received information, forms the basic currency of exchange, through education, for later social goods. It is important to note that these changes were brought on partly by the printing press, in its role not so much as an agent of literacy, but rather as an agent of the standardization of knowledge, now able to be recast as an object that can be reproduced by students with demonstrably differing degrees of accuracy and completeness.

Within the corporatized school system that grew rapidly in the century following Gutenberg's invention, and that became legally mandated generally in the late 19th century, it is grades that are surrogates, promissory notes for the social goods that lie in wait for the achieving student. In this way the immediately social nature of education is presented as always-deferred, postponed, "what our students will be ready for when we have done with them." As well, the logics of the Social are always projected through the normative moral relationships within the institution of schooling via students' demonstrated dispositions for accessing and substantiating valued knowledge. Where knowledge comes from, how it is substantiated, how it is subjected to various forms of procedural truth-, use-, or value-criteria, all become the moral systems that drive the uses to which education puts knowledge, now recast as "information." Here we can see the reciprocal relationship between cognitive psychology's equation of knowledge with information, the use to which the public administration one of schooling puts this information in a controlled print-based knowledge environment, and the local, institutional morality of currency exchange by which that reciprocation is legitimated.

BEYOND BARTERING AND COERCION

> Each mind keeps its own thoughts to itself. There is no giving or bartering between them.... The altogether unique kind of interest which each human mind feels in those parts of creation which it can call me or mine may be a moral riddle, but it is a fundamental psychological fact. (James, 1890/1950, p. 226)

> We have ... succeeded in delineating for ourselves the exact field of sociology. It embraces one single, well defined group of phenomena. A social fact is identifiable through the power of external coercion which it exerts or is capable of exerting upon individuals. (Durkheim, 1895, p. 51)

The statements of James and Durkheim announce a century-old division of labor among those studying human behavior: On the one hand, the "fact" under investigation is the private and unshareable thought of individual human beings; on the other hand, the "fact" under investigation is the impersonal force of the social, coercing individuals until they accept and internalize the scope and direction of that force. This division is also a division of interest, intellectual and professional investment, and methodological disposition.

It is not coincidental that the development and divergence of these disciplines matured in developed European countries at almost the precise moment at which governments were undertaking responsibility for universal education and passing laws that mandated schooling experiences for all members of their societies. Not only did such legal intervention into the lives of ordinary people create immediate need for a large welfare profession, it also created a need for concepts and techniques in which such professionals could be trained, and which at the same time afforded a certain administrative convenience, as governments attempted to manage and legitimate universal schooling, often across extensive territories, and often applied to people of considerable cultural, linguistic, and social diversity.

A theme for this chapter has been to wonder aloud: To what extent this division of labor, interest, and methodological and intellectual disposition has been bridged by the branding of approaches that yoke two key terms together: "socio-cognitive"? and how has the necessity of engaging specifically educational issues given shape to that bridge?

It is in our interest to consider how it is that the application of such a term as apparently oxymoronic as "socio-cognitive" to educational practice has shaped the ways in which that term itself is understood. It has long been considered that education, rather than a discipline, is a professional enterprise that applies understandings and reasoning practices from its so-called core, informing, foundational disciplines, typically psychology, sociology, economics, history, philosophy, and, more recently, anthropology. Less evident, or at least less recognized, have been the important ways in which the study of education, along with other major institutionalized operations such as health care and welfare provision, have put pressure on the theoretical and methodological directions of the foundational disciplines themselves, and given some lines of work more prominence, energy, and urgency than others.

For instance, at the simplest level, it is not clear what William James would say were he to visit a classroom full of talk and texts in all sorts of formations and interactions—if these people cannot "give or barter thoughts" then what on earth could they possibly be doing? Clearly, educational events are sites in which the participants at least believe themselves to be "giving and or bartering thoughts." Their practices are predicated on that fundamental belief, and some pretty fancy theoretical machinery would need to be installed to ironize that belief in the service of the fundamental Jamesian psychological fact.

On the other hand, Durkheim might well find educational events in which there is evident coercion, both direct, from teachers and parents, and more indirect, from principals, curriculum policies, imported projections of "what you need for a job," and so on. But again, the participants in some educational events seem, oddly enough, to behave as if they've chosen to be there, even in the face of what they see as external pressures. Unless the possible "reasonableness" of that choice is preemptively dismissed as a fond epiphenomenon, then the "rule-governed machine" logic will not work. Like James, Durkheim would need to retreat to an ironized redefinition of the participants' evident practice, and claim a specialist ownership of these everyday words that would strike us as almost perverse.

In these senses, sociocognitive approaches cannot occupy a space "between" the psychological and sociological "facts," because we don't need one, and anyway, there isn't one. Rather, they may offer a distinct way of describing and explaining everyday educational practice, in terms that do not ironize people's everyday practices and understandings, and that may give some theoretical purchase on educational events—on what is distinctive about them, and on how they may fall short of educators' and stakeholders' aspirations. While there may not be a space between the facts of psychology, cognitive science, and sociology, there is, right there in front of us, the data of everyday negotiation in the spaces between individuals and social formations—the data that trace—materially, interactionally, institutionally, and ideologically—the practices that transform private considerations into public matters, and vice versa. Studying these practices in culturally high-stakes sites such as educational settings can enable us to see better how it is that some societies can (and some might even want to) achieve the acculturation of individuals whose minds keep their own thoughts to themselves, neither giving or bartering with other minds, or, in contrast, of individuals whose most relevant attribute is the efficacy of the external coercions that their society is capable of exerting upon them.

The case made here is that the proper study of education is via naturalistic studies of language as it is used in everyday educational practices (Macbeth, 2003), organized by natural language, through the public use and manipulation of common sense as ways of accessing new forms of knowledge, and through educational events that themselves have potentially dialectic relationships to commonsensical ways of reasoning about the world.

There are three senses with which any serious cognitive account of educational practice can develop its commitment to the Social. First, the *material* sources and tasks that make up educational practice are deeply and inextricably "social." They, and the puzzles they afford teachers and learners, are the direct products of histories of victories and defeats. Some of these victories and defeats were earned hundreds, maybe thousands of years ago in far-off places, and some in this classroom last week. The alphabet, the numeric representational system, historical accounts, atomic theory, bunsen burners, Venn diagrams, multiplication tables, pens, cuisinaire rods, calculators, and all the rest, are not arbitrary, but neither are they ontological, predetermined, ahistorical or uncontested. Second, the *interactional* patterns that make up educational practice are deeply and inextricably "social" in that epistemological victories and defeats are recast, reenacted (reproduced, adapted, revised, and contested) in the currency of exchangeable knowledge-commodities in and through everyday educational events in classrooms, in the propositions and procedures made available through teachers' talk, and in the prevalent organizational characteristics of that talk. The more standardized that organization, the more thoroughly submerged and unremarked will be both the consensual and contested histories of school-knowledge. Finally, the *epistemological* organizations that make up educational practice are deeply and inextricably "social" in that their histories of victories and defeats are revisited, reproduced, adapted, revised, and contested anew each day in the management of the bodies of propositions and procedures that make up the canons or disciplines underwriting the legitimacy of the curriculum formations used in schools.

"Knowing" and "acting knowingly" are central concepts in education as a field both of practice and of study. Managing and producing communications, often print communications, afford the resources through which knowledge problems are solved, knowledge is made and organized, and "learning"—adding, refining, recalibrating, and restructuring thought and activity—happens. A critical point here, acknowledged in some domains of study but, again, almost never in educational psychology, cognitive or otherwise, is that there is variation among disciplines—logical as well as dispositional—at fundamental epistemological levels, including how they identify, name, and inquire into their central problems. MacDonald (1994, p. 22ff), for instance, has identified and exemplified four such critical continua on which disciplines organize their propositional and procedural knowledge:

- compact versus diffuse;
- explanatory versus interpretive;
- driving the relationship between the general and the particular through conceptual versus textual methods; and
- high versus low in the extent to which they are "epistemically self-conscious."

I discuss each of these briefly to emphasize the fundamentally social nature of the key phenomena of educational inquiry, but it is important initially to recognize the professional sense of the Social, and the reciprocal and always-changing nature of the relationship between the organization of forms of inquiry and professional communities; in MacDonald's (p. 24) terms: "Knowledge problems shape the map of discourse communities, and discourse communities shape the map of knowledge problems."

Adapting Toulmin's (1975) terminology, MacDonald has pointed to variations in the compactness and diffusion of knowledge fields, specifically in terms of how generally agreed upon is the "reservoir of problems" that make up a discipline. At one end of this continuum, we find disciplines inhabited by large groups of researchers focusing on a small number of commonly recognized and understood problems ("urban" disciplines); at the other end, we find many problems and many different kinds of problems, each receiving the attention of a relatively small number of researchers ("rural" disciplines), with considerable debate about what topics constitute the key puzzles of the discipline. The differences at these extremes of the continuum are absolute, but they soften in the middle ranges: definitions of what a knowledge problem is, why it matters, how it is approached and presented for inquiry, how problems are solved, and, crucially for learners, what counts as evidence and procedures for challenging conclusions.

Disciplines differ as well in terms of their key discursive goals on a continuum from explaining to interpreting. This relates to the distinctive ways in which disciplines relate the general to the particular. MacDonald makes this distinction in terms of the interpretation of data as being "driven" by recourse to the materials themselves (generally text) rather than to the central conceptual questions. In the latter instance, cases are examined and enquiries and interventions designed all in the service of the key conceptual question at stake. That is, the "case" at hand is selected to be, and taken to be, a representative of the concept or category framed in the question. In the instance of more textually driven procedures for relating the particular to the general, it is the distinctiveness of the particular that draws descriptive and interpretive attention. The key here is the differing interplays between abstraction and description in the building of knowledge, in the maintenance of attention to the major knowledge problems, and in the defining force of the community of scholars in the field as adjudicators of these issues.

MacDonald (1994) has drawn attention to the importance of textual and conceptual procedures for managing abstraction and particularism in the development of knowledge domains. She described the potential liabilities of textually driven fields:

> That laissez-faire community practices would allow the movement from particular to abstraction to be made without any rational controls on the diverg-

> ing results, that the community would lack either well-defined problems to address or well-defined criteria for adjudicating contributions to the field, and that, in consequence, staking out "new" interpretative positions would fill the vacuum where rational negotiation or communal problem solving might conceivably operate. (p. 47)

MacDonald (1994) elaborated on three contrasts in the ways in which problems and solutions are presented. First, in conceptually driven disciplinary work, presentations begin immediately with a focus on the key problem, whereas textually driven disciplinary work is characterized by an initial need to motivate the reader and to advocate a particular orientation to the problem. Second, conceptually driven work tends to display a focus on the current state of knowledge and debate surrounding the topic at hand, whereas textually driven work tends to focus on the phenomena as they are readily evident to common sense as the starting point for discussion. Finally, conceptually driven work readily employs standard terminology and concepts taken to be known and known-in-common within the field; textually driven work usually necessitates the stipulation of definitions to be used on particular occasions, "new takes" on the topic or "reframings" of the problem.

To draw out these distinctions, MacDonald used a discourse-structure perspective, but the distinctions are reminiscent of those made in instructional science between "well-structured" and "ill-structured" knowledge and task domains. This distinction has been discussed by, for example, Jonassen (1997, 2000) in these terms: Well-structured problems are those that are constrained, having convergent solutions that engage the application of a limited number of rules and principles, and with well-defined parameters, right answers, and right ways to them. Ill-structured problems entail potentially multiple solutions and solution pathways, fewer set parameters that are less manipulable, and uncertainties about which concepts, rules, sequences, and principles may be necessary for the solution, how these issues may be organized, and which solution is best for the particular task at hand:

> Conceptually, ill-structured problem-solving may be thought of as a design process, not a systematic search for problem solutions.... It is a frame experiment in which the problem solver engages in a reflective conversation with the elements of the problem situation.... [T]he problem solvers must frame the design problem, recognize the divergent perspectives, collect evidence to support or reject the alternative proposals and ultimately synthesize their own understanding of the situation rather than find a solution for a prescribed problem. (Jonassen, 1997, p. 78)

Both knowledge domains and tasks can be thought of as lying on this continuum—and the structure of a task may not align with the structure of the relevant knowledge domain on which it relies.

This distinction between knowledge and task is the focus of much of the work developed by Spiro and colleagues on the topic of "cognitive flexibility" (Spiro, Feltovich, Jacobson, & Coulson, 1992; Spiro & Jehng, 1990). One of the key points here is that the social organizational features of schools and classrooms are such that teachers often inappropriately simplify important knowledge as a way of protecting students from excessive effort or failure, or maybe as a way of simply getting through a cluttered syllabus. This is a move, it is argued, that has serious and negative consequences for students. The main point, however, relates to the inextricable connection between curriculum, teaching, and learning—tasks, social practice, and cognitive processes. It is this connection that gives the work its distinctive value as both educational and cognitive science. Again, the social-individual divide shows itself to be inadequate in helping us understand everyday educational phenomena.

The subtitle of this chapter refers to the quip of John Dewey (1916, Ch. 2) that there is something extraordinary, perhaps even vaguely miraculous about education. That comment is worth revisiting in its context:

> By various agencies, unintentional and designed, a society transforms uninitiated and seemingly alien beings into robust trustees of its own resources and ideals ... in general it may be said that the things which we take for granted without inquiry or reflection are just the things which determine our conscious thinking and decide our conclusions. And these habitudes which lie below the level of reflection are just those which have been formed in the constant give and take of relationship with others.

For educators and educational psychologists alike, the urgency is for closer theoretical and empirical attention to this "constant give and take of relationship with others," as that give and take animates and orders, everyday, a society's valued resources and ideals, in unremarked but immensely significant ways.

Education can be considered a collection of valued practices and processes. Engaging in educational activities constitutes a significant interruption to daily practice, not only in the lives of young people through legally-mandated formal schooling, but also in the ways in which valued knowledge is systematized, partitioned, sequenced, and generally recast from the means of its original production into exchangeable packages suitable for use in publicly administered institutions such as schools. Similarly, the scientific studies of human behavior, psychology as much as the other social sciences, themselves represent a collection of valued practices and processes. These studies are, more significantly, also an interruption to the seamless ways in which we common-sensically account for one another's behavior, and the ways in which we routinely display those accounts in our dealings with people day-to-day.

Humans interact both to conduct the daily social and business of their lives and, simultaneously and through the very same resources, to display the normativity, accountability, and, in general, the understandability, appropriateness, and functionality of their actions (Garfinkel, 1967, p. 1). This is a key aspect of individual and social reflexivity, both as ordinary people enact it in their daily activities, and as social scientists attempt to document its occurrence and consequences for ongoing social order. Interaction is where individuals put their cognitive contents and processes on display in order to do the particular work at hand. That is to say, human interaction is "built for inspection" (Macbeth, 2003). In educational situations, in teaching and learning, this purpose-built interaction is doubly available for inspection, both in the normal sense of getting certain social practices accomplished and doing so evidently accountably, but also in the sense that learners are meant to pay the interaction going on around them in lessons a special kind of attention. So it has a very particular kind of inspect-ability. It is this reflexive orchestration of the accomplishment of daily social practice, the presentation of the self as acting accountably in a given social circumstance, and the particular demands of attention and inspection of the contributions of others and of oneself, that together become the key work of cognitive processes in learning situations. That is to say, while it is clear that language is the domain and resource through which the private is made public, and the individual made communal, it is also the case that interaction is the phenomenon through which cognitive practices—motivations, dispositions, habits, strategies, processes, whatever—are made available in teaching and learning situations for inspection and modification, by the self, by peers, or by a teacher or expert.

We ask a lot of our theories of teaching and learning, cognitive or otherwise. They are expected to provide an adequate description of the practices we observe as people conduct everyday educational activities. They have to offer some plausible hypotheses about where these practices "come from" in the first place, how they are optimally modeled, transmitted, and acquired, and the practical and cognitive consequences of not quite getting it right. They need to inform our everyday work at selecting and framing curriculum, at conducting face-to-face teaching, and evaluating students' progress. They are often used to guide and implement educational policy, central funding, and everyday administrative decisions about the actual running of schools. More broadly, they are used to influence, shape, and motivate the mobilization of ideas around such cultural archetypes as the active and informed citizen, the critical thinker, and the lifelong learner.

Taken together, these are tall orders. Perhaps the only thing more evident than the impossibility of the task is educational psychologists' ready, almost wistful acknowledgment of that impossibility. As researchers, teachers, administrators, and bureaucrats expand the range of practical, social, scientific demands on theories and models of teaching and learning, so too does the scrutiny and stress placed on our theories of teaching and learning increase.

The orientation outlined here aims to reemphasize the significance of the study of naturally occurring educational events as the primordial empirical activity for educational psychology and as the prime theoretical affordance of cognitive approaches to that broader tradition. This reemphasis harks back to the unfinished business of our forebears such as Binet and Huey and, using the powerful theoretical and analytic lenses that have been developed through the "revolutionary" activities of recent generations, to take up the disciplined documentation of educational activity—as itself a primordial social formation that shapes individuals and communities, and individuals into communities—that these forebears tried to push to new levels a century ago. So we can arrive at an even more assertive, not to say overstated way of turning the question for this chapter around: Would there be a "sociocognitive perspective" worth discussing had it not been for the pressure exerted by our need to study and theorize educational practices? Would cognitive science itself be anything more than a fond, entertaining cottage industry but for the urgency of our need to revisit and reinvigorate our educational practices in the persistent interests of our economic, cultural, intellectual, and moral well-being?

REFERENCES

Bakhurst, D., & Shanker, S. (Eds.). (2001). *Jerome Bruner: language, culture, and self.* London: Sage.

Barnes, D., Britton, J., & Rosen, H. (1969). *Language, the learner and the school.* Harmondsworth, UK: Penguin.

Bransford, J. D., Brown, A. L., & Cocking, R. R. (1999). *How people learn: Brain, mind, experience and school.* Washington, DC: National Academy Press.

Bruner, J. S. (1960). *The process of education.* Cambridge, MA: Harvard University Press.

Bruner, J. S. (1986). The *cognitive revolution in psychology.* New York: Guilford Press.

Bruner, J. S. (1990). *Acts of meaning.* Cambridge, MA: Harvard University Press.

Cobb, P., & Bowers, J. (1999). Cognitive and situated learning perspectives in theory and practice. *Educational Researcher, 28,* 2.

Desrosieres, A. (1998). *The politics of large numbers: A history of statistical reasoning.* Cambridge, MA: Harvard University Press.

Dewey, J. (1916). *Democracy and education.* New York: Macmillan.

Dreeben, R. (1968). *On what is learned in school.* Reading, MA: Addison-Wesley.

Drew, P., & Heritage, J. (1992). *Talk at work.* Cambridge, UK: Cambridge University Press.

Durkheim, E. (1982). *The rules of the sociological method* (Steven Lukes, Ed.; W. D. Halls, Trans.). New York: Free Press. (Original work published 1895)

Edwards, A., & Westgate, D. (1994). *Investigating classroom talk.* London: Falmer Press.

Freebody, P. (2003). Qualitative research in education: Interaction and practice. London: Sage.

Goodwin, C., & Duranti, A. (1992). *Rethinking context: Language as an interactive phenomenon* Cambridge, UK: Cambridge University Press.

Heap, J. L. (1982, March). *Word recognition in theory and in classroom practice.* Paper presented at the Ethnography in Education Conference, Philadelphia.

Heap, J. L. (1985, August). Applied ethnomethodology: Looking for the local rationality of reading activities. In *Proceedings of the Seventh International Institute for Ethno and CA,* pp. 39–74.

Huey, E. B. (1908). *The psychology and pedagogy of reading,* New York: Macmillan.

James, W. (1950). *Principles of psychology* (Vol. 1). New York: Dover. (Original work published 1890)

Jonassen, D. H. (1997). Instructional design models for well-structured and ill-structured problem-solving learning outcomes. *Educational Technology Research and Development, 45,* 65–95.

Jonassen, D. H. (2000). Toward a design theory of problem solving. *Educational Technology Research and Development, 48,* 63–85.

Luke, A. (1988). *Literacy, textbooks and ideology.* London: Falmer Press.

Macbeth, D. (2003). Hugh Mehan's "Learning Lessons" reconsidered: On the differences between the naturalistic and critical analysis of classroom discourse. *American Education Research Journal, 40,* 239–280.

MacDonald, S. P. (1994). *Professional academic writing in the humanities and social sciences.* Carbondale: Southern Illinois University Press.

Mead, G. H. (1934). *Mind, self, and society from the standpoint of a social behaviorist.* Chicago: University of Chicago Press.

Miller, G. A. (2003). The cognitive revolution: A historical perspective. *Trends in Cognitive Sciences,* 7, 141–144.

Miller, G., Galanter, E., & Pribram, K. H. (1960). *Plans and the structure of behavior.* New York: Holt, Rinehart & Winston.

Ong, W. (1958). *Ramus, method and the decay of dialogue.* Cambridge, MA: Harvard University Press.

Pollack, B. (1995). *The experimental psychology of Alfred Binet, selected papers.* New York: Springer.

Resnick, L.B., Saljo, R., Pontecorvo, C., & Berge, B. (Eds.). (1997). *Discourse, tools and reasoning: essays on situated cognition.* Berlin: Springer-Verlag.

Rogoff, B., & Lave, J. (Eds.). (1984). *Everyday cognition: its development in social context.* Cambridge, MA: Harvard University Press.

Saloman, G. (Ed.). (1993). *Distributed cognition: Psychological and educational considerations.* New York: Cambridge University Press.

Scheffler, I., & Howard, V. (1995). *Work, education, and leadership.* New York: Peter Lang.

Searle, J. (1994). *The rediscovery of the mind.* Boston: MIT Press.

Shotter, J. (2001). Toward a third revolution in Psychology: From inner mental representations to dialogical social practices. In D. Bakhurst & S. Shanker (Eds.), *Jerome Bruner: Language, culture, and self.* London: Sage.

Spiro, R. J., Feltovich, P. J., Jacobson, M. J., & Coulson, R. L. (1992). Cognitive flexibility, constructivism, and hypertext: Random access instruction for advanced knowledge acquisition in ill-structured domains. In T. M. Duffy & D. H. Jonassen (Eds.), *Constructivism and the technology of instruction: A conversation.* Hillsdale, NJ: Erlbaum.

Spiro, R. J., & Jehng, J. (1990). Cognitive Flexibility And Hypertext: Theory and technology for the non-linear and multidimensional traversal of complex subject matter. In D. Nix & R. J. Spiro (Eds.), *Cognition, education, and multimedia.* Hillsdale, NJ: Erlbaum.

Wenger, E. (1998). *Communities of practice: Learning, meaning, and identity.* Cambridge, UK: Cambridge University Press.

Wortham, S. (1994). *Acting out participant examples in the classroom.* Philadelphia: John Benjamins.

Wortham, S. (2001). *Narratives in action: A strategy for research and analysis.* New York: Teachers College Press.

APPENDIX
Transcription Conventions Used in This Paper

O::open	Extended vowel sound
=yes	Latched turns, connecting immediately with prior turns
//	Interruption
^	Upward intonation
V	Downward intonation
[talk]	Overlapping talk
(n)	Pause of *n* seconds
((transcriber's comment))	

Printed in the United States
38063LVS00002B/78

9 781593 111625